International Review of Science

# Physical Chemistry
# Series Two

Consultant Editor

**A. D. Buckingham, F.R.S.**

# Publisher's Note

The International Review of Science is an important venture in scientific publishing presented by Butterworths. The basic concept of the Review is to provide regular authoritative reviews of entire disciplines. Chemistry was taken first as the problems of literature survey are probably more acute in this subject than in any other. Biochemistry and Physiology followed naturally. As a matter of policy, the authorship of the Review of Science is international and distinguished, the subject coverage is extensive, systematic and critical.

The Review has been conceived within a carefully organised editorial framework. The overall plan was drawn up, and the volume editors appointed by seven consultant editors. In turn, each volume editor planned the coverage of his field and appointed authors to write on subjects which were within the area of their own research experience. No geographical restriction was imposed. Hence the 500 or so contributions to the Review of Science come from many countries of the world and provide an authoritative account of progress.

The publication of Physical Chemistry Series One was completed in 1973 with thirteen text volumes and one index volume; in accordance with the stated policy of issuing regular reviews to keep the series up to date, volumes of Series Two will be published between the middle of 1975 and early 1976. Series Two of Organic Chemistry will be published at the same time, while Inorganic Chemistry Series Two was published during the first half of 1975. Volume titles are the same as in Series One but the articles themselves either cover recent advances in the same subject or deal with a different aspect of the main theme of the volume. In Series Two an index is incorporated in each volume and there is no separate index volume.

Butterworth & Co. (Publishers) Ltd.

## PHYSICAL CHEMISTRY SERIES TWO

Consultant Editor
A. D. Buckingham, F.R.S., *Department of Chemistry, University of Cambridge*

*Volume titles and Editors*

1 **THEORETICAL CHEMISTRY**
Professor A. D. Buckingham, F.R.S., *University of Cambridge* and Professor C. A. Coulson, F.R.S., *University of Oxford*

2 **MOLECULAR STRUCTURE AND PROPERTIES**
Professor A. D. Buckingham, F.R.S., *University of Cambridge*

3 **SPECTROSCOPY**
Dr. D. A. Ramsay, F.R.S.C., *National Research Council of Canada*

4 **MAGNETIC RESONANCE**
Professor C. A. McDowell, F.R.S.C., *University of British Columbia*

5 **MASS SPECTROMETRY**
Professor A. Maccoll, *University College, University of London*

6 **ELECTROCHEMISTRY**
Professor J. O'M. Bockris, *The Flinders University of S. Australia*

7 **SURFACE CHEMISTRY AND COLLOIDS**
Professor M. Kerker, *Clarkson College of Technology, New York*

8 **MACROMOLECULAR SCIENCE**
Professor C.E.H. Bawn, C.B.E., F.R.S., *formerly of the University of Liverpool*

9 **CHEMICAL KINETICS**
Professor D. R. Herschbach, *Harvard University*

10 **THERMOCHEMISTRY AND THERMO-DYNAMICS**
Dr. H. A. Skinner, *University of Manchester*

11 **CHEMICAL CRYSTALLOGRAPHY**
Professor J. M. Robertson, C.B.E., F.R.S., *formerly of the University of Glasgow*

12 **ANALYTICAL CHEMISTRY—PART 1**
Professor T. S. West, *Macaulay Institute for Soil Research, Aberdeen*

13 **ANALYTICAL CHEMISTRY—PART 2**
Professor T. S. West, *Macaulay Institute for Soil Research, Aberdeen*

---

## INORGANIC CHEMISTRY SERIES TWO

Consultant Editor
H. J. Emeléus, F.R.S., *Department of Chemistry, University of Cambridge*

*Volume titles and Editors*

1 **MAIN GROUP ELEMENTS—HYDROGEN AND GROUPS I–III**
Professor M. F. Lappert, *University of Sussex*

2 **MAIN GROUP ELEMENTS—GROUPS IV AND V**
Dr. D. B. Sowerby, *University of Nottingham*

3 **MAIN GROUP ELEMENTS—GROUPS VI AND VII**
Professor V. Gutmann, *Technical University of Vienna*

4 **ORGANOMETALLIC DERIVATIVES OF THE MAIN GROUP ELEMENTS**
Professor B. J. Aylett, *Westfield College, University of London*

5 **TRANSITION METALS—PART 1**
Professor D. W. A. Sharp, *University of Glasgow*

6 **TRANSITION METALS—PART 2**
Dr. M. J. Mays, *University of Cambridge*

7 **LANTHANIDES AND ACTINIDES**
Professor K. W. Bagnall, *University of Manchester*

8 **RADIOCHEMISTRY**
Dr. A. G. Maddock, *University of Cambridge*

9 **REACTION MECHANISMS IN INORGANIC CHEMISTRY**
Professor M. L. Tobe, *University College, University of London*

10 **SOLID STATE CHEMISTRY**
Dr. L. E. J. Roberts, *Atomic Energy Research Establishment, Harwell*

---

## ORGANIC CHEMISTRY SERIES TWO

Consultant Editor
D. H. Hey, F.R.S., *formerly of the Department of Chemistry, King's College, University of London*

*Volume titles and Editors*

1 **STRUCTURE DETERMINATION IN ORGANIC CHEMISTRY**
Professor L. M. Jackman, *Pennsylvania State University*

2 **ALIPHATIC COMPOUNDS**
Professor N. B. Chapman, *University of Hull*

3 **AROMATIC COMPOUNDS**
Professor H. Zollinger, *Eidgenössische Technische Hochschule, Zürich*

4 **HETEROCYCLIC COMPOUNDS**
Professor K. Schofield, *University of Exeter*

5 **ALICYCLIC COMPOUNDS**
Professor D. Ginsburg, *Technion-Israel Institute of Technology, Haifa*

6 **AMINO ACIDS, PEPTIDES AND RELATED COMPOUNDS**
Professor H. N. Rydon *University of Exeter*

7 **CARBOHYDRATES**
Professor G. O. Aspinall, *York University, Ontario*

8 **STEROIDS**
Dr. W. F. Johns, *G. D. Searle & Co., Chicago*

9 **ALKALOIDS**
Professor K. Wiesner, F.R.S., *University of New Brunswick*

10 **FREE RADICAL REACTIONS**
Professor W. A. Waters, F.R.S., *University of Oxford*

## BIOCHEMISTRY SERIES ONE

Consultant Editors
H. L. Kornberg, F.R.S.,
*Department of Biochemistry University of Leicester* and
D. C. Phillips, F.R.S., *Department of Zoology, University of Oxford*

*Volume titles and Editors*

1 **CHEMISTRY OF MACRO-MOLECULES**
Professor H. Gutfreund, *University of Bristol*

2 **BIOCHEMISTRY OF CELL WALLS AND MEMBRANES**
Dr. C. F. Fox, *University of California*

3 **ENERGY TRANSDUCING MECHANISMS**
Professor E. Racker, *Cornell University, New York*

4 **BIOCHEMISTRY OF LIPIDS**
Professor T. W. Goodwin, F.R.S., *University of Liverpool*

5 **BIOCHEMISTRY OF CARBO-HYDRATES**
Professor W. J. Whelan, *University of Miami*

6 **BIOCHEMISTRY OF NUCLEIC ACIDS**
Professor K. Burton, F.R.S., *University of Newcastle upon Tyne*

7 **SYNTHESIS OF AMINO ACIDS AND PROTEINS**
Professor H. R. V. Arnstein, *King's College, University of London*

8 **BIOCHEMISTRY OF HORMONES**
Professor H. V. Rickenberg, *National Jewish Hospital & Research Center, Colorado*

9 **BIOCHEMISTRY OF CELL DIFFER-ENTIATION**
Professor J. Paul, *The Beatson Institute for Cancer Research, Glasgow*

10 **DEFENCE AND RECOGNITION**
Professor R. R. Porter, F.R.S., *University of Oxford*

11 **PLANT BIOCHEMISTRY**
Professor D. H. Northcote, F.R.S., *University of Cambridge*

12 **PHYSIOLOGICAL AND PHARMACO-LOGICAL BIOCHEMISTRY**
Dr. H. K. F. Blaschko, F.R.S., *University of Oxford*

---

## PHYSIOLOGY SERIES ONE

Consultant Editors
A. C. Guyton,
*Department of Physiology and Biophysics, University of Mississippi Medical Center* and
D. F. Horrobin,
*Department of Physiology, University of Newcastle upon Tyne*

*Volumes titles and Editors*

1 **CARDIOVASCULAR PHYSIOLOGY**
Professor A. C. Guyton and Dr. C. E. Jones, *University of Mississippi Medical Center*

2 **RESPIRATORY PHYSIOLOGY**
Professor J. G. Widdicombe, *St George's Hospital, London*

3 **NEUROPHYSIOLOGY**
Professor C. C. Hunt, *Washington University School of Medicine, St. Louis*

4 **GASTROINTESTINAL PHYSIOLOGY**
Professor E. D. Jacobson and Dr. L. L. Shanbour, *University of Texas Medical School*

5 **ENDOCRINE PHYSIOLOGY**
Professor S. M. McCann, *University of Texas*

6 **KIDNEY AND URINARY TRACT PHYSIOLOGY**
Professor K. Thurau, *University of Munich*

7 **ENVIRONMENTAL PHYSIOLOGY**
Professor D. Robertshaw, *University of Nairobi*

8 **REPRODUCTIVE PHYSIOLOGY**
Professor R. O. Greep, *Harvard Medical School*

International Review of Science

# Physical Chemistry
# Series Two

## Volume 9
## Chemical Kinetics

Edited by **D. R. Herschbach**
Harvard University

**BUTTERWORTHS**
LONDON - BOSTON
Sydney - Wellington - Durban - Toronto

THE BUTTERWORTH GROUP

ENGLAND
Butterworth & Co (Publishers) Ltd
London: 88 Kingsway, WC2B 6AB

AUSTRALIA
Butterworths Pty Ltd
Sydney: 586 Pacific Highway, NSW 2067
Also at Melbourne, Brisbane, Adelaide and Perth

CANADA
Butterworth & Co (Canada) Ltd
Toronto: 2265 Midland Avenue,
Scarborough, Ontario M1P 4S1

NEW ZEALAND
Butterworths of New Zealand Ltd
Wellington: 26–28 Waring Taylor Street, 1

SOUTH AFRICA
Butterworth & Co (South Africa) (Pty) Ltd
Durban: 152–154 Gale Street

USA
Butterworths (Publishers) Inc
Boston: 19 Cummings Park, Woburn, Mass. 01801

**Library of Congress Cataloging in Publication Data**
Main entry under title:

Chemical kinetics.

(Physical chemistry, series two; v. 9) (International review of science)
Includes index.
1. Chemical reaction, Rate of. I. Herschbach, D. R. II. Series. III. Series: International review of science.
QD450.2.P59 vol. 9 [QD503] 541′.3′08s [541′.39]
ISBN 0 408 70608 2 75–22304



Typeset by Amos Typesetters, Hockley, Essex
Printed and bound in Great Britain by
REDWOOD BURN LIMITED, Trowbridge and Esher

# Consultant Editor's Note

The International Review of Science was conceived as a comprehensive, critical and continuing survey of progress in research. The difficult problem of keeping up with advances on a reasonably broad front makes the idea of the Review especially appealing, and I was grateful to be given the opportunity of helping to plan it.

Physical Chemistry Series One was published in 1972–73. Its success was assured by the very great distinction of its editors and authors. Our need for critical reviews at a high level has not diminished. In the rather difficult times being experienced in most parts of the world, research workers should seek to broaden the range of their expertise; it is hoped that this Series will be of use in this connection.

Like its forerunner, Series Two consists of thirteen volumes covering Physical and Theoretical Chemistry, Chemical Crystallography and Analytical Chemistry. Each volume has been edited by a distinguished chemist; in several cases the person responsible for Series One has acted again as editor. The editors have assembled a strong team of authors, each of whom has assessed and interpreted recent progress in a specialised field in terms of his own experience. I believe that their efforts have again produced useful and timely articles which will help us in our efforts to keep abreast of progress in research.

It is my pleasure to thank all those who have collaborated in this venture—the volume editors, the authors and the publishers.

Cambridge A. D. Buckingham

# Preface

In his classic essay, *The Republic of Science*, Michael Polanyi describes a race to assemble a giant jigsaw puzzle. The competing teams have exactly the same talent but some are organised in the hierarchial ways traditional in practical affairs whereas one enjoys the chaotic freedom of science. Polanyi shows that the chaotic team is by far the most effective. It is in fact coordinated by 'an invisible hand,' provided that each independent worker has the opportunity to observe and apply the results found by the others. This theme deserves emphasis in the recurrent discussions of 'planning' for scientific progress. It is also an apt leitmotiv for the present volume of reviews devoted to a field which owes much to Polanyi's own pioneering work on reaction rates.

Chemical Kinetics is a notoriously eclectic science. It seeks to map the molecular pathways for chemical transformations in a great variety of systems and must invoke many disparate techniques and levels of approximation. This gives Kinetics a lively character that often erupts in unexpected discovery, controversy, or drastic changes in perspective. For the most active parts of Kinetics, which may appear almost kaleidoscopic, workers in kindred fields are hard put to 'observe and apply.' I hope the reviews given here will make more tangible the present and future course of 'the invisible hand.'

The topics covered in this volume include: unimolecular reactions, molecular dynamics in single collisions, enzyme kinetics, reactions in planetary atmospheres, and chemiluminescence in solution. The first three areas were reviewed four years ago in the predecessor volume of Series One, edited by J. C. Polanyi. The last two are major new areas and each is treated in two complementary chapters. The authors are among the most evangelical leaders in their respective fields. They have put particular effort into outlining for non-specialists the main themes and prospects. I should like to renew my congratulations and earnest thanks.

Harvard D. R. Herschbach

# Contents

# 1
# Unimolecular Reactions

**J. TROE**
Institut de Chimie-Physique de l'Ecole Polytechnique Fédérale, Lausanne



## 1.1 INTRODUCTION

The term *unimolecular reaction* comprises many aspects. Initially it was used for thermal dissociations and isomerisations near their first-order limit.

Later on the pressure-dependent fall-off range of these reactions was also included. A generalisation of the term was obtained when reactions with different types of activation were compared. Statistical theories of thermal unimolecular reactions and of mass spectrometer fragmentations for a long time had been developed almost independently. With a better communication between these two fields, several specific problems could have been avoided. Today one tries to apply unimolecular reaction rate theory to reactions in which activated particles are produced by thermal collisions, by chemical and optical activations, by electron impact, by field-, charge exchange- and photo-ionisation, by radiolysis, by hot-atom reactions, by primary exothermic processes, by bimolecular reactions, etc. As far as the dissociation (or isomerisation) lifetime of the activated particle is concerned, the description of all systems will certainly profit from a common point of view. Besides the dissociative process, other elementary processes such as collisional energy transfer, radiationless transitions between electronic states, etc. may be similar in the different mechanisms. Nevertheless, in practice the unimolecular aspect is often only of minor importance; the specific mechanism and the different ways of preparation of excited species may be of primary interest.

Whenever the unimolecular aspect of a reaction system is considered, one has to distinguish between the description of the total reaction composed of a mechanism of different elementary steps, and the description of the elementary process of activated single molecules. In unimolecular reactions — this might serve as a vague definition — the latter process corresponds to the dynamics on bound potential surfaces with relatively deep potential minima between the reacting parts. Bimolecular processes involving such potential surfaces, correspondingly, should be discussed as a special case of unimolecular processes.

Dissociations at non-bonding surfaces, on the other hand, should be discussed together with bimolecular reactions without complex formation. This distinction with respect to the properties of the potential surfaces involved corresponds to the different success of applying statistical reaction rate theories. Whereas such theories have been most successful in unimolecular reactions and bimolecular reactions involving bound intermediates, the non-statistical character of many direct bimolecular reactions has often become quite evident[1].

Several inconsistencies in its application to unimolecular processes and to bimolecular reactions involving bound intermediates have to be removed. Two important versions, transition state theory and phase space theory, often use different statistical postulates; conservation of total angular momentum is often introduced in a problematic way. After the discussion of unimolecular processes, some composed mechanisms of unimolecular reactions, with competition of collisional energy transfer and reaction, are considered. Thermal and optical unimolecular reactions are analysed here, whereas examples for chemical activation have been extensively discussed in *Series One* of this Review[2]. Within this chapter, bimolecular reactions involving long-lived intermediates are also considered. One may hope that such reactions in the future will be studied for systems where unimolecular dissociations or recombinations with the same intermediate complex are also accessible. With different processes on the same potential one certainly has the best chance to arrive at a meaningful experimental characterisation of the basic quantity of a reaction, i.e. its potential surface.

## 1.2 STATISTICAL MODELS OF UNIMOLECULAR PROCESSES

### 1.2.1 General formalism

#### *1.2.1.1 Specific rate constants*

We consider activated molecules of energy $E$, total angular momentum $J$ and certain symmetry properties. These molecules can dissociate or isomerise via channels $a'$ to products. In the statistical theory, one assumes that the channels $a'$ are populated in a random way, and that activation and decay into the channels $a'$ are independent processes. If the statistical assumption does not hold, the general description of different unimolecular reactions by one common formalism loses much of its significance.

Specific rate constants $k(E,J,a')$ for decay of activated molecules into a channel $a'$ are defined by

$$\left(\frac{\mathrm{d}\, n(E,J)}{\mathrm{d}t}\right)_{a'} = -k(E,J,a')\, n(E,J) \tag{1.1}$$

where $n(E,J)$ is the number of activated molecules in state $(E,J)$. If the formation of products in levels $a$ is considered, one has to take into account the degeneracy of the fragment levels $a$ at infinite separation with respect to orbital angular momentum $l$ and with respect to the relative orientations of their rotational angular momenta. Assuming independence of the different reaction channels $a'$, one obtains for reaction from activated molecule levels $(E,J)$ to product levels $(E,a)$, by summation

$$\frac{\mathrm{d}\, p(E,a)}{\mathrm{d}t} = +\sum{}' k(E,J,a')\, n(E,J) \tag{1.2}$$

$p(E,a)$ is the number of products in level $(E,a)$; the sum extends over all channels $a'$ leading to levels $a$; in the summation the known rules for coupling of orbital and rotational angular momenta of the product levels $a$ are taken into account.

The specific rate constants in the statistical theory are given by

$$k(E,J,a') = \frac{\gamma(E,J,a')}{h\,\rho(E,J)} \tag{1.3}$$

$\rho(E,J)$ is the density of states of activated molecules with angular momentum $J$, i.e. their number of states per unit energy, $\gamma(E,J,a')$ is a transmission coefficient for passage through channel $a'$. Equation (1.3) has amply been used in the theory of compound nucleus decompositions[3,4]. Its factors can easily be interpreted: $h\rho(E,J)$ is the mean recurrence time for a specific configuration within a set of strongly coupled states, and $\gamma(E,J,a')$ is the number of successful attempts to escape through channel $a'$ divided by the number of attempts to escape[3]. The general form of (1.3) is conserved by more general quantum scattering theories[5–9], the main problem, of course, being the explicit calculation of $\gamma(E,J,a')$. A different interpretation of (1.3) is often used in transition state theory[10]. Assuming equilibrium populations for all states of the activated molecule up to a certain activated complex ‡, the flow towards products in a channel $a'$ is given by the flow velocity in length per unit time, multiplied by an equilibrium density of molecules at ‡ in number per unit length, and multiplied by a transmission coefficient

$$\left(\frac{\mathrm{d}\, n(E,J)}{\mathrm{d}t}\right)_{a'} = -\left[\left(\frac{\mathrm{d}q}{\mathrm{d}t}\right)\left(\frac{\mathrm{d}p\,\mathrm{d}q/h}{\mathrm{d}q\,\rho(E,J)\mathrm{d}E}\,n(E,J)\right)\gamma(E,J,a')\right]_{\ddagger} = -\frac{\gamma_{\ddagger}(E,J,a')}{h\,\rho(E,J)}\,n(E,J) \quad (1.4)$$

Whenever the same activated complex is used for *all* product channels $a'$, the latter interpretation may lead to problems (see below).

#### *1.2.1.2 Cross sections*

The reverse process of a fragmentation of activated molecules is complex formation by bimolecular association of the fragments. Both directions of the reaction are coupled by microscopic reversibility, and should be treated in a consistent manner. According to microscopic reversibility, $\gamma(E,J,a')$ is the same for forward and backward reaction in the channel $a'$. As the complex is only specified by its energy $E$, total angular momentum $J$, and eventually symmetry properties conserved during reaction, the fragment levels $a$ correlating with complex levels in general are highly degenerate. The total cross section $\sigma_{ca}(E,J,a)$ for formation of the complex in the level $(E,J)$ from fragment states $a$, therefore, is given by the sum

$$\sigma_{ca}(E,J,a) = \sum(2l+1)\,\pi\lambda\!\!\!^{-2}(a)\,\gamma(E,J,a') \quad (1.5)$$

$\lambda\!\!\!^{-}(a)$ is the de Broglie wavelength $\hbar/[2\mu_a E_{\mathrm{trans}}(a)]^{\frac{1}{2}}$ with the reduced mass $\mu_a$ and the initial relative translational energy $E_{\mathrm{trans}}(a)$ of the fragments in level $a$; the sum extends over all orbital angular momenta $l$ and relative orientations of orbital and rotational angular momenta of the products leading to a total angular momentum $J$ of the complex with arbitrary orientations of $J$. If the fragment level $a$ is characterised by rotational angular momenta $j_{a1}$ and $j_{a2}$ of the two fragments, there are $(2l+1)(2j_{a1}+1)(2j_{a2}+1)$ possibilities to combine these angular momenta with the orbital angular momentum $l$. $(2J+1)$ of these possibilities correspond to the different orientations of the total angular momentum $J$ of the complex. Therefore

$$\sigma_{ca}(E,J,a) = \frac{(2J+1)\,\pi\lambda\!\!\!^{-2}(a)}{(2j_{a1}+1)(2j_{a2}+1)}\sum{}'\gamma(E,J,a') \quad (1.6)$$

with the sum extending over all channels $a'$ leading to fragment levels $a$.

If complex formation with subsequent stabilisation prior to redissociation is not considered, but reactive or inelastic scattering with only intermediate complex formation, the statistical theory postulates independence of complex formation and complex dissociation. The probability of dissociation into a particular channel $b'$ is then given by

$$P(E,J,b') = \frac{k(E,J,b')}{\sum\limits_{a',b'}\sum{}'\,k(E,J,a')} = \frac{\gamma(E,J,b')}{\sum\limits_{a',b'}\sum{}'\,\gamma(E,J,a')} \quad (1.7)$$

The degeneracy of the product channels with respect to the final orbital angular momentum and to the relative orientations of final rotational angular momenta, and the contributions from different total angular momenta $J$, have to be taken into account. The total cross section for scattering from $a$ into $b$ is thus given by

$$\sigma_{ba}(E) = \sum_J \sigma_{ca}(E,J,a) \left\{\sum{}' P(E,J,b')\right\}$$
$$= \frac{\pi \lambda\!\!\!^{-2}(a)}{(2j_{a1}+1)(2j_{a2}+1)} \sum_J (2J+1) \frac{\sum' \gamma(E,J,a') \sum' \gamma(E,J,b')}{\sum\limits_{a',b'} \sum' \gamma(E,J,a')} \tag{1.8}$$

Equation (1.8) has been successfully applied to nuclear reactions[3,4].

### 1.2.2 Adiabatic channel models

#### *1.2.2.1 Adiabatic approximation*

The basic problem of statistical theory is the characterisation of the different channels $a'$ and the counting of numbers of open channels. In general, simplified transmission coefficients $\gamma(E,J,a')$ are used.

One may, for instance, construct a set of adiabatic channel states $a'(q)$ along a curvilinear reaction coordinate $q$. In a dissociation reaction these channel states at $q \to \infty$ become identical with the product states $a$. Approximately one might separate[6,11] the motions along and perpendicular to $q$ at intermediate values of $q$. Each channel can be characterised by an 'adiabatic channel potential curve' $V_a(q)$ which is the solution of a suitable Schrödinger equation at a given $q$. These potential curves may be understood as effective potentials similar to the well-known effective potentials of rotating molecules. If the electronic potential energy of the molecule along the reaction coordinate $q$ is denoted by $V(q)$, the adiabatic channel potential curves $V_a(q)$ may also be expressed by

$$V_a(q) = V(q) + E_a(q) \tag{1.9}$$

with some 'channel eigenvalue' $E_a(q)$. The adiabatic potential curves $V_a(q)$ in the range $q \gtrsim q_e$ ($q_e$ = equilibrium position of $q$ in the activated molecule) can have their maxima $V_{a\,\max}$ at $q \approx q_e$, $q \to \infty$ or at intermediate values of $q$.

The motion in the activated molecule is characterised by strong coupling and frequent transitions between the adiabatic channel states. At $q \to \infty$ this strong coupling disappears. In the adiabatic approximation one assumes that the decay in one channel and the corresponding $\gamma(E,J,a')$ are determined by the properties of the single adiabatic channel state $a'$ only. The strongly non-adiabatic coupling in the activated molecule is thus assumed to be followed by an adiabatic decay in the individual channel states. (If non-adiabatic transitions between adiabatic channel states are to be included, this could be done in extreme cases by changing the set of adiabatic channel states, or by including this effect in the transmission coefficients as indicated, for example, in Refs. 6 and 12.) For an allowed reaction the approximation

$$\gamma(E,J,a') \approx \begin{cases} 0 \text{ at } E \leqslant V_{a\,\max} \\ 1 \text{ at } E > V_{a\,\max} \end{cases} \tag{1.10}$$

is used. Equation (1.10) neglects tunnelling and reflections at barriers of $V_a(q)$. For forbidden reactions, suitable semiclassical transmission coefficients $\gamma(E,J,a')$ can be derived from the Landau–Zener–Stückelberg–Nikitin formalism[13] or the corresponding multi-dimensional formulae[14].

On the basis of (1.10) at a given $E$ and $J$, the channels $a'$ can be grouped into open channels with $\gamma \approx 1$ and closed channels with $\gamma \approx 0$. Abbreviating the number of open channels $\sum' \gamma(E,J,a')$ by $W(E,J,a)$, with equations (1.2), (1.3), (1.6) and (1.8) the specific rate constant for dissociation into channels $a$ becomes

$$k(E,J,a) \approx \frac{W(E,J,a)}{h\,\rho\,(E,J)} \tag{1.11}$$

the specific rate constant for dissociation into arbitrary channels

$$k(E,J) \approx \frac{\sum_a W(E,J,a)}{h\,\rho(E,J)} = \frac{W(E,J)}{h\,\rho(E,J)} \tag{1.12}$$

the cross section for complex formation

$$\sigma_{ca}(E,J,a) \approx \frac{2J+1}{(2j_{a1}+1)(2j_{a2}+1)}\,\pi\lambda^2(a)\,W(E,J,a) \tag{1.13}$$

and the cross section for the scattering $a \rightarrow b$

$$\sigma_{ba}(E) = \frac{\pi\lambda^2(a)}{(2j_{a1}+1)(2j_{a2}+1)} \sum_J (2J+1)\,\frac{W(E,J,a)\,W(E,J,b)}{W(E,J)} \tag{1.14}$$

Equation (1.12) has the usual form given by transition state theory[15,16], (1.14) that given by phase space theory[17]. Nevertheless, the numbers of open channels $W(E,J,a)$ are not directly the same as the numbers of states usually counted at localised activated complexes or at critical surfaces (see below). Again, one has to insist in a consistent handling of (1.11)–(1.14).

### *1.2.2.2 Construction of adiabatic channel potential curves*

The real difficulty is certainly not the formulation of the basic formulae but their explicit evaluation. In general, large numbers of channels are contributing: for a known potential surface even the detailed specification of one channel is complicated; in reality the potential surfaces are unknown in many details. As for all averaged quantities, one has therefore to ask for the essential parameters which survive averaging, and which independently of simplifications have to appear correctly in the final rate constant or cross section. In the following, an empirical, approximative model will be discussed[18] which allows for a study of this question.

With bonding potential surfaces, often the potential parameters of the separated fragments and of the complex (at $q \approx q_e$) are known from spectroscopy. The intermediate range is poorly known, and suitable interpolations are required. For the interaction potential between the two fragments, $V(q)$ in (1.9), one may often start with a Morse potential. The transformation of reactant into product eigenvalues $E_a(q)$ to a first approximation can be described by empirical rules such as Pauling's rules, Badger's rules or the bond energy–bond order method[19]. As a starting point one may use one 'Pauling parameter' to interpolate all eigenvalues $E_a(q)$. More parameters might be required in reality, although these should be introduced only when a comparison with quantum-chemical potential surfaces becomes possible.

With increasing interfragment distances, Pauling's rules suggest an exponential relation between force constants or energies and bond lengths; this dependence is described by empirical 'Pauling parameters' such as listed in Ref. 19. Analogous to such rules, for a non-linear triatomic molecule, which is approximately a symmetric top, the channel eigenvalues $E_a(q)$ in Ref. 18 have been interpolated between reactant and product states by the following equation

$$\begin{aligned} E_a(q) \approx {} & [\,(v_n + \tfrac{1}{2})\varepsilon_n + (v_{bn} + \tfrac{1}{2})\varepsilon_{bn} + \{A_n(q_e) - B_n(q_e)\}K^2]\exp[-\alpha(q - q_e)] \\ & + [(v_p + \tfrac{1}{2})\varepsilon_p + A_p j(j+1)][1 - \exp\{-\alpha(q - q_e)\}] \\ & + B_s(q)P(P+1) \end{aligned} \tag{1.15}$$

with

$$P = J\exp\{-\alpha(q - q_e)\} + l[1 - \exp\{-\alpha(q - q_e)\}]$$

$\alpha$ is the 'Pauling parameter'. One stretching mode of the reactant molecule is identified with the reaction coordinate $q$; $\varepsilon_n$ and $\varepsilon_{bn}$ are the other stretching and bending quanta of the reactant molecule, $A_n(q_e)$ and $B_n(q_e)$ its rotational constants; $\varepsilon_p$ and $A_p$ are the vibrational quantum and the rotational constant of the diatomic fragment; $B_s(q)$ is the small rotational constant of the reactant at intermediate $q$; $v_n$ and $v_{bn}$ are the vibrational quantum numbers, $J$ and $K$ the rotational quantum number of the reactant molecule; $v_p$ and $j$ are the vibrational and rotational quantum numbers of the diatomic fragment, $l$ the quantum number of orbital angular momentum; the centrifugal term with the pseudo-quantum number $P$ interpolates between the corresponding rotations of the reactant and the products. Similar equations have been discussed for more complicated cases.

In order to use $E_a(q)$ one has to specify which reactant and product states most reasonably are correlated via adiabatic channel states. Here, much the same questions arise as in correlation diagrams in spectroscopy. In particular one has to discuss the question which crossings, from a dynamical point of view, are allowed or avoided. It appears reasonable to assume, for instance, a conservation of the quantum number of the stretching mode in (1.15), i.e. $v_n = v_p$. On the other hand, bending and rotation ($v_{bn}$ and $K$ in the reactant, $j$ and $l$ in the fragments) cannot be separated, and their terms should simply be correlated in the order of increasing energies avoiding crossing. In more complicated molecules such correlations cannot be specified uniquely without explicit calculations; however, the essential channels associated to the disappearing bending motions coupled to rotations can be handled in a similar way. Fortunately enough in most cases not the properties of single channels but only the general pattern of channel eigenvalues is of importance; in the averaged rate constants and cross sections inadequate correlations are often averaged out.

Considering the individual channel states has a further advantage. One may rigorously include symmetry restrictions in the correlation diagram. The individual nuclear spin statistical weights can be taken into account, and symmetry numbers are introduced in a most natural way. The ambiguity of the term 'reaction path degeneracy' is easily removed[18].

### 1.2.3 Activated complexes

#### *1.2.3.1 Minimum local entropy criteria*

As indicated in the general formulation of the statistical theory, Section 1.2.1.1, the statistical expression of the specific rate constant for a single channel $k(E,J,a')$ can also be understood in terms of transition state theory: the positions $q^{\ddagger}$ of the maxima $V_{a\,\max}$ of the adiabatic channel potential curves $V_a(q)$ are related to the ordinary 'activated complexes'. The basic postulate of the general statistical theory, nevertheless, often appears less restrictive than the different versions of the equilibrium hypothesis formulated in transition state theory.

In conventional transition state theory, the same activated complex is used for all channels $a$, or at least large groups of channels. $W(E,J,a)$ or $W(E,J)$ is identified with the number of channels passing at an energy $\leqslant E$ through this activated complex, or the 'number of states of the activated complex'. Whenever a number of channels is counted in this way, in general some closed channels are included. For potential surfaces with pronounced barriers at intermediate distances between reactant and products, this problem is not too serious: practically all channel states have their maxima $V_{a\,\max}$ at the place of the maxima of $V(q)$. For surfaces without such barriers, like in simple bond fission reactions, however, this simplification is questionable.

In the following, we shall consider energies $E > E_0(J)$, where $E_0(J)$ is the threshold energy for reaction at a total angular momentum $J$, and denote by $W_c(E,J,q)$ the total number of open and closed channels below an energy $E$ counted at a fixed value of the reaction coordinate $q$. $q^{\ddagger}$ shall denote the value of $q$ in the range $q_e \leqslant q^{\ddagger} \leqslant \infty$ where $W_c(E,J,q)$ has its minimum. Obviously, $W_c(E,J,q^{\ddagger})$ includes a minimum number of closed channels and presents an upper limit to the number of open channels $W(E,J)$. Often it is possible to estimate also a lower bound of $W(E,J)$. $W_c'(E,J,q^{\ddagger})$ shall denote the number of states at the activated complex $q^{\ddagger}$ for which, at a total energy $\leqslant E$, there is an energy larger than the threshold energy contained in the reaction coordinate. In general there exist some open channels with strongly decreasing eigenvalues $E_a(q)$, which contribute part of their energy to the reaction coordinate only at $q > q^{\ddagger}$. These channels are not counted in $W_c'(E,J,q^{\ddagger})$. The opposite behaviour, increase of $E_a(q)$ with increasing $q$, is also possible but generally less frequent. Therefore, $W_c'(E,J,q^{\ddagger})$ in most cases is a lower limit of $W(E,J)$, and

$$W_c'(E,J,q^{\ddagger}) \lesssim W(E,J) \leqslant W_c(E,J,q^{\ddagger}) \tag{1.16}$$

Equation (1.16) is the basis of minimum local entropy criteria. In such criteria the position of the activated complex is identified with $q^{\ddagger}$ obtained from the condition

$$\frac{\partial W_c(E,J,q)}{\partial q} = 0 \text{ at } q = q^{\ddagger} \tag{1.17}$$

$W_c(E,J,q)$ corresponds directly to a local entropy at fixed $q$, i.e. $S_c(E, J,q) = k \ln W_c(E, J,q)$; its minimum thus corresponds to the minimum of the local entropy. Originally[20], instead of (1.17) a 'minimum density of states criterion' had been formulated, in which the local density of states included the translational states in the reaction coordinate. As indicated above, a minimum number of states criterion excluding translational states probably better approximates the general statistical theory.

### *1.2.3.2 Properties of activated complexes*

In the general statistical theory, i.e. equations (1.3), (1.6) and (1.8), each channel $a'$ is characterised by the suitable transmission coefficient $\gamma(E, J,a')$. In the language of transition state theory, each channel $a'$ has its own activated complex. By the minimum number of states criterion (1.17) an 'average activated complex' is determined for all channels at a given total energy $E$ and total angular momentum $J$. An approximate value of the number of open channels is obtained from the number of states of this average activated complex in (1.16). Conventional transition state theory goes one step further in assuming one activated complex for the whole reaction, or at least for all energies at a given $J$. In order to study the adequateness of this assumption, one has to consider the positions of the average activated complexes $q^{\ddagger}(E, J)$, determined from (1.17), as a function of $E$ and $J$.

As shown in Refs. 18 and 21, $q^{\ddagger}$ may vary strongly with $E$ and $J$. In general, $q^{\ddagger}$ is quite different from the positions of centrifugal barriers[18,20,21] often used as activated complexes in conventional transition state theory. Furthermore, there is never a precisely determined value of $q^{\ddagger}$ but only a range of values, because $W_c(E, J,q)$ takes only integer values. This range can extend up to $q \rightarrow \infty$. The ambiguity in $q^{\ddagger}$ is of no significance, because not $q^{\ddagger}$ but $W_c(E, J,q^{\ddagger})$ interests; this is uniquely defined. An energy dependence of $q^{\ddagger}$, in the language of transition state theory, corresponds to energy dependent 'rigidity' or 'looseness' of activated complexes. Often $q^{\ddagger}$ decreases with increasing energy[18,21]. This is mainly due to the relative contribution of loose coordinates (like bending vibrations being transformed during fragmentation into orbital motions) which decreases at increasing energies. Compared with conventional transition state theory this results in a comparably slow increase of $k(E, J)$ at high energies. Thermally averaged rate constants at low temperatures, therefore, have looser average activated complexes than at high temperatures, and comparably low apparent activation energies result.

The difference between the statistical adiabatic channel model, minimum local number of states calculations and conventional transition state theory are illustrated in Figure 1.1 for the $NO_2 \rightleftharpoons NO + O$ model reaction. The figure shows $k(E, J = 0)$. Curve 1 gives the result from the statistical adiabatic channel model; the channel eigenvalues $E_a(q)$ have been determined with (1.15) and (1.16); a Pauling parameter $\alpha$ of 1.3 Å$^{-1}$ was chosen after fitting of the thermally averaged rate constant to experimental data (Pauling's rules would have predicted $\alpha \approx 2$ Å$^{-1}$ for stretching force constants). Curve 2 corresponds to the upper limit $W_c(E, J,q^{\ddagger})$, curve 3 to the lower

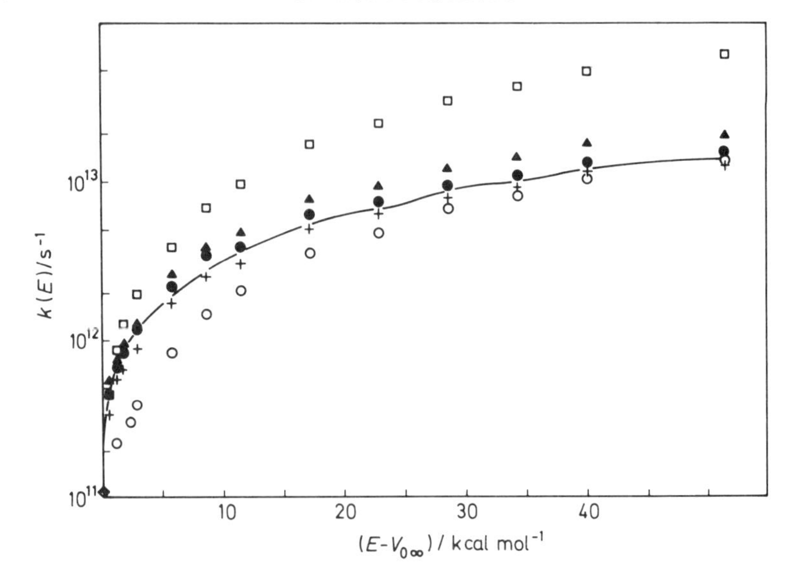


**Figure 1.1** Specific rate constants $k(E, J = 0)$ for the reaction $NO_2 \rightleftharpoons NO + O$ (——, adiabatic channel model (curve 1); ●, minimum local entropy model, upper limit (curve 2); +, minimum local entropy model, lower limit (curve 3); ▲, as curve 2, but harmonic oscillator approximation (curve 4); ○, □, best RRKM calculations (curves 5 and 6); $V_{0\infty}$ corresponds to $E_0(J = 0)$; for details see text and Ref. 18

limit $W_c'(E, J, q^{\ddagger})$ [see (1.16)] of a minimum local number of states calculation. The agreement with curve 1 is remarkably good, although larger differences between $W_c(E, J, q^{\ddagger})$ and $W_c'(E, J, q^{\ddagger})$ are found for larger molecules. Curve 4 is the same as curve 2 but with the frequency of the bending motion decaying towards 0 at $q^{\ddagger} \rightarrow \infty$, instead of the correct transformation into rotational motion. Curves 5 and 6 are conventional RRKM calculations, the former with a transition state chosen to fit the thermally averaged rate constant, the latter to fit curve 1 at *ca.* 50 kcal mol$^{-1}$.

The comparison of the different curves in Figure 1.1 suggests that conventional transition state calculations with a suitable choice of one activated complex for the whole reaction (at a fixed $J$) can only locally reproduce the results of the general statistical theory. Whenever details are needed, such as the energy dependence of $k(E, J)$ over large energy intervals or temperature coefficients of thermally averaged rate constants in large temperature ranges, the simplification of one activated complex is inadequate. Minimum number of states calculations are much nearer to the statistical theory, and offer convenient upper and lower bounds for the number of open channels.

The influence of rotation is illustrated in Figure 1.2. $k(E, J = 0)$, $k(E, J = 30)$ and $k(E, J = 60)$ have been calculated with the statistical adiabatic channel model for the same reaction as in Figure 1.1. The $J$ dependence of the threshold rate constants $[h\,\rho(E, J)]^{-1}$ is governed by two effects: the appearance of centrifugal barriers, of 0.24 kcal mol$^{-1}$ at $J = 30$ and 1.07 kcal mol$^{-1}$ at $J = 60$, and the increase of the density of states $\rho(E, J)$

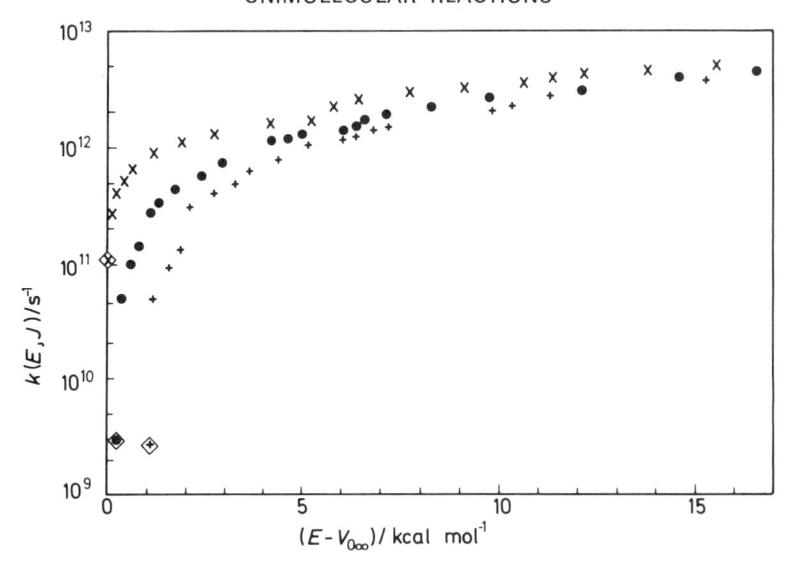


**Figure 1.2** Specific rate constants $k(E, J)$ for the reaction $NO_2 \rightleftharpoons NO + O$ adiabatic channel model: $\times$, $J = 0$; ●, $J = 30$; $+$, $J = 60$; $V_{0\infty}$ corresponds to $E_0(J = 0)$; ◇ indicates the threshold rate constants for transmission coefficients from equation (1.10); for details see text and Ref. 18

at $J \neq 0$, because of the levels with quantum numbers $K \leqslant J$. The latter rotational states are fully coupled with the vibrational states at large amplitudes. Above the threshold energy the number of open channels $W(E, J)$ includes those channels which on the molecule side correspond to $K$ levels with $K \leqslant J$. Again a separation of vibration and rotation is impossible. Because of the small energy range shown, one can recognise in Figure 1.2 the opening of new groups of channels with NO ($v = 1$) and NO ($v = 2$) at the bends of the $k(E, J)$ curves at *ca.* 5 and 10 kcal mol$^{-1}$.

### 1.2.4 Further approximations

#### *1.2.4.1 Loose and rigid coordinates*

The number of open channels $W(E, J, a)$ and the density of states $\rho(E, J)$, for triatomic systems can be obtained by direct counting without excessive use of computer time. For larger molecules, numbers of channels and densities of states increase so markedly that simplified counting procedures often have to be used. First of all one may try to distinguish between rigid coordinates — which correspond to vibrations in the reactant and in the fragments — and loose coordinates — which correspond to deformation vibrations and rotation in the reactant, and rotational and orbital motion of the fragments. The eigenvalues of the rigid coordinates often change relatively little along $q$, whereas marked changes occur for loose coordinates. One may try, therefore, to use an approximate partial density of open channels $\rho_{ri}(E - \varepsilon, J, a)$ for rigid coordinates, as given by an average between reactant and product numbers of states. Then the loose coordinates are treated separately, channel after channel as indicated above, giving a partial number of open channels $W(\varepsilon, J, a)$. Finally, the total $W(E, J, a)$ is obtained by convolution

$$W(E,J,a) = \int_0^E \rho_{ri}(E-\varepsilon,J,a)\, W(\varepsilon,J,\alpha)\, d\varepsilon \tag{1.18}$$

### *1.2.4.2 Densities of states*

$\rho(E, J)$ may be determined from a vibrational contribution $\rho_{vib}[E_{vib}(E, J, K)]$ and a summation over all quantum numbers $K$ (for a symmetrical top, degeneracy $g_K = 1$, for $K = 0$, $g_K = 2$ for $K \neq 0$, $K \leqslant J$)

$$\rho(E,J) = \sum_{K=0}^{J} g_K\, \rho_{vib}\,[E - BJ(J+1) - (A-B)K^2 - E_Z] \tag{1.19}$$

In semiclassical approximation, neglecting anharmonicity effects, the density of vibrational states $\rho_{vib}(E_{vib})$ is given by

$$\rho_{vib}(E_{vib}) = \frac{[E_{vib} + a(E_{vib})E_Z]^{s-1}}{(s-1)!\prod_i^s \varepsilon_{ni}} \tag{1.20}$$

with the quanta $\varepsilon_{ni}$, the zero point energy $E_Z = \frac{1}{2}\sum_i \varepsilon_{ni}$ and a suitable factor $0 < a(E_{vib}) < 1$ [22]. Internal rotations of the reactant molecule couple with external rotation; both motions cannot be separated and have to be treated similar to (1.19). In semiclassical approximation, the partial density of states $\rho_K(E_K)$, which corresponds to the quantum number $K$ [with $E_K = (A-B)K^2$], is given by

$$\rho_K(E_K) \approx \frac{1}{[(A-B)E_K]^{\frac{1}{2}}} \text{ with } E_K \leqslant (A-B)J^2 \tag{1.21}$$

The convolution of $\rho_K$ and $\rho_{vib}$, equations (1.20) and (1.21), leads to

$$\rho(E,J) \approx \int \rho_K(E_K)\, \rho_{vib}(E - BJ(J+1) - E_K - E_Z)\, dE_K$$

$$\approx \int_0^{\mathrm{Min}\{(A-B)J^2, E-BJ(J+1)-E_z\}} \frac{1}{[(A-B)E_K]^{\frac{1}{2}}} \frac{[E - BJ(J+1) - E_K - (1-a)E_Z]^{s-1}}{(s-1)!\prod_i^s \varepsilon_{ni}}\, dE_K \tag{1.22}$$

At small $J$, (1.19) and (1.22) give approximately

$$\rho(E,J) \approx (2J+1)\frac{[E-(1-a)E_Z]^{s-1}}{(s-1)!\prod_i^s \varepsilon_{ni}} \tag{1.23}$$

The increase of $\rho(E, J)$ with increasing $J$ is limited by the approximation of (1.22) for large $J$:

$$\rho(E,J) \approx \frac{[E-(1-a)E_Z - BJ(J+1)]^{s-1}}{(s-1)!\prod_i^s \varepsilon_{ni}} \frac{(\frac{1}{2})!(s-1)!}{(s+\frac{1}{2})!}\left[\frac{E-(1-a)E_Z - BJ(J+1)}{A-B}\right]^{\frac{1}{2}} \tag{1.24}$$

which is obtained by allowing $0 \leqslant E_K \leqslant E - BJ(J+1) - (1-a)E_Z$.

The described contribution of $K$ to $\rho(E, J)$ and $k(E, J)$ has been neglected in most applications. Semi-classical formulae like (1.20) and (1.22), of course, should be considered only as approximations. Exact counting should be preferred, whenever this is possible. Recent advances in calculational methods for $\rho(E)$ and its integral $W(E)$ have been described[23].

### *1.2.4.3 Number of channels and specific rate constants*

When the use of one activated complex for all open channels (at energy $E$ and total angular momentum $J$) appears possible, according to (1.16) the lower and upper limits $W_c'(E, J, q^\ddagger)$ and $W_c(E, J, q^\ddagger)$ can be used to obtain an estimate of the number of open channels $W(E, J)$. Without a channel correlation diagram in this case, a partitioning of $W(E, J)$ into the contributions of different product channels $a$ often is difficult. As a rough approximation, one may try to estimate the quanta $\varepsilon_i(q^\ddagger)$ of the different coordinates at the position $q^\ddagger$ of the activated complex, and count $W_c'$ and $W_c$. Applying again semiclassical approximations, this gives, similar to (1.22):

$$W_c(E,J,q^\ddagger) \approx \int_0^{\mathrm{Min}\{(A-B)^\ddagger J^2,\, E - V(q^\ddagger) - B^\ddagger J(J+1) - E_z^\ddagger\}} \frac{1}{[(A-B)^\ddagger E_K]^{\frac{1}{2}}} \frac{[E - V(q^\ddagger) - B^\ddagger J(J+1) - E_K - (1-a^\ddagger)E_Z^\ddagger]^{s-1}}{(s-1)! \prod_i^{s-1} \varepsilon_i(q^\ddagger)} \, \mathrm{d}E_K \tag{1.25}$$

For $W_c'(E, J, q^\ddagger)$, $V(q^\ddagger)$ is replaced by $V[q_c(J)]$ where $q_c(J)$ denotes the position of the centrifugal maximum. It is evident that direct counting instead of the semi-classical approximations is much more important for $W_c(E, J, q^\ddagger)$ than for $\rho(E, J)$, because of the smaller amount of energy to be distributed.

Two important limiting cases are 'rigid' or 'loose' activated complexes. Rigid complexes correspond to $q^\ddagger \to q_e$, loose complexes to $q^\ddagger \to \infty$. It is worthwhile to consider these extreme cases.

(a) *Rigid complexes* — Equation (1.25) has already been formulated for a reactant-like state pattern of the activated complex with the harmonic oscillator approximation for all disappearing deformation vibrations. It can immediately be used for rigid complexes with $q^\ddagger \to q_e$. Normally, the lower limit $W_c'(E, J, q^\ddagger)$ is used, rotation and vibration are separated, and some non-rigidity is allowed for. Neglecting the contribution from $K$, one obtains from (1.25) the conventional formula of RRKM or quasiequilibrium theory

$$k(E,J) \approx \frac{\prod_{i=1}^{s} \varepsilon_{ni}}{\prod_{i=1}^{s-1} \varepsilon_i(q^\ddagger)} \left( \frac{E - E_0(J) - (1-a^\ddagger)E_Z^\ddagger}{E - (1-a)E_Z} \right)^{s-1} \tag{1.26}$$

It should be emphasised that (1.26) involves many rough approximations and inconsistencies, and should be used as a first orientation only. Furthermore, (1.26) can hardly be used to obtain a distinction between different product channels $a$.

(b) *Loose complexes* — If a product-like state pattern of the activated complex is used, a distinction between the different product channels $a$ is much easier. However, the harmonic oscillator approximation of (1.25) cannot be employed any more. Instead, one has to consider the different rotational motions of the fragments. Depending on the type of the products (different tops, linear fragments and/or atoms), one obtains different state patterns. We shall illustrate this for a loose activated complex ($q^{\ddagger} \rightarrow \infty$) in the dissociation of a symmetric top into a linear fragment and an atom. The product channel $a$ is characterised by vibrational quantum numbers $v_p$ and a rotational quantum number $j$. $W(E, J, v_p, j)$ is given by the rotational degeneracy $g_{rot}(j)$. Because of the various constraints $\boldsymbol{j} + \boldsymbol{l} = \boldsymbol{J}$, $m_j + m_l = M_J$, $-j \leqslant m_j \leqslant j$, $-l \leqslant m_l \leqslant l$ and $-J \leqslant M_J \leqslant J$, $g_{rot}(j)$ is the number of orbital angular momenta $l$ which are in the range $|J - j| \leqslant l < J + j$. Therefore

$$g_{rot}(j) = \begin{cases} 2j + 1 & \text{for } 0 \leqslant j \leqslant J \\ 2J + 1 & \text{for } J \leqslant j \end{cases} \tag{1.27}$$

and

$$W_c(E,J,a) = g_{rot}(j) \tag{1.28}$$

If $W(E, J)$ is required, one has to sum (1.28) over all $a$, with the result

$$W_c(E,J) = \sum_j W_{vib}[E - E_0 - E_{rot}(j)] g_{rot}(j) \tag{1.29}$$

The sum over $j$ according to (1.27) consists of two parts with $0 \leqslant E_{rot}(j) \leqslant A_p J(J + 1)$ and $A_p J(J + 1) \leqslant E_{rot}(j) \leqslant E - E_0$. If $J$ is small and the first part is neglected, one obtains using a classical description for the rotation:

$$W_c(E,J) \approx (2J+1) \int_0^{E-E_0} W_{vib}\,[E - E_0 - E_{rot}(j)] \frac{\mathrm{d}\,E_{rot}(j)}{2[A_p\,E_{rot}(j)]^{\frac{1}{2}}} \tag{1.30}$$

By partial integration [with $W_{vib}(0) \approx 0$], again for small $J$, with (1.23) this gives

$$k(E,J) \approx [h\,\rho_{vib}(E)]^{-1} \int_0^{E-E_0} \rho_{vib}(x) \left(\frac{E - E_0 - x}{A_p}\right)^{\frac{1}{2}} \mathrm{d}(E - E_0 - x) \tag{1.31}$$

Similar formulae have been derived for several different types of fragments[24]. It should be emphasised that (1.31) is only valid for small $J$. It gives a threshold rate constant of 0, which is an artifact due to the classical description of rotation, and to the use of $W_{vib}(0) = 0$. These approximations can easily be avoided by the direct application of (1.28) and (1.29).

As for the non-rigidity introduced in (1.26), some non-looseness can be introduced in the derivation of (1.31). In general, this is done by considering the reaction from the fragment side, i.e. by combining (1.11) and (1.13) to give

$$k(E,J,a) = [h\,\rho(E,J)]^{-1} \frac{(2j_{a_1} + 1)(2j_{a_2} + 1)}{(2J+1)} \frac{\sigma_{ca}(E,J,a)}{\pi \lambda^2(a)} \tag{1.32}$$

Then, additional constraints for the orbital angular momenta $l$ contributing to the cross section $\sigma_{ca}(E, J, a)$ of formation of a 'strongly coupled complex' are derived. For instance, the Langevin cross section, based on the effective potential of orbital motion and the long-range ion–molecule potential, has been used for ion decompositions[24]. Similar methods have been employed in the phase space theory of bimolecular reactions[17]. The adiabatic channel calculations of Sections 1.2.2 and 1.2.3. indicate that for strongly coupled reactants generally the long-range potential is not sufficient. Furthermore, introduction of non-rigidity in rigid complexes or non-looseness in loose complexes, because of partial separation of vibration and rotation, always leads to some internal inconsistency. Finally, too liberal use of semi-classical approximations may easily lead to wrong conclusions.

### *1.2.4.4 Product distributions and thermally averaged rate constants*

The specification of individual channels $a$ and $b$ in the adiabatic channel model, (1.11)–(1.14), allows for a direct calculation of product distributions. In decomposition processes, a product distribution is simply given by

$$P(a) = \frac{k(E,J,a)}{\sum_a k(E,J,a)} = \frac{W(E,J,a)}{\sum_a W(E,J,a)} \tag{1.33}$$

In bimolecular processes, a product distribution for instance is given by

$$P(b,a,E) = \frac{\sigma_{ba}(E)}{\sum_b \sigma_{ba}(E)} \tag{1.34}$$

The same calculational methods for $W(E, J, a)$ as described before, and the same approximations for rigid or loose coordinates or activated complexes, can be used. Many examples have been given in the literature, e.g. product distributions in photolysis[25], in mass spectrometer fragmentations[24,26], in molecular beam scattering involving long-lived complexes[27], and in vibrational energy exchange involving sticky complexes[28,29]. In most cases rough approximations for the numbers of channels and for the potential surfaces have been used. Often conservation and constraints of angular momenta have not been taken into account properly. At the present state, some conclusions against the validity of the statistical theory should be taken with precautions.

Many details of specific rate constants or cross sections are lost during thermal averaging. In thermal unimolecular reactions at high pressures, $k(E, J)$ is averaged over an equilibrium population $P(E, J)$ of $E$ and $J$:

$$\begin{aligned} k_\infty &= \sum_J \int_{E_0}^{\infty} k(E,J)\,P(E,J)\,\mathrm{d}E \\ &= \sum_J g_J \int_{E_0}^{\infty} \frac{W(E,J)}{Q} \exp\left(-\frac{E}{kT}\right) \frac{\mathrm{d}E}{h} \end{aligned} \tag{1.35}$$

This may be represented in the form of transition state theory

$$k = \frac{kT}{h}\frac{Q^\ddagger}{Q} \exp\left(-\frac{E_0}{kT}\right) \tag{1.36}$$

with

$$Q^\ddagger = \sum_J g_J \int_{E_0}^{\infty} W(E,J) \exp\left(-\frac{E-E_0}{kT}\right) \frac{\mathrm{d}E}{kT} \tag{1.37}$$

The 'partition function of the activated complex' $Q^\ddagger$ in the statistical theory is determined by the contributions from all different channels, and can in general not be attributed to one particular configuration. If the reactant or the fragments have some symmetry, one has to study which states and channels really are possible for a specified nuclear spin function. If densities of states and numbers of channels are calculated without considering symmetry restrictions, and are denoted by the index $m$, on average one has $Q = Q_m/\sigma$ and $Q^{\ddagger} = Q^{\ddagger}{}_m/\sigma_1\sigma_2$, where $\sigma$ = symmetry number of the reactant and $\sigma_1$, $\sigma_2$ = symmetry numbers of the fragments (see Refs. 18, 30).

Cross sections of bimolecular reactions upon thermal averaging give rate constants of the form[31]

$$k_{ba}(T) = \frac{1}{kT}\left(\frac{8}{\pi\mu_a kT}\right)^{\frac{1}{2}} \int_0^{\infty} E_{\text{trans}}(a)\sigma_{ba}\left[E_{\text{trans}}(a)\right] \exp\left(-\frac{E_{\text{trans}}(a)}{kT}\right) \mathrm{d}E_{\text{trans}}(a) \tag{1.38}$$

and can easily be calculated from (1.14) with suitable $W(E, J, a)$. Rate constants for vibrational relaxation $O + NO(v = 1) \rightarrow O + NO(v = 0)$ or $O + O_2(v = 1) \rightarrow O + O_2(v = 0)$, and for isotope exchange $^{18}O + N^{16}O \rightarrow {}^{16}O + N^{18}O$ or $^{18}O + O_2 \rightarrow {}^{16}O + {}^{16}O^{18}O$ have been shown[18,32,33] to be consistent with rate constants for thermal dissociation and recombination of $NO_2$ and $O_3$ at high pressures.

## 1.3 THERMAL UNIMOLECULAR REACTIONS

### 1.3.1 Limiting low-pressure rate constants

In the following some unimolecular reactions, composed of different elementary steps including a unimolecular process, are described. Only a few aspects can be considered; for more detailed discussions the reader is referred elsewhere[2,33,35,38]. Thermal unimolecular reactions (or the reverse recombinations) are governed by the competition between collisional activations and deactivations, and unimolecular dissociation or isomerisation of activated molecules. In the low-pressure limit, only collisional processes are rate determining, and the specific rate constants $k(E, J)$ described in Section 1.2 do not enter into the overall rate constant. We shall give a brief analysis of this reaction range.

The collisional activation–deactivation sequence in the low-pressure limit is described by a master equation

$$\frac{\mathrm{d}n_i}{\mathrm{d}t} = -\sum_{j, E_j < \infty} k_{ji}[M]n_i + \sum_{j, E_j < E_0} k_{ij}[M]n_j \tag{1.39}$$

$k_{ji}$ denotes the rate constants for collisional energy transfer from state $i$ to state $j$, and $[M]$ is the concentration of collision partners. The number $n_j$ of molecules in state $j$, because of the comparably rapid unimolecular process at $E_j \geqslant E_0$, is negligible at $E_i \geqslant E_0$. Therefore, the second sum extends only up to $E_0$. For polyatomic molecules many different states contribute to the reaction. A complete specification of the set of $k_{ji}$ is rarely practicable. Therefore, average rate constants between groups of states are used, either with a small system of differential equations of the type of (1.39)[34], or with one smoothed integral equation instead of (1.39)[28,33,35].

The energy transfer rate constants $k_{ji}$ may be presented by a total energy transfer rate constant $Z_i$ and a collisional transition probability $P_{ji}$:

$$k_{ji} = P_{ji}\,Z_i = \left(\frac{k_{ji}}{\sum\limits_j k_{ji}}\right)\left(\sum_j k_{ji}\right) \tag{1.40}$$

The total energy transfer rate constant in most cases is probably very near to the gas kinetic collision rate constant. The collisional transition probabilities $P_{ji}$ of highly excited molecules are difficult to calculate. They are generally represented by parametrised model functions. Detailed balancing requires

$$P_{ji}\,Z_i = \left(\frac{n_j}{n_i}\right)_{\text{eq}} P_{ij}\,Z_j \tag{1.41}$$

(index eq = equilibrium). Analytical solutions of the master equation (1.39) are available for exponential cusp models of $P_{ji}$ [28]. Figure 1.3 shows the

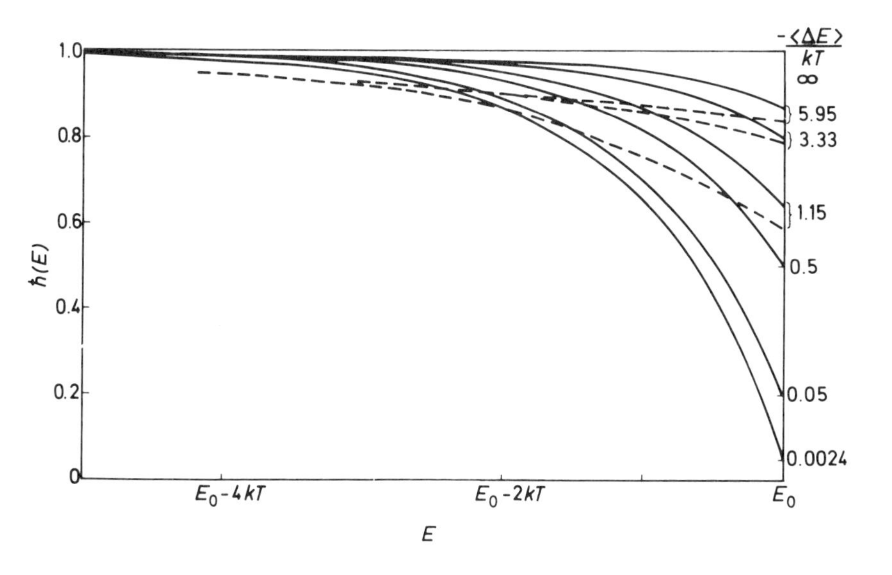


**Figure 1.3** Non-equilibrium populations in low-pressure thermal unimolecular reactions (——, $P_{ji}$ peaked at $E_j = E_i$, - - - -$P_{ji}$ peaked at $E_j < E_i$; for details see text and Ref. 28

non-equilibrium populations of molecular states below $E_0$ which have been obtained for the steady state of the reaction $h(E) = n(E)/[n(E)]_{\text{eq}}$. Two models of $P_{ji}$ have been used: one with $P_{ji}$ peaked at $E_j = E_i$ (full lines) and a parametrised statistical model with $P_{ji}$ peaked at $E_j < E_i$ (dotted lines). For decreasing amounts of the average energy transferred per collision $\langle \Delta E \rangle$, one finds increasing non-equilibrium effects.

The first-order low-pressure rate constant of the overall reaction, $k_0$, is given by

$$k_0 = [M]\,\beta_c Z \int_{E_0}^{\infty} f(E)\,\mathrm{d}E \tag{1.42}$$

with $f(E) = \rho(E)\exp(-E/kT)/Q$, an average total energy transfer rate constant $Z$ from (1.40) and a 'collision efficiency' $\beta_c < 1$; $\beta_c$ has been calculated for the mentioned exponential methods of transition probabilities $P_{ji}$[28]. Figure 1.4 shows the dependence of $\beta_c$ on $\langle \Delta E \rangle$, again for the two models described in Figure 1.3. The dependence on the particular shape of

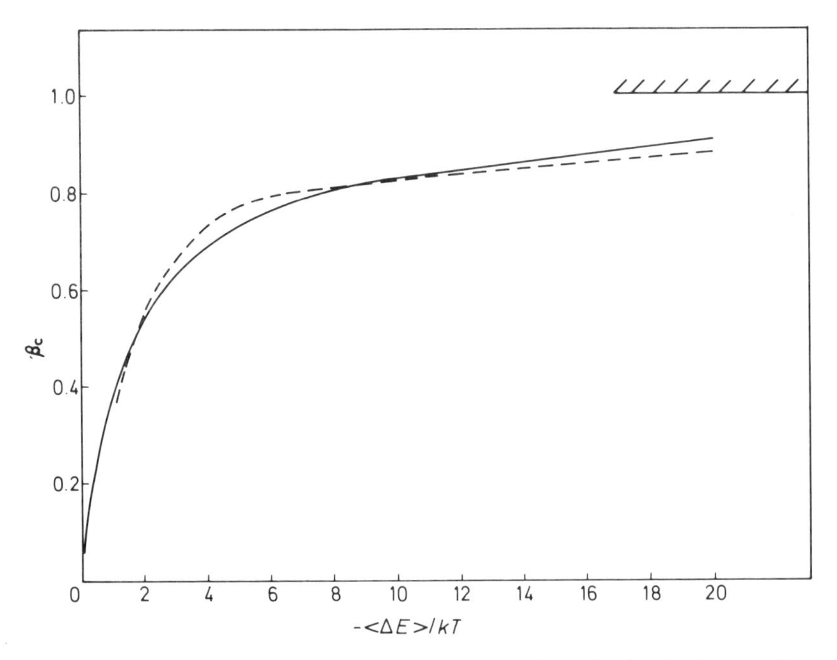


**Figure 1.4** Collision efficiencies $\beta_c$ in low pressure thermal unimolecular reactions (——, $P_{ji}$ peaked at $E_j = E_i$; - - -, $P_{ji}$ peaked at $E_j < E_i$; for details see text and Ref. 28

$P_{ji}$ is remarkably small; therefore $\beta_c$ offers a direct access to $\langle \Delta E \rangle$. For moderately complex reacting molecules

$$-\frac{\langle \Delta E \rangle}{kT} \approx \frac{\beta_c}{1 - \sqrt{\beta_c}} \tag{1.43}$$

Comprehensive studies of collision efficiencies $\beta_c$ are now available for several thermal unimolecular reactions, such as the recombinations $O + NO + M \rightarrow NO_2 + M$ and $O + NO_2 + M \rightarrow NO_3 + M$, the dissociation $NO_2Cl + M \rightarrow NO_2 + Cl + M$, and the isomerisations $CH_3NC + M \rightarrow CH_3CN + M$ and $C_2H_5NC + M \rightarrow C_2H_5CN + M$ (see analysis[28,33]). $\beta_c$ may become very small at high $T$; the old 'strong collision assumption' used in the RRKM theory, i.e. $\beta_c \approx 1$, appears invalid in many cases.

Further discussions of the strong collision part

$$k_0{}^{\text{s.c.}} = [M]Z\int_{E_0}^{\infty} f(E)\,\mathrm{d}E \tag{1.44}$$

of the rate constant $k_0$ are based on expressions of the densities of states $\rho(E)$ as described in Section 1.2.4.2. Anharmonicity effects have to be taken into account. The role of rotation has to be considered. One may only hope that averaging of $k_0{}^{\text{s.c.}}$ and $\beta_c$ over an equilibrium rotational distribution of activated molecules, with $J$-dependent threshold energies $E_0(J)$ (see Section 1.2), is adequate. In spite of these uncertainties, with our present knowledge on $\beta_c$ and $k_0{}^{\text{s.c.}}$, today a prediction of low-pressure rate constants within about a factor of 2 appears possible, whenever the molecular constants are known.

### 1.3.2 Fall-off curves

In the high-pressure limit of thermal unimolecular reactions an equilibrium population of states below and above $E_0$ is established by rapid collisional energy transfer. The unimolecular process is rate determining. The overall rate constant

$$k_\infty = \int_{E_0}^{\infty} k(E)\, f(E)\, \mathrm{d}E \tag{1.45}$$

has been considered in (1.35)–(1.37). In the intermediate fall-off range between the low and the high pressure limits, properties of collisional energy transfer ($k_{ji}$) and of the unimolecular process [$k(E, J)$] enter into the rate constant. Nevertheless, it can be shown that the information contained in $k_0$ and $k_\infty$ is sufficient to describe the intermediate fall-off range with adequate precision. For this purpose, fall-off curves should be represented in a ($k_0$,$k_\infty$)-reduced form. Computer calculated, non-strong collision RRKM curves have been described[36]. For practical application the use of reduced Kassel integrals leads to equivalent results[37]. ($k_0$,$k_\infty$)-reduced Kassel integrals have the form

$$\frac{k}{k_\infty} = \frac{1}{\Gamma(S_K)}\int_0^{\infty} \frac{x^{S_K-1}\exp(-x)}{1 + \dfrac{k_\infty}{k_0} I_0(S_K,B_K)\left(\dfrac{x}{x+B_K}\right)^{S_K-1}}\,\mathrm{d}x \tag{1.46}$$

with

$$I_0(S_K,B_K) = \sum_{v=0}^{S_K-1} \frac{B_K{}^v}{v!} \tag{1.47}$$

The two Kassel parameters $S_K$ and $B_K$ can be easily estimated by means of the relations

$$S_K \approx \frac{U_{\text{vib}} + E_{a\infty} - E_0}{kT} \tag{1.48}$$

and

$$F(S_K,B_K) \approx F(s,B') \tag{1.49}$$

$s$ is the number of oscillators of the reactant, $U_{\text{vib}}$ denotes the vibrational part of the internal energy of the reactant, $E_{a\infty}$ is the measurable activation energy in the high-pressure limit, $B' = (E_0 + a(E_0)E_Z/kT$ and the function $F(S,B)$ is given by

$$F(S,B) = \frac{\Gamma(S)}{B^{S-1}} \sum_{v=0}^{S-1} \frac{B^v}{v!} \tag{1.50}$$

The comparison of the $(k_0, k_\infty)$-reduced fall-off curves from RRKM non-strong collision calculations and from (1.46)–(1.50) showed very good agreement. Reduced fall-off curves can therefore easily be obtained from tabulated reduced Kassel integrals[37].

## 1.4 PHOTOCHEMICAL UNIMOLECULAR REACTIONS

### 1.4.1 Energy transfer studies

The mechanism of photochemical unimolecular reactions in general is much more complicated than that of thermal unimolecular reactions: light absorption as an activation process prepares non-statistical initial populations; dissociation or isomerisation often is coupled with non-radiative processes such as intersystem crossing, internal conversion or strong mixing of states, with collisional energy transfer and with fluorescence. Only few aspects can be mentioned in the following (for detailed discussions, see Ref. 39).

If the photochemical mechanism is sufficiently simple, one may investigate details of collisional energy transfer. Different methods have been employed: one can use measurements of fluorescence lifetimes and quantum yields, measurements of photolysis and photoisomerisation quantum yields, etc. Some representative studies are those on the pressure dependence of fluorescence in $I_2$ [40], $NO_2$ [41] or $\beta$-naphthylamine[29], and of the pressure dependence of the photoisomerisation of cycloheptatriene[42,43]. Fluorescence experiments allow for an absolute calibration of the energy transfer rate via the zero-pressure lifetimes; analysis of the fluorescence spectrum permits a determination of average energies transferred per collision. For diatomic molecules even the detailed transition probabilities can be analysed[40]. Measurements of photolysis or isomerisation quantum yields like those with chemical activation provide only data relative to the estimated rate constants $k(E, J)$ of the unimolecular process. In all cases, master equations for energy transfer, radiative and/or unimolecular processes similar to (1.39) have to be solved. Remarkably enough, the shape of collisional transition probabilities $P_{ji}$ for $I_2$ has been found to depend nearly exponentially on the difference between initial and final energies $E_i - E_j$ [40]. Also the average energies $\langle \Delta E \rangle$ transferred per collision have been found to be very similar for similar colliders after thermal, chemical or optical activation[43].

### 1.4.2 Specific rate constants

Photochemical unimolecular reactions have the advantage that the initial excitation energy can be varied over fairly large ranges. Single line and broad band excitations are both possible. Therefore, one may try to measure directly specific rate constants of unimolecular processes and their energy dependences. Again, several methods are available; either fluorescence can be used to monitor populations of excited states, or pressure dependences of photolysis quantum yields can be interpreted in terms of lifetimes and of (hopefully known) collisional deactivation rates. Fluorescence lifetime measurements up to now have mainly been performed in the $1$–$10^4$ ns range; an extension down to the ps range has been achieved recently with new laser techniques. Pressure dependences of photolysis quantum yields at high pressures permit indirect measurements down to the $10^{-12}$–$10^{-13}$ s range, although some doubt about the quantitative calibration of the collisional reference rate persists.

Again, only a few studies can be mentioned. Several sets of data on specific rate constants of intersystem crossing or other non-radiative processes are available, such as for $\beta$-naphthylamine[44], aniline[45], hexafluoroacetone[46], formaldehyde[47], chloro- and bromo-acetylene[48], etc. Specific rate constants have generally been found to increase with increasing excitation energy; in some cases marked fluctuations with strong dependence on the excitation

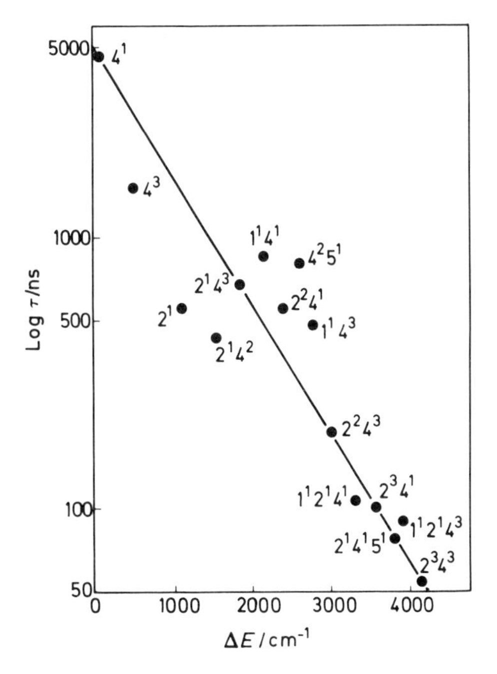


**Figure 1.5** Single vibronic level lifetimes for $D_2CO$ at zero pressure ($\Delta E$ = vibrational energy relative to the origin of the $S_1$ electronic state, the numbers characterise different bands; see Ref. 47

level could be resolved. Figure 1.5 shows single vibronic level lifetimes of $D_2CO$ at zero pressure. Whereas the lifetime of the $4^1$ level is approximately equal to the radiative lifetime, the shorter lifetimes correspond to radiationless decays. The attribution of such lifetimes to an elementary process often is doubtful. Statistical estimates with (1.12) and (1.23) could suggest that the lifetimes observed with $D_2CO$ correspond to a direct dissociation from the electronic ground state, even though $S_1$–$S_0$ coupling can also be the rate determining process. Figure 1.6 illustrates another example

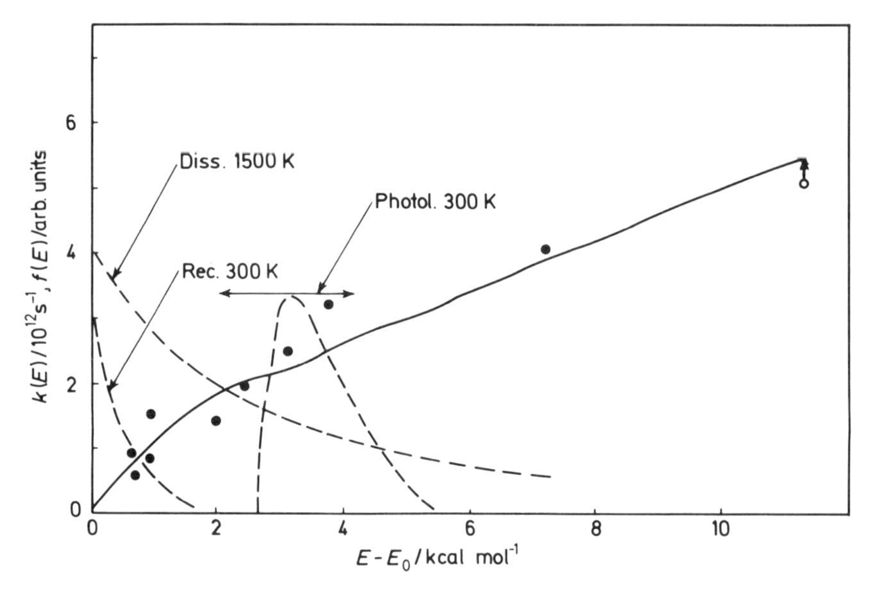


**Figure 1.6** *J*-averaged specific rate constants of $NO_2$ decomposition (●, experimental data from high-pressure photolysis[49]; ○, from zero-pressure photofragment spectroscopy; ——, minimum density of states calculation[21]; - - -, populations in thermal recombination and dissociation at high pressures and photolysis; see Refs. 49 and 50

which is related to Figures 1.1 and 1.2. $NO_2$ is known to show a strong coupling between electronic ground state and excited states. Photolysis at $\lambda \leqslant$ 410 nm is pressure dependent at inert gas pressures larger than 1 atm [49]. This can be interpreted in terms of collisional stabilisation and photolysis lifetimes in the range $10^{-11}$–$10^{-13}$ s. Approximative calibration of specific rate constants $k(E, J)$ is possible with information on collisional energy transfer from fluorescence studies[41]. The specific rate constants obtained agree within the experimental accuracy with the statistical calculations for the electronic ground state as given in Figures 1.1 and 1.2. In this case it is possible that thermal dissociation and recombination at high pressures, and high pressure photolysis, all involve essentially the same excited states[49], only with the different populations shown in Figure 1.6. Nevertheless, optical single band excitation in collision-free cases may give different behaviour. New information on specific rate constants of unimolecular processes is certainly to be expected from optical excitation studies over extended time ranges. The general statistical theory of unimolecular processes can serve at least as a first guideline for an interpretation.

## Acknowledgement

Many helpful discussions with Dr. M. Quack and generous financial support from the Schweizerischer Nationalfonds are gratefully acknowledged.

## References

1. George, T. F. and Ross, J. (1973). *Ann. Rev. Phys. Chem.*, **24**, 263
2. Setser, D. W. (1972). *MTP International Review of Science, Physical Chemistry Series One*, Vol. 9, *Chemical Kinetics*, 1 (J. C. Polanyi, editor) (London: Butterworths)
3. Blatt, J. and Weisskopf, V. (1952). *Theoretical Nuclear Physics*, Chap. VIII (New York: Wiley)
4. Hauser, W. and Feshbach, H. (1952). *Phys. Rev.*, **87**, 366
5. Mies, F. H. and Krauss, M. (1966). *J. Chem. Phys.*, **45**, 4455
6. Mies, F. H. (1969). *J. Chem. Phys.*, **51**, 787, 798
7. Van Santen, R. A. (1972). *J. Chem. Phys.*, **57**, 5418
8. Levine, R. D. (1969). *Quantum Mechanics of Molecular Rate Processes* (Oxford: Clarendon)
9. Smith, F. T. (1969). *Kinetic Processes in Gases and Plasma*, 321 (A. R. Hochstim, editor) (New York: Academic Press)
10. Slater, N. B. (1959). *Theory of Unimolecular Reactions* (London: Methuen)
11. Hofacker, G. L. (1963). *Z. Naturforsch.*, **18a**, 607; Marcus, R. A. (1964). *J. Chem. Phys.*, **41**, 2614; (1965). *J. Chem. Phys.*, **43**, 1598
12. Fischer, S. F., Hofacker, G. L. and Seiler, R. (1969). *J. Chem. Phys.*, **51**, 3951
13. Landau, L. D. and Lifshitz, E. M. (1965). *Quantum Mechanics*, 2nd ed. (Oxford: Pergamon); Nikitin, E. E. (1968). *Chemische Elementarprozesse*, 43 (H. Hartmann, editor) (Berlin: Springer)
14. Miller, W. H. and George, T. F. (1972). *J. Chem. Phys.*, **56**, 5637
15. Rosenstock, H. M., Wallenstein, M. B., Wahrhaftig, A. L. and Eyring, H. (1952). *Proc. Nat. Acad. Sci. USA*, **38**, 662
16. Marcus, R. A. and Rice, O. K. (1951). *J. Phys. Colloid Chem.*, **55**, 894
17. Light, J. C. (1967). *Discuss. Faraday Soc.*, **44**, 14; Keck, J. C. (1958). *J. Chem. Phys.*, **29**, 410; Nikitin, E. E. (1965). *Theor. Exp. Chem. USSR*, **1**, 144
18. Quack, M. and Troe, J. (1974). *Ber. Bunsenges. Phys. Chem.*, **78**, 240; (1975). *ibid.*, **79**, 170, 469
19. Johnston, H. S. (1966). *Gas Phase Reaction Rate Theory* (New York: Ronald)
20. Bunker, D. L. and Pattengill, M. (1968). *J. Chem. Phys.*, **48**, 772
21. Gaedtke, H. and Troe, J. (1973). *Ber. Bunsenges. Phys. Chem.*, **77**, 24
22. Forst, W. (1971). *Chem. Rev.*, **71**, 339
23. Stein, S. E. and Rabinovitch, B. S. (1973). *J. Chem. Phys.*, **58**, 2438; Hoare, M. R. and Pal, P. (1971). *Mol. Phys.*, **20**, 695
24. Klots, C. E. (1971). *J. Phys. Chem.*, **75**, 1526; (1972). *Z. Naturforsch.*, **27a**, 553
25. Campbell, R. J. and Schlag, E. W. (1967). *J. Amer. Chem. Soc.*, **89**, 5103
26. Le Roy, R. L. (1971). *J. Chem. Phys.*, **55**, 1476; Vestal, M. L. (1965). *J. Chem. Phys.*, **43**, 1356
27. Shobatake, K., Parson, J. M., Lee, Y. T. and Rice, S. A. (1973). *J. Chem. Phys.*, **59**, 1402, 1416, 1435, 2483, 6104; Miller, W. B., Safron, S.A. and Herschbach, D. R. (1967). *Discuss. Faraday Soc.*, **44**, 108; Safron, S. A., Weinstein, N. D., Herschbach, D. R. and Tully, J. C. (1972). *Chem. Phys. Lett.*, **12**, 564
28. Troe, J. (1973). *Ber. Bunsenges. Phys. Chem.*, **77**, 665
29. Von Weyssenhoff, H. and Schlag, E. W. (1973). *J. Chem. Phys.*, **59**, 729
30. Schlag, E. W. and Haller, G. L. (1965). *J. Chem. Phys.*, **42**, 584
31. Light, J. C., Ross, J. and Shuler, K. E. (1969). *Kinetic Processes in Gases and Plasma*, 281 (A. R. Hochstim, editor) (New York: Academic Press)
32. Hippler, H. and Troe, J. (1971). *Ber. Bunsenges. Phys. Chem.*, **75**, 27
33. Troe, J. (1974). *Fifteenth Symp. on Combustion*, 667 (Pittsburgh: Combustion Institute)
34. Tardy, D. C. and Rabinovitch, B. S. (1968). *J. Chem. Phys.*, **48**, 5194
35. Troe, J. and Wagner, H. Gg. (1967). *Ber. Bunsenges. Phys. Chem.*, **71**, 937

36. Tardy, D. C. and Rabinovitch, B. S. (1968). *J. Chem. Phys.*, **48,** 1282
37. Troe, J. (1974). *Ber. Bunsenges. Phys. Chem.*, **78,** 478
38. Benson, S. W. and O'Neal, H. E. (1970). *Kinetic Data on Gas Phase Unimolecular Reactions* (NSRDS-NBS 21) (Washington: National Bureau of Standards); Spicer, L. D. and Rabinovitch, B. S. (1970). *Ann. Rev. Phys. Chem.*, **21,** 349; Troe, J. and Wagner, H. Gg. (1972). *Ann. Rev. Phys. Chem.*, **23,** 311; Robinson, P. J. and Holbrook, K. A. (1972). *Unimolecular Reactions* (New York: Wiley); Troe, J. (1975) In *Physical Chemistry, An Advanced Treatise*, Vol. VIb, *Gas Kinetics* (W. Jost, editor) (New York: Academic Press); Forst, W. (1973). *Unimolecular Reactions* (New York: Academic Press)
39. Rice, S. A. (1976). To be published; Jortner, J., Rice, S. A. and Hochstrasser, R. M. (1969). *Adv. Photochem.*, **7,** 149
40. Kurzel, R. B. and Steinfeld, J. I. (1970). *J. Chem. Phys.*, **53,** 3293
41. Keyser, L. F., Levine, S. Z. and Kaufman, F. (1971). *J. Chem. Phys.*, **54,** 355
42. Atkinson, R. and Thrush, B. A. (1970). *Proc. Roy. Soc. (London)*, **A316,** 123, 131, 143; Orchard, S. W. and Thrush, B. A. (1972). *Proc. Roy. Soc. (London)*, **A329,** 233
43. Luu, S. H. and Troe, J. (1973). *Ber. Bunsenges. Phys. Chem.*, **77,** 325; (1974). *Ber. Bunsenges. Phys. Chem.*, **78,** 766
44. Von Weyssenhoff, H. and Schlag, E. W. (1969). *J. Chem. Phys.*, **51,** 2508
45. Von Weyssenhoff, H. and Kraus, F. (1971). *J. Chem. Phys.*, **54,** 2387
46. Halpern, A. M. and Ware, W. R. (1970). *J. Chem. Phys.*, **53,** 1969
47. Yeung, E. S. and Moore, C. B. (1973). *J. Chem. Phys.*, **58,** 3988
48. Evans, K., Schelps, R., Rice, S. A. and Heller, D. (1973). *J. Chem. Soc. Faraday Trans. II*, **69,** 856
49. Gaedtke, H., Hippler, H. and Troe, J. (1972). *Chem. Phys. Lett.*, **16,** 174
50. Gaedtke, H. and Troe, J. (1975). *Ber. Bunsenges. Phys. Chem.*, **79,** 184

# 2
# Ion–Molecule Collision Phenomena

**B. H. MAHAN**
University of California, Berkeley

---



---

## 2.1 INTRODUCTION

This review deals with recent developments in the study of elastic, inelastic and reactive ion–molecule scattering phenomena. The topics have been chosen on the basis that they contribute directly to our general understanding of the detailed dynamics of chemical reactions. The presence of sections devoted to elastic and inelastic scattering is prompted by the fact that these experiments are the sources of information on the intermolecular potentials, non-adiabatic phenomena and molecular mechanics which can help to elucidate the more complicated process of chemical reaction. No attempt has been made to catalogue the literature comprehensively, since a number of excellent reviews of both the broad and specific aspects of ion–molecule reaction phenomena have recently appeared[1–11].

## 2.2 RECENT ACCOMPLISHMENTS

### 2.2.1 Elastic scattering

Investigations of the elastic scattering of ions by atoms and simple molecules have been reviewed recently by Weise[12] in an article which contains a concise and edifying summary of the theory of molecular elastic scattering. At the risk of redundancy we present the following summary of accomplishments in this area in order to emphasise the relation of the work to problems of reactive scattering, and to similar experiments performed on uncharged systems.

#### *2.2.1.1 The proton–noble gas systems*

The ions $HA^+$, where A is a noble gas atom, have long been known as products of ion–molecule reactions. However, only relatively recently have there been quantitative determinations of the proton affinities of the noble gas atoms, either from *ab initio* quantum mechanical calculations, reaction threshold energies, or the temperature dependence of equilibria. The relatively large de Broglie wavelength of the proton even at energies of several electron volts, and the substantial potential energy well depth expected for all the proton–noble gas systems, have made them particularly appealing subjects for elastic scattering studies.

The groups of Henglein[13–15] and of Doverspike and Champion[16,17] have determined the angular distribution of protons scattered from noble gas atoms at a number of different initial relative energies. The distributions display considerable oscillatory structure. The relatively low-frequency rainbow and supernumerary rainbow oscillations resulting from interferences between trajectories which have turning points close to, but on either side of, the potential energy minimum appear clearly, and permit accurate determination of the potential energy well depth. In addition, higher-frequency oscillations of the differential cross-section resulting from the interference of trajectories which turn, respectively, in the region of the attractive tail and the repulsive wall of the potential have been resolved for all proton–noble gas pairs. Potential energy curves have been determined either by adjusting the parameters of analytical potentials so that the calculated scattering fitted the experimental data[14], or by inversion procedures[16–18] which do not involve an assumed form for the potential. The locations and depths of the potential energy wells found by the groups of Henglein and Doverspike and

**Table 2.1 Potential energy well-depths and locations**

| *System* | $\varepsilon$/eV | | $r_m$/nm | |
|---|---|---|---|---|
| $H^+$–He | 2.00* | 2.18† | 0.077* | 0.075† |
| $H^+$–Ne | 2.28 | 2.27 | 0.099 | 0.100 |
| $H^+$–Ar | 4.04 | 4.22 | 0.131 | 0.124 |
| $H^+$–Kr | 4.45 | 4.6 | 0.147 | 0.153 |
| $H^+$–Xe | 6.75 | — | 0.174 | — |

*Ref. 14
†Ref. 17

Champion are compared in Table 2.1. The agreement seems highly satisfying. In addition, the full potential energy curves determined for $H^+$–He, $H^+$–Ne and $H^+$–Ar are very close to those determined by *ab initio* calculations.

The values of the noble gas proton affinities derived from scattering studies establish the overall energetics of the much studied reactions

$$A^+ + H_2 \rightarrow AH^+ + H \qquad (A = Ar, Kr, Xe)$$
$$A + H_2^+ \rightarrow AH^+ + H \qquad (A = He, Ne, Ar, Kr, Xe)$$

and many others in which protonated rare gas atoms are formed. In addition, the determination of the full potential curves has provided a valuable set of tests of the validity of the much used ion–induced dipole expression

$$V(r) = -\frac{\alpha e^2}{2r^4} \qquad (2.1)$$

where $\alpha$ is the polarisability and $e$ is the fundamental charge. For protons interacting with Ne, Ar and Kr, deviations of equation (2.1) from the experimental potentials become substantial (>50%) at distances smaller than three times the equilibrium separation, with the experimental potential lying below the value given by equation (2.1). The failure of the ion–induced dipole expression in these cases should not be surprising, since it is derived using the point dipole approximation in the perturbation theory limit. Both these approximations are of doubtful validity at distances which approach the equilibrium internuclear separation, and energies greater than 0.1 eV. For the $H^+$–He system the approximations are better justified, and in this extraordinary case equation (2.1) is within 0.15 eV of the experimental curve for distances greater than 10 nm, and energies less than 1 eV. In general, however, it appears[19] that the ion–induced dipole potential is very likely to be in serious error for chemically interacting systems when the potential energy is greater than 0.1 eV.

#### 2.2.1.2 *The* $H_2^+$*–noble gas systems*

Scattering of $H_2^+$ from Ar and Kr has also been investigated[15]. Because of the non-spherical nature of the potential in these systems, the amplitude of the rainbow oscillations is considerably diminished compared with the data from the atom–atom cases. Nevertheless, approximate values for potential energy well depths have been deduced. The exact significance of these numbers is somewhat obscure, since the intermolecular potential is certainly a function of the $H_2^+$ internuclear separation and orientation. It is not clear how closely the well depth derived from the scattering experiments approaches the true minimum, which would correspond to relaxation of all internal coordinates to their equilibrium values. For the closely related case of $H^+$ scattered by $H_2$, a well depth of 4.04 eV is deduced, which is 0.51 eV less than the value of 4.56 eV obtained from *ab initio* calculations. Thus the well depths deduced from scattering of $H_2^+$ from Ar and Kr almost certainly provide an upper limit to the true minimum energy of $ArH_2^+$ and $KrH_2^+$. The results suggest[9,12,15] that for the reactions $Ar^+(H_2,H)ArH^+$ and $H_2^+(Ar,H)ArH^+$ there may be a shallow energy basin ($\sim$0.1 eV) in the exit channel, but it is not deep enough to have an appreciable effect on the reaction dynamics. The well depth deduced from $H_2^+$–Kr scattering suggests that for $Kr^+(^2P_{3/2})$ reacting with $H_2$ there may be a small activation energy ($\sim$0.33 eV). These conclusions are qualitatively consistent with diatomics-in-molecules calculations for these systems by Kuntz and Roach[20].

#### 2.2.1.3 *The alkali ion–noble gas systems*

There have now been a number of investigations[21–27] of the energy dependence of the total elastic scattering cross-section of alkali ion–noble gas atom pairs. These experiments are performed at an angular resolution low enough and at relative energies high enough so that the apparent cross-section is determined by repulsive interaction potentials in the range of a few tenths to a few electron volts. It is this range that is very important in inelastic and reactive scattering phenomena at moderate to high energies, so the activity in high-energy elastic scattering is most welcome. Potentials have been determined in four different laboratories, and while there is some measure of agreement, certain consistent discrepancies remain.

### 2.2.2 Inelastic scattering

The ease with which ion beams of comparatively high kinetic energy can be produced makes them attractive vehicles for the study of inelastic scattering. Recently, Doering[28] has reviewed the progress in this field, with particular emphasis on phenomena which take place at ion energies (in the laboratory frame) in the 0.1–1 keV range. Here, we again choose to summarise developments that are closely related to chemical reaction phenomena.

### *2.2.2.1 Electronic inelasticity*

The study of the electronic transitions in molecules induced by collisions with simple ions has led to the development of the field of ion impact spectroscopy[29–33]. The value of this type of experiment is that the electronic properties of the impacting ion can influence which transitions are induced in the target molecule. For example[29], 500 eV protons incident on $N_2$ show at zero scattering angle a peak at 9.3 eV in the energy-loss spectrum, which corresponds to the X $^1\Sigma^+_g \rightarrow$ a $^1\Pi_g$ transition of $N_2$. The same type of transition is observed when the isoelectronic molecules CO and $C_2H_2$ are used as targets. However, when the experiments are executed using $H_2^+$ as the projectile ion, additional features appear which for $N_2$ correspond to the X $^1\Sigma^+_g \rightarrow$ B $^3\Pi_g$ transition, for CO to the X $^1\Sigma^+ \rightarrow$ a $^3\Pi$ transition, and for $C_2H_2$ correspond to excitation of two low-lying triplet states. Thus, in agreement with the spin conservation rule, these singlet–triplet excitations have relatively large transition probabilities when the impacting ion is in a doublet state, and have small probabilities when the ion is an electronic singlet.

The energy loss spectra of $H^+$ and $H_2^+$ incident on ethylene[29] provide a particularly spectacular example of the importance of the electron spin of the projectile ion. A very intense singlet–triplet transition at 4.5 eV is induced by $H_2^+$ collisions, but is completely absent from the proton scattering spectrum. Other examples of this type have been given by Moore[30,32] and Doering and Moore[31].

Besides its potential for providing valuable supplementary information on the spin multiplicity of electronic states, this type of experiment can be used to test the validity of the electron spin conservation rule. In collisions of $N^+(^1D)$ with He, Ne, Ar and $N_2$, Moore[33] found that transitions of $N^+$ in which spin was not conserved were roughly 0.002 times as probable as comparable spin conserving transitions. In the $N^+$–$O_2$ system, however, spin conserving transitions were favoured by only one order of magnitude.

There have been very few studies of electronic excitation produced by collisions in which the relative energy was in the range of a few electron volts. At Berkeley we have observed[107] the $^3P \rightarrow {}^3D$ and $^3P \rightarrow {}^5S$ transitions of $N^+$ induced by collision with He. In the former case the excitation energy is 11.4 eV, and yet the transition appears to be very probable even when the relative energy of collision is as low as 13 eV. Such highly probable transitions must involve the near intersection of two or more adiabatic potential energy curves. The experimental study of these phenomena combined with analysis in terms of accurate *ab initio* calculations of potential energy curves and transition probabilities may provide a valuable background for the understanding of the effects of potential surface intersections in reactive polyatomic systems. In a limited sense, such investigations have begun. Miller, Schaefer and co-workers[34] have calculated potential curves for all valence states of the $HeO^+$ system, and evaluated the collision cross section for the spin forbidden $^4S \rightarrow {}^2D$ transition of $O^+$ to be approximately $8.6 \times 10^{-5}$ nm$^2$ at 6 eV relative energy. Gillen, Mahan and Winn[35] found no evidence for this transition in experimental studies of the scattering of $O^+$ by He, which is consistent with the small value of the calculated cross section.

### 2.2.2.2 *Vibrational inelasticity*

Ion beam scattering techniques have been notably effective in elucidating the nature of vibrationally inelastic molecular collisions. In the earliest published experiment of this type, Gentry *et al.*[36] measured the complete angular distribution of the vibrational–rotational inelasticity of collisions of $N_2^+$ with He. Subsequent experiments from the same laboratory[37] gave complete velocity vector distributions for the inelastic scattering of $O_2^+$ and $NO^+$ by He at a number of initial energies in the range 0.4–24 eV. No individual vibrational transitions were resolved in these experiments, but the data showed clearly that in these systems the inelasticity increases as the barycentric scattering angle increases. Also, the magnitude of the inelasticity measured at a scattering angle of 180° increases as the relative collision energy increases; in the higher-energy range, the ratio of the most probable inelasticity to the initial relative energy $\Delta E/E_r$ is roughly constant. This behaviour is qualitatively consistent with an impulse model which pictures the collision as a hard-sphere interaction between the helium atom and one of the atoms of the diatomic, although such a fully impulsive model overestimates the inelasticity at all scattering angles. Interpretation of the experiments led to the refined impulse approximation[38] as the correct classical expression for the vibrational inelasticity in collinear atom–diatomic collisions in the near impulse limit.

Moore and Doering[39] were the first to resolve the excitation of discrete vibrational levels in their experiments on small-angle, high-energy scattering of $H^+$ and $H_2^+$ from $H_2$, $D_2$ and $N_2$. A more extensive investigation by Herrero and Doering[40] of the $H^+$–$H_2$ system has since appeared, and these authors have also reported[41] the observation of superelastic collisions of vibrationally excited $H_2^+$ with several targets.

Udseth, Giese and Gentry[42,127] have carried out an extensive investigation of the inelastic scattering of $H^+$ by $H_2$, HD and $D_2$ in the relative energy range 4–16 eV, and at scattering angles smaller than the rainbow angle. The results are of particular interest, since the $H^+$–$H_2$ system has a potential energy well 4.56 eV deep in the equilateral triangle conformation. Scattering at angles smaller than the rainbow angle comes principally from trajectories that explore the outer attractive wall of the potential energy well, although trajectories which reach the low-energy portions of the repulsive wall also contribute. Thus the mechanism of the vibrational excitation of $H_2$ is rather different from what must operate in systems[37] like $O_2^+$–He and $NO^+$–He, where the potential energy well depths are much smaller, and the strong repulsive forces which operate in small impact parameter collisions produce the vibrational excitation. In the $H^+$–$H_2$ system a grazing collision must induce a force which tends to expand the $H_2$ internuclear separation, since the equilibrium distance in $H_3^+$ is greater than in $H_2$.

Figure 2.1. shows some of the results of the investigation of Udseth *et al.*[127]. For $H^+$–$H_2$ scattering the vibrational transition probability increases with increasing scattering angle, and decreases as the energy of the transition increases. When the transition probability to each of the first three excited levels of $H_2$, HD and $D_2$ is plotted as a function of the initial collision energy divided by the magnitude of the vibrational quantum, the data for all

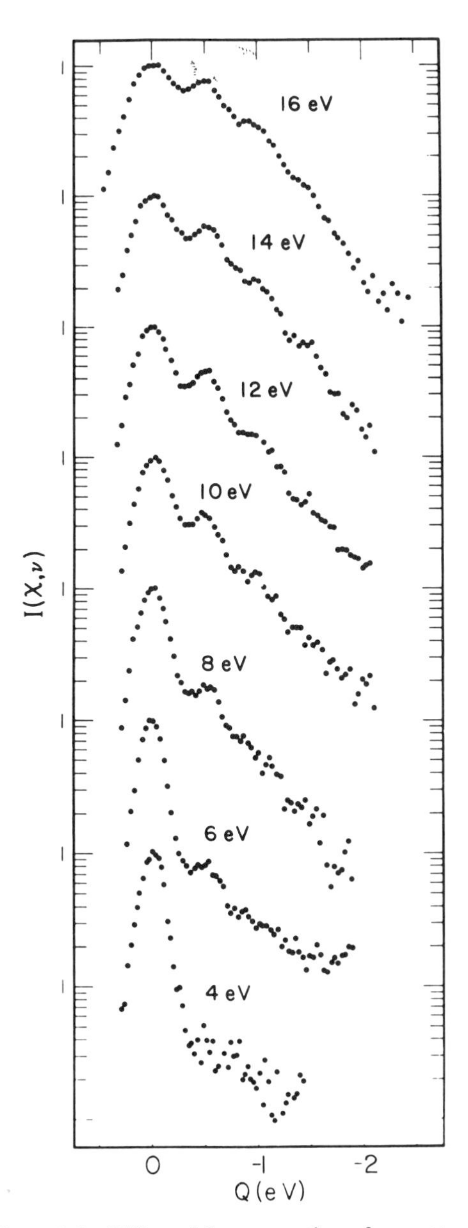


**Figure 2.1** Differential cross-sections for scattering of $H^+$ from $H_2$ as a function of the translational exoergicity $Q$. The centre-of-mass scattering angle is 11° and the primary kinetic energies range from 4 to 16 eV. Each curve is normalised to unity at the elastic ($Q = 0$) peak. Note the development of peaks corresponding to excitations to the $v = 1, 2$ and 3 vibrational levels of $H_2$ as the initial relative energy increases (From Udseth *et al.*[127], by courtesy of the American Institute of Physics)

isotopic targets fall on common curves, each characteristic of a given change in vibrational quantum number. Thus, the entire isotope effect is expressible in terms of the size of the vibrational quantum.

Other recent experiments in which individual vibrational excitations have been resolved have been reported by Cosby and Moran[43], and Petty and Moran[44]. In the latter work, transitions to several of the lower vibrational levels of $CO^+$ induced by collisions with Ar were studied as a function of scattering angle and initial translational energy. At small scattering angles and low energies, transitions to the lowest vibrational levels predominate, but at barycentric angles of roughly 15° or greater, and for relative energies in excess of approximately 8 eV, transitions in which the vibrational quantum number changes by more than unity become the most probable.

These experiments were analysed in terms of the version of the forced oscillator worked out by Shin[45]. In this model, a static orientation and internuclear separation are assumed for the diatomic molecule. However, for the mass combination in which the free atom is heavier than the atoms in the diatomic, it is very unlikely that a static orientation will be maintained in an appreciable fraction of the collisions (see Section 2.2.2.3) and thus the model, whatever its virtues, seems inappropriate for the data. These very elegant experiments deserve a much more detailed theoretical treatment, although the problem of an appropriate potential energy surface is formidable.

The alkali metal ions in collision with $H_2$, $D_2$ and HD are particularly attractive systems for the study of collisional inelasticity. Both collision partners have closed electronic shells, and the possible electronic excitations and chemical reactions which might introduce complications are quite endoergic. Experiments on the large-angle scattering of $K^+$ by $H_2$ and $D_2$ were reported by Dittner and Datz[46] in 1968, and a more complete description of these experiments, together with some results on the scattering of $Na^+$ from $H_2$ and $D_2$, were published in a second paper[47]. Van Dop, Boerboom and Los[48] also have studied the scattering of $K^+$ from $H_2$ and $D_2$, with particular emphasis on high relative energies. Toennies and co-workers[49–51] have made a series of investigations of the large-angle scattering of $Li^+$ by $H_2$ and, in the most recent experiments, have resolved the individual transitions of $H_2$ to its first three excited vibrational levels. These results are of particular interest and importance, since an extensive *ab initio* (SCF) potential energy surface for the $Li^+$–$H_2$ system has been determined by Lester[52], and comparisons of the scattering calculated using this surface with the experimental results of Toennies and co-workers are now being made.

The large-angle scattering of $Na^+$ by $D_2$ has recently been investigated by Dimpfl and Mahan[108] by Schöttler[53]. The resolution in these experiments was insufficient to detect individual vibrational transitions, so the most probable inelasticity and the shape of the envelope of the scattered ion energy spectrum were determined as a function of initial relative energy. Dimpfl and Mahan reached the conclusion that the collisions mainly responsible for the most probable inelasticity at barycentric angles near 180° are those with very small impact parameters, and with the axis of the diatomic oriented perpendicularly (or nearly so) to the initial relative velocity vector. Classical trajectory calculations were performed using a potential composed of exponential repulsions between $Na^+$ and the individual deuterium atoms, combined with a Morse potential for the oscillator. The parameters of the repulsive terms were varied until agreement with experimental inelasticities was obtained. The trajectories showed that exactly collinear collisions lead to very small inelasticities, while conformations slightly off collinearity give large inelasticities. However, these collisions occur rather rarely, since the probability of finding a diatomic molecule oriented with its axis at an angle $\alpha$ with respect to the relative velocity vector is proportional to $\sin \alpha$. The nearly perpendicular type of collision is somewhat less inelastic, but occurs much more frequently. Such collisions are therefore responsible for the most probable feature of the inelastic scattering spectrum. Similar conclusions have been reached by Faubel and Toennies[54] as a result of their analysis of the data of Schöttler[53] for $Na^+$–$D_2$ scattering.

Dimpfl and Mahan[52] deduced an intermolecular potential for the $Na^+$–$D_2$ system by fitting the results of classical trajectory calculations to the measured inelasticities. In the range of validity ($5 < V < 12$ eV, $0.9 > r > 0.14$ nm) the $Na^+$–$D_2$ potential falls between the potential curves for $Li^+$–$H_2$ and $K^+$–$D_2$ determined from elastic scattering by Inouye and Kita[55]. This, and a similar analysis by Faubel and Toennies[54], seems to be the first instance in which vibrational inelastic scattering has been used to deduce an intermolecular potential.

### *2.2.2.3 Impulse models for vibrational inelasticity*

In analysing vibrationally inelastic scattering, we have often found it valuable to use a very primitive model in which a hard sphere atom A makes a collinear collision with the atom B of a diatomic molecule BC which is bound by a square well potential[56]. If one uses the coordinates[57]

$$x = r_{AB} + \gamma r_{BC}, \quad y = r_{BC}/a$$

with

$$\gamma = C/(B + C), \quad a^2 = A(B + C)^2/BCM$$

where $r_{AB}$ and $r_{BC}$ are the internuclear separations, and $A$, $B$ and $C$ stand for the atomic masses, then the kinetic energy has the diagonal form

$$T = \tfrac{1}{2}\,\frac{A\,(B + C)}{(A + B + C)}\,(\dot{x}^2 + \dot{y}^2) \tag{2.2}$$

In addition, if one uses $x$ and $y$ as Cartesian coordinates, then lines of constant $r_{AB}$ and $r_{BC}$ intersect with an internal angle $\beta$, where

$$\tan^2\beta = \frac{BM}{AC} \tag{2.3}$$

and $M$ is the total mass of the system. The results of plotting the potential energy of the simple hard sphere–square well oscillator system and some of its variants is shown in Figure 2.2.

The simple form of the kinetic energy expression which appears in equation (2.2) means that a mass point sliding without friction on the potential energy surface will carry out a motion which exactly represents the full three-atom

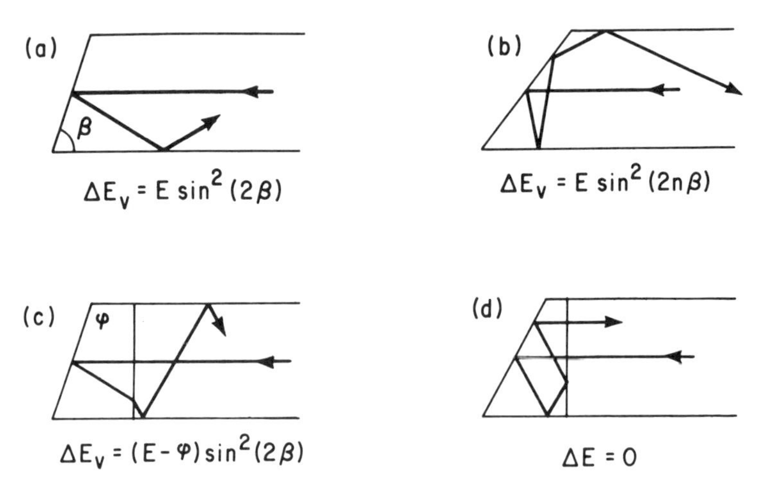


**Figure 2.2** Representation of the collision of a hard-sphere atom A with a square-well diatomic oscillator BC in Cartesian coordinates which diagonalise the kinetic energy. (a) The kinematic angle $\beta = \tan^{-1}\left(\frac{\mathrm{BM}}{\mathrm{AC}}\right)^{\frac{1}{2}}$ is large enough so that a single A–B collision occurs. (b) The angle $\beta$ is small enough so that a second A–B collision occurs on the outgoing leg of the trajectory. (c) In this case there is an attractive potential energy $\phi$ between A and the centre-of-mass of BC, but only one A–B collision occurs. (d) Again, a short-range attraction acts, and A is temporarily trapped as it attempts to leave BC

collinear system in its two internal dimensions. If the total energy of the system is small enough so that the oscillator remains bound, the trajectories on the surfaces of Figure 2.2 consist of a series of straight lines which follow the laws of specular reflection at the potential walls, and refraction at potential steps. In a collision of an atom with a non-vibrating oscillator, the initial trajectory is parallel to the $x$-axis. If, as is true in Figure 2.2a, the angle $\beta$ is large enough, there is only one collision between A and B followed by separation of the atom and the vibrating oscillator. From the geometry of the trajectory we obtain directly

$$\frac{\Delta E}{E_r} = \sin^2(2\beta) = \frac{4ABC\,M}{(A+B)^2(B+C)^2} \tag{2.4}$$

for the ratio of the excitation energy of the oscillator to the initial relative energy, which is exactly the impulse approximation result. In order to include the effects of a more realistic potential energy surface, in the refined impulse approximation[38] for vibrational inelasticity, equation (2.4) is multiplied by a function of $\omega L/v$, where $\omega$ is the oscillator frequency, and $v$ is the relative collision velocity. However, when $\beta$ is small ($\lesssim 50°$), a second A–B encounter occurs, and equation (2.4) is no longer correct. Straightforward analysis shows that it must be replaced with the more general form

$$\frac{\Delta E}{E_r} = \sin^2(2n\beta) \tag{2.5}$$

where $n$ is the number of times A hits B.

The value of this primitive approach is that it reveals clearly the limits of validity of theories of vibrational excitation that are based explicitly or implicitly on the static oscillator or impulse approximation. When $\beta$ is large (which implies B > A or B > C), the outgoing trajectory is nearly parallel to the incoming trajectory, and the conditions for the impulse or static oscillator approximation are satisfied. When B is a light atom, $\beta$ is small, and a large displacement of the oscillator occurs before the atom A has departed. This leads to exploration of parts of the potential energy surface which are not included in an impulse approximation, or even to multiple encounters between A and B. These effects are evident in the exact classical trajectories for the $Na^+$–$D_2$ system ($\beta = 47°$) by Dimpfl and Mahan[52]. In general, good results cannot be expected from approximate theories based on the Landau–Teller model[58] in cases where $\beta$ is small ($\leqslant 60°$). This has also been clearly demonstrated by the exact trajectory calculations of Kelley and Wolfsberg[59], and in the analysis of this problem by Marcus and Attermeyer[60].

Certain other useful qualitative conclusions can be drawn from these primitive collinear collision models. If a small potential step $\phi$ is introduced perpendicular to the $x$-axis, corresponding to an attraction or repulsion between the centres of mass of A and BC, then the inelasticity is given by

$$\Delta E = (E_r - \phi) \sin^2 (2\beta) \tag{2.6}$$

where $\phi$, the potential energy increment, is negative for attractive potentials and positive for repulsive potentials. Equation (2.6), which holds only for 'simple' trajectories (one A–B encounter, no reflection at $\phi$) indicates that attractive forces increase the inelasticity, and repulsions between the centres-of-mass do the opposite. However, the orientation of the step also matters: a step oriented at an angle $\beta$ has no effect on the inelasticity of most types of simple trajectories. In addition, for certain combinations of $E$, $\beta$ and $\phi$, the atom A will be temporarily trapped in a potential well, as Figure (2.2d) shows. This trapping can affect the magnitude of the energy transfer greatly. Thus theories of vibrational energy transfer based on the impulse approximation which have been amended to take account of attractive potentials should be used very cautiously. In effect, they are most likely to be invalid under those conditions where the attractive part of the potential is important.

It should be realised that the restriction of theories based on the impulse approximation to large values of $\beta$ also applies to collisions which are not collinear. An indication of why this is so can be gleaned from the velocity vector construction for the impulse model shown in Figure 2.3. There the atom A (particle 1) with laboratory velocity $V_1$ approaches a stationary BC molecule (particles 2 and 3, respectively) and makes an elastic collision with B. The possible laboratory velocity vectors $V_1'$ and $V_2'$ for A and B which can result from this collision lie on circles of radii $V_1B/(A + B)$ and $V_1A/(A + B)$ with centres at the velocity of the centre-of-mass of the A–B system, $V_1A/(A + B)$. The velocity $V_2'$ is also the velocity of B relative to C after the A–B interaction. Depending on the initial orientation of the BC axis with respect to the direction of $V_2'$, this motion may correspond to pure vibration of BC ($V_2'$ and $r_{BC}$ collinear), or pure rotation ($V_2'$ and $r_{BC}$ perpendicular), or to a mixture of vibration and rotation. However, if A $\ll$ B ($\beta$ large), the maximum velocity of B will be small compared with that of A,

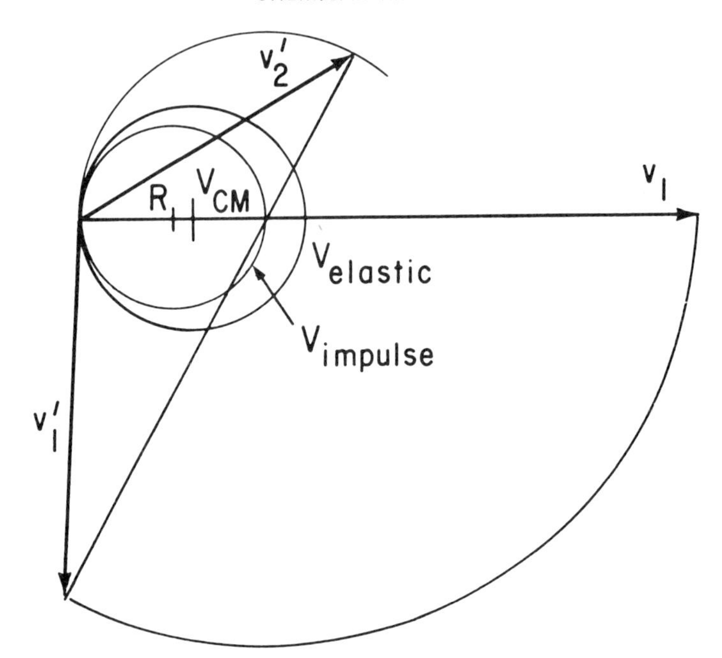


**Figure 2.3** Velocity vector diagram for the impulsive collision of an atom with initial laboratory velocity $V_1$ with particle 2 of a stationary homonuclear diatomic. The resulting laboratory velocity of particle 2 is $V'_2$ and the centre-of-mass velocity of the diatomic lies on the circle labelled $V_{\text{impulse}}$ which is centred at $R$. An elastic collision between the atom and the diatom would leave the latter with a velocity on the circle labelled $V_{\text{elastic}}$, which is centred at $V_{\text{CM}}$, the velocity of the centre-of-mass of the system. The masses have been chosen to represent the $^6Li^+$–$N_2$ system

and A will be able to enter and exit before B has moved appreciably. Thus the conditions for the impulse approximation will be met. However, if A $\gg$ B ($\beta$ small), A will induce a large velocity in B, and the orientation or internuclear distance of the oscillator will change before A exits. Thus the outgoing leg of the trajectory will differ from the incoming leg, and theories based on the static oscillator approximation can be expected to fail.

For completeness and subsequent use, it is well to note that the eventual motion of the centre-of-mass of BC resulting from the A–B collision can be found easily, and is indicated in Figure 2.3. The velocity vector of the centre-of-mass of BC must lie on a 'centroid circle' (actually a sphere in three dimensions) of radius $R$, where

$$R = \frac{A}{A+B}\,\frac{B}{B+C}V_1$$

The centre of this sphere is located on the vector $V_1$ at a distance $R$ from the origin. In contrast, if the A–BC collision were elastic, the centroid vector would lie on a circle of radius $AV_1/M$ whose centre is located on $V_1$ at this distance from the origin. Thus the inelasticity predicted by the impulse model is a function of the barycentric scattering angle $\theta$. Straightforward analysis shows that it is given by

$$\frac{\Delta E}{E_r} = 1 - \left(\frac{g'}{g}\right)^2$$

where $g'/g$, the ratio of the final to the initial relative velocity vector, can be most compactly expressed as

$$\frac{g'}{g} = \cos^2 \beta \, [\cos \theta + (\tan^4 \beta - \sin^2 \theta)^{\frac{1}{2}}]$$

Combination of these expressions gives

$$\frac{\Delta E}{E_r} = 2\cos^4 \beta \, [\tan^2 \beta + \sin^2 \theta - \cos \theta(\tan^4 \beta - \sin^2 \theta)^{\frac{1}{2}}] \qquad (2.7)$$

an equation first given in a slightly different form by Cheng *et al.*[37]. When $\theta = \pi$, equation (2.7) gives the same result as equation (2.4).

Gillen, Mahan and Winn[61] found that at energies of 20 eV or higher, equation (2.7) describes prominent features of the non-reactive scattering of $O^+$ by $H_2$, $D_2$ and HD fairly closely. For these systems, $\beta$ is not large, and evidently the fact that impulsive scattering is observed clearly in the non-reactive channel is to some degree a consequence of the possibility that non-impulsive collisions can lead to chemical reaction.

### 2.2.3 Reactive scattering

#### *2.2.3.1 The* $H^+$–$H_2$ *system*

A number of investigations[62–67] have dealt with the simplest of all ion–molecule systems, $H^+ + H_2$. Early experimental work led to the determination of the reaction cross-section for the atom exchange and charge transfer reactions. More recently, measurements of the energy and angular distributions of the products of the reactions of $D^+$ with HD have been made by Krenos and co-workers[62,67]. The fact that both atom rearrangement and charge transfer occur means that departures from adiabatic behaviour are important, and to take account of these effects Tully and Preston[68,69] developed the trajectory surface hopping (TSH) model. Calculations with this model have been carried out at five initial energies for the five product channels (including dissociation and dissociative charge transfer) of the reaction of $D^+$ with HD. The analogous reactions of $H^+$ with $D_2$ have also been studied. Tully[70] has recently reviewed the published and forthcoming work on these systems, and we shall confine ourselves to a few of the salient points.

The energy dependence of the total cross-sections calculated[67,70] with the TSH model using surfaces generated by the diatomics-in-molecules method are in fairly close agreement with experiment for the reactions of $H^+$ with $D_2$ to give $D^+ + HD$, $H^+ + D$ and $D_2{}^+ + H$. Similar agreement is found for the reactions of $D^+$ with HD to give $H^+ + D_2$, $D_2{}^+ + H$ and, to a lesser degree, $HD^+ + D$. In fact, the discrepancies between the various experimental values are of the same general magnitude as the deviation of the TSH calculations from the experiments.

Similarly, there is good agreement between experimental and calculated product relative energy distributions for the reactions of $D^+$ with HD to give $H^+ + D_2$, $D_2^+ + H$ and $HD^+ + D$ at initial relative energies of 3.0, 4.0 and 5.5 eV. Comparisons of complete calculated product velocity vector distributions with experimental results have also been made for the reactions of $D^+$ with HD to give $H^+ + D_2$, $D_2^+ + H$ and $HD^+ + D$ at 5.5 eV initial relative energy. Certain of the calculated features are indeed found in the experimental results. However, because of the difficulty associated with detecting slow-moving ions, some important features are inaccessible and do not appear in the experimental results. Thus the predictions of the TSH model are fairly consistent with the experimental results insofar as they are available, but more accurate and extensive experimental determination of the product velocity vector distributions would be desirable. Nevertheless, the work of Krenos *et al.*[66,67] stands as the first example of the comparison of detailed product energy and angular distributions with calculations which employed a realistic potential energy surface determined by the diatomics-in-molecule method from *ab initio* calculations.

The trajectory calculations[67,70] for this system illuminate the transition from the persistent collision complex to the direct interaction mode of reaction as the initial relative energy is increased. At relative energies below 3 eV, multiple encounters between the nuclei occur, and the product distributions are approximately symmetric with respect to inversion through the centre-of-mass velocity. At relative energies above 4 eV, most of the trajectories show reaction by direct interaction. Thus in this system, where energy exchange among nuclei of comparable mass should be very facile, the initial relative energy must be less than the well depth if persistent complex behaviour is to be observed.

### *2.2.3.2 The* $H_2^+$–He *system*

The reaction

$$H_2^+ + He \rightarrow HeH^+ + H \quad \Delta H^\circ_0 = 0.803 \text{ eV}$$

is of interest because the simplicity of the reactants and products suggests that a profitable comparison of experimental results with *ab initio* calculations will eventually be made. In addition, in the stimulating experiments of Chupka and co-workers[71,72], reaction cross-sections as a function of translational energy were measured for individual vibrational states of $H_2^+$ selected by photoionisation. At low relative energies ($<3$ eV), the reaction cross-section increases very rapidly with increasing vibrational energy, and is much less sensitive to increases in the translational energy. This effect of vibrational energy diminishes as the translational energy is increased.

Rutherford and Vroom have also studied the energy dependence of the total cross-section by using a crossed-beam apparatus and $H_2^+$ ions prepared by electron impact on $H_2$. The effect of variation of the energy of the ionising electrons, which alters the vibrational energy distribution of $H_2^+$, was consistent with the findings of Chupka and co-workers[71,72]. Rutherford and Vroom found that, at relative energies above 5 eV, the reaction cross-section is quite small and depends primarily on the translational energy. These workers also determined the energy dependence of the cross-section of the reverse reaction, and found that it follows an $E_r^{-\frac{1}{2}}$ dependence at relative energies below 0.6 eV. However, the cross-section is smaller in magnitude than the values predicted from the expression based on the ion–induced dipole model.

It is clear that studies of the energy and angular distribution of the products of this reaction would be valuable, and such experiments have been attempted by at least three groups. Before discussing the results, we should point out that because it is the projectile ion which breaks up to form an ion product which is heavier than the neutral product, there is potential for confusion over the definition and interpretation of the terms 'forward' and 'backward' scattering. We shall use the definition which has become conventional; that is, the *ion* product is forward scattered if it appears in the same direction in the centre-of-mass system as the reaction projectile ion velocity. However, with this definition, the $HeH^+$ formed by spectator stripping or a similar process will appear in the backward region ($\theta \approx 180°$) of barycentric velocity space rather than in the forward direction, as is much more common. This point of interpretation seems to have eluded one investigator. A similar behaviour for the $O_2^+(H_2, O)H_2O^+$ reaction was noted earlier by Chiang *et al.*[74].

Leventhal[75] has measured the energy distribution of $HeH^+$ formed at a laboratory angle of 0° from collisions of $H_2^+$ with a stationary scattering gas. Thus in some experiments two product peaks were detected which represent forward (high laboratory energy) and backward (low laboratory energy) scattering of $HeH^+$. The low-velocity peaks were at first dismissed as spurious. However, this interpretation is surely incorrect, and was later retracted[76]. It is clear that the low-velocity peaks correspond to formation of $HeH^+$ by grazing collisions which may be of the spectator stripping type. Because the laboratory energy of the backscattered peaks in the Leventhal experiments is quite small ($\lesssim 1$ eV), they are undoubtedly very seriously distorted by detector collection efficiency factors which diminish rapidly at low energies. Thus neither the intensity nor the energy of these peaks can be awarded much significance. The high-velocity forward scattered peaks detected by Leventhal correspond to formation of $HeH^+$ by small impact parameter collisions. In such processes the $HeH^+$ recoils forward and the H atom recoils backward with respect to the original $H_2^+$ direction.

The reaction $H_2^+(He, H)HeH^+$ was studied by Neynaber and Magnuson[77] using a merged beam apparatus which permitted energy analysis of the ion product. Because both reactants move with kilovolt energies in this type of experiment, the problems associated with the collection efficiency of low-velocity ions are avoided, and it is possible to obtain the full translational energy spectrum of the product ions. However, the product energy and angular distributions are folded together in these experiments, and the interpretation of the product laboratory energy distribution is complicated by this fact.

At low initial relative energies ($<1$ eV), Neynaber and Magnuson[77] can resolve only a single peak in the product energy spectrum, which corresponds to formation of back-scattered (in their convention, forward-scattered) $HeH^+$ by a grazing collision mechanism. As the initial relative energy is increased, a tail appears on this peak which corresponds to forward scattering of $HeH^+$ by a rebound mechanism. At initial relative energies above 3.5 eV, this forward-scattered peak is fully resolved, but remains of lesser intensity than the backward peak up to the highest relative energy investigated, 12.0 eV.

Although the intrinsic nature of an experiment in which only product energy distributions can be measured imposes some limitations on detailed interpretation, the experiments of Neynaber and Magnuson indicate clearly that $HeH^+$ is formed both by small impact parameter collisions of the rebound type and, more importantly, by grazing collisions. The authors assert that the spectator stripping mechanism is not consistent with the latter events, but it is very difficult to tell from their graphical presentation of the primary data just how important this departure is. At relative energies above 3.8 eV, the spectator stripping mechanism cannot apply, since it would produce molecules unstable with respect to dissociation.

Herman and co-workers[78] have studied this reaction using a crossed-beam apparatus which allowed them to determine complete velocity vector distributions of $HeH^+$ for initial relative energies in the 0.35–3.5 eV range. The product ions were found scattered both forward and backward about the centre-of-mass, with greater intensity in the backward direction. The authors state that the backward peak corresponds closely to the stripping mechanism over the entire range of relative energies studied. Thus there is a qualitative consistency between the results of Herman and co-workers[78] and Neynaber and Magnuson[77], although the two groups differ over whether the back-scattered peak is well represented by the spectator stripping mechanism. It is to be hoped that in the future we shall see experiments that display the effects of internal excitation of $H_2^+$ on the energy and angular distributions of $HeH^+$.

### *2.2.3.3 The* $O^+$–$H_2$ *system*

The reaction $O^+(H_2,H)OH^+$ or its isotopic variants have been investigated by Harris and Leventhal[79], and by Gillen, Mahan and Winn[80,81]. Harris and Leventhal measured product velocity profiles at a fixed laboratory scattering angle of 0° and were able to show that the product distributions were asymmetric about the centroid velocity even when the initial relative energy was as low as 0.76 eV. Thus the reaction proceeds by a direct, short-lived interaction mechanism.

Gillen, Mahan and Winn[80,81] measured complete velocity vector distributions for the reaction of $O^+$ with $H_2$, HD and $D_2$ at relative energies which ranged from 3 to 50 eV. Some of their results are shown in Figure 2.4. At low relative energies the spectator stripping peak is the most prominent feature of the distribution. However, at relative energies which are high enough so that $OH^+(^3\Sigma)$ formed by spectator stripping would be unstable with respect to dissociation to O and $H^+$, the spectator stripping peak is lost.

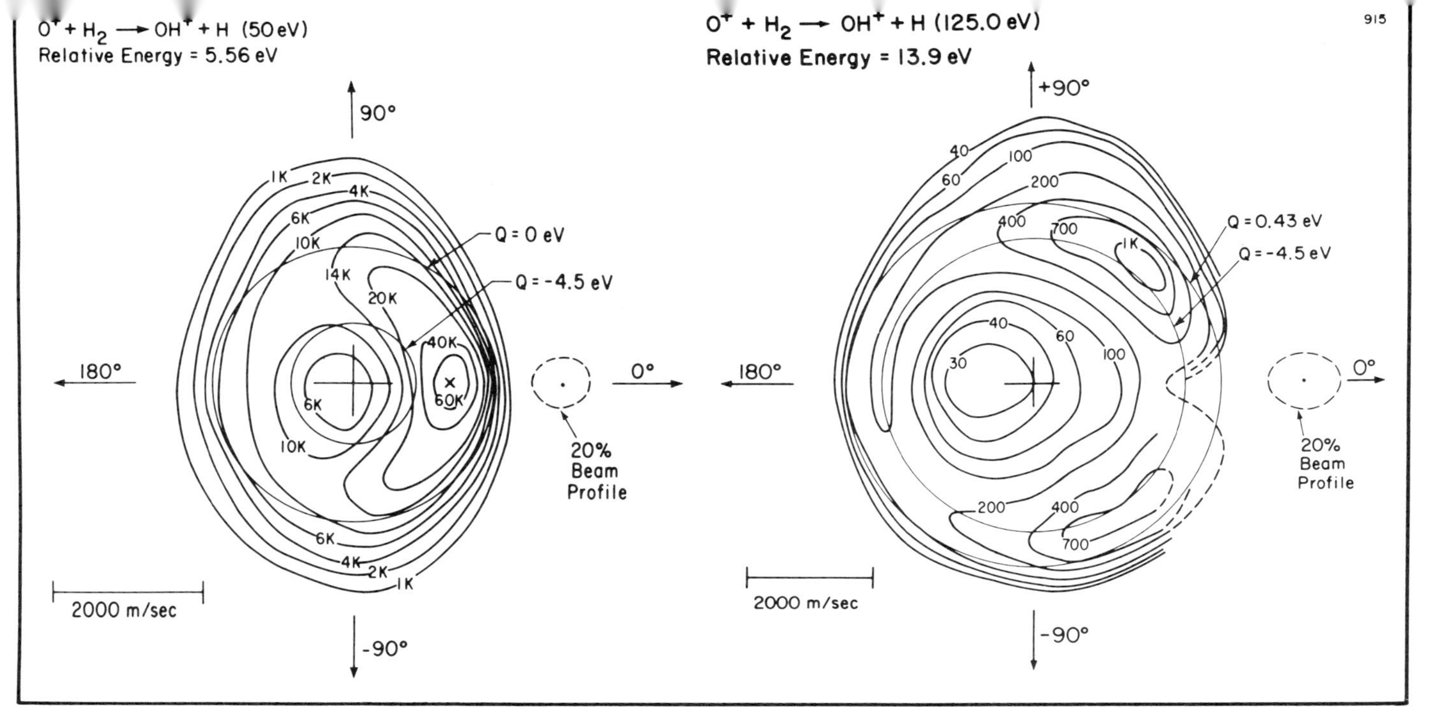


**Figure 2.4** Contour maps of the specific intensity of $OH^+$ from the $O^+$–$H_2$ reaction at 5.56 eV (left) and 13.9 eV (right) initial relative energy. In each case the radial coordinate is the speed of $OH^+$ relative to the centre-of-mass, and the angular coordinate is the barycentric scattering angle, relative to the original direction of the $O^+$ beam. The circle marked $Q = -4.5$ eV is the locus of the smallest speed that $OH^+(^3\Sigma^-)$ can have and be stable with respect to dissociation. Note that there is a prominent peak at the spectator stripping velocity (marked by a small cross) at low initial relative energy, but that this peak at zero angle is not present at high initial relative energy. In the latter case, the spectator stripping velocity falls at $Q = -7.35$ eV, well into the region where $OH^+(^3\Sigma^-)$ is unstable (Data of Gillen *et al.*, unpublished, but part of work reported in Ref. 80)

This behaviour is experimentally unprecedented, although it was asserted to be the expected general behaviour in early discussions[82] of the mechanism of spectator stripping. The experimental facts[82,83] are that the reactions $Ar^+(D_2,D)ArD^+$, $N_2^+(D_2,D)N_2D^+$ and $CO^+(D_2,D)COD^+$ all display a spectator stripping peak when the initial relative energy is low. An intensity peak in the forward direction is still observed even when the relative energy is so high that the products formed by spectator stripping would be unstable. However, these forward-scattered product peaks fall at velocities high enough that the product internal energy does not exceed the dissociation limit. In other words, the potential energy surfaces produce, in these cases, forward recoil of the products into a velocity zone where they are stable. This product stabilisation by forward recoil does not occur in the $O^+(H_2,H)OH^+$ reaction.

Elementary molecular mechanics indicates that in order to have substantial forward recoil, there must be a repulsive interaction between the products as they leave the collision scene. For reactions without an activation energy barrier, this product repulsion will occur, if it occurs at all, as a result of a late release of the exoergicity of reaction. If this exoergicity is small, the product repulsion is apt to be small, and forward recoil of the products will not occur. The reaction $O^+(H_2,H)OH^+$, which does not display forward recoil, is only 0.43 eV exoergic; the other three reactions cited, which do display forward recoil, are all exoergic by approximately 1.5 eV.

The reaction $O^+(H_2,H)OH^+$ proceeds for the most part on a potential energy surface of $^4\Sigma^-$–$^4A''$–$^4A_2$ symmetry to form $OH^+$ in its ground electronic state, $^3\Sigma^-$. However, Gillen, Mahan and Winn[80,81] found evidence that in some collisions $OH^+(^1\Delta)$ is formed. This requires a switch from the $^4\Sigma^-$–$^4A''$–$^4A_2$ surface to a $^2\Delta$–$^2A''$–$^2B_2$ surface, which correlates to electronically excited $OH^+(^1\Delta)$. This point is discussed more fully in Section 2.3.2 in connection with the electronic state correlation diagram for the system.

At relative energies of 6 eV or greater, Gillen, Mahan and Winn[80,81] found that the angular distributions of $OH^+$ and $OD^+$ from the $O^+$–HD reaction were quite different. Examples are shown in Figure 2.5. The $OH^+$ is found principally at barycentric angles smaller than 90°, although the stripping peak at zero degrees is absent if the collision energy is high enough. In contrast, the $OD^+$ product is found at barycentric angles greater than 50°. A similar angular dependence of the isotope effect was found by Chiang *et al.*[84] for the reaction of $O_2^+$ with HD (to form $O_2D^+$ and $O_2H^+$) at high relative energies. The form of the angular distributions are determined at these high relative energies principally by the necessity to stabilise the product molecules with respect to dissociation. Detailed consideration[81] of the mechanics of the atom transfer process shows that this stabilisation can be accomplished in the $O^+$–HD system events in which $OH^+$ is scattered into small angles or $OD^+$ is scattered into larger angles.

Gillen, Mahan and Winn[81] were able to reproduce the major features of the product angular distributions observed for the $O^+$–HD reaction at high relative energies by performing exact trajectory calculations on a potential energy surface made up of hard-sphere repulsions between atoms. The attractive parts of the intermolecular potential served only to hold the

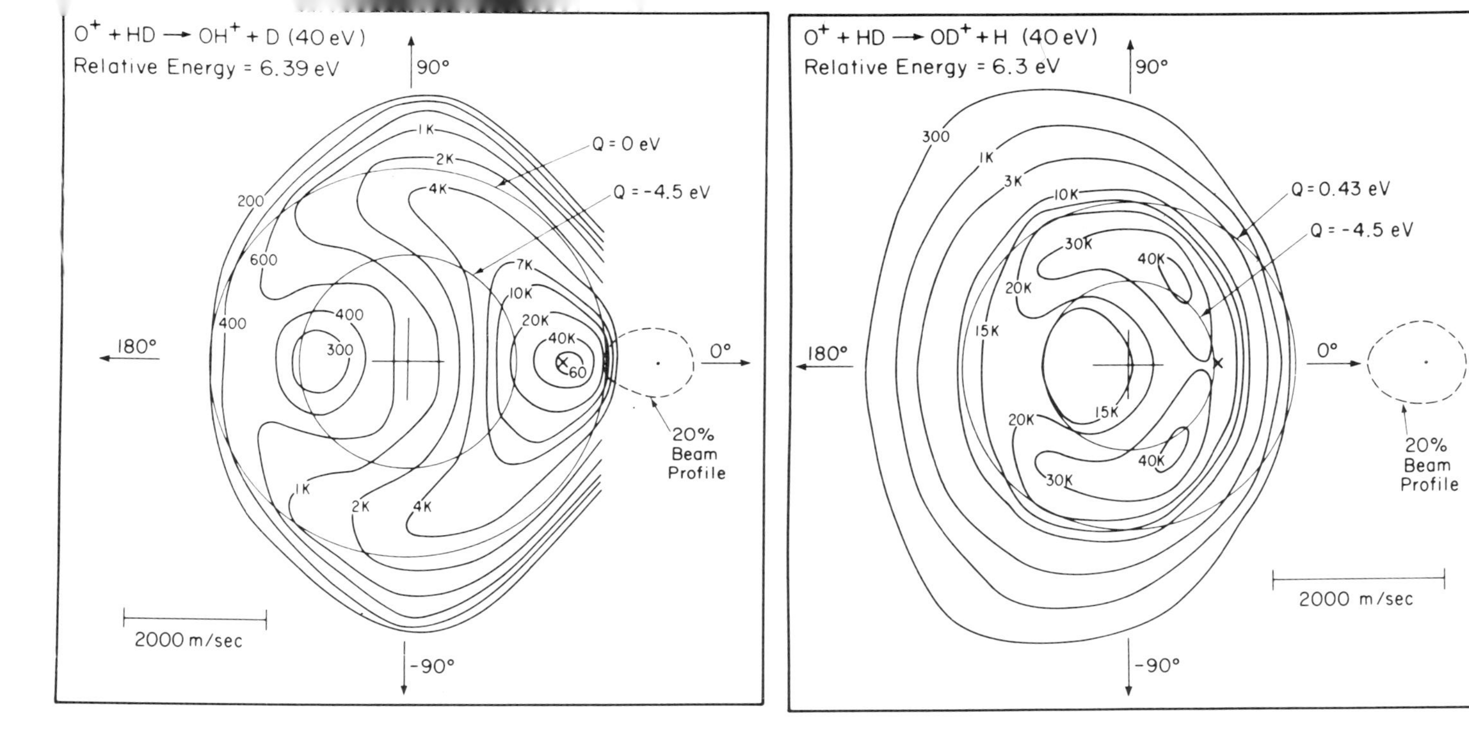


**Figure 2.5** Contour maps of the specific intensity of $OH^+$ (left) and $OD^+$ (right) from $O^+$–HD collisions at an initial relative energy of 6.3 eV. Note that for the $OH^+$ product there is a strong intensity peak at the spectator stripping velocity, which is marked by a small cross. For the $OD^+$, the stripping velocity lies at the border of the zone of velocities (marked by the $Q = -4.5$ eV circle) where $OD^+(^3\Sigma^-)$ is stable to dissociation. No stripping peak is visible, but some intensity for $Q < -4.5$ eV persists, and indicates that some $OD^+(^1\Delta)$ is being formed (From Gillen *et al.*[80], by courtesy of the American Institute of Physics)

reactant and product diatomic molecules together. The agreement between the experimental and the calculated product angular distributions is not perfect, nor is it expected to be. However, the agreement is close enough to be convincing evidence that the major features of the high-energy reaction dynamics are consequences of the repulsive forces between atoms, to which hard-sphere impulsive interactions are a satisfactory approximation (see Section 2.4).

#### *2.2.3.4 The* $C^+$–$H_2$ *system*

Product energy and angular distributions for the reaction $C^+(D_2,D)CD^+$ have been determined by Koski and co-workers[85,86] at six initial relative energies between 3.5 and 9.1 eV. At low relative energies ($\lesssim$4.4 eV) the product velocity vector distributions appeared to be highly symmetric about the $\pm 90°$ axis in the barycentric system. This was interpreted as an indication that the reaction proceeds through a persistent $CD_2^+$ complex mechanism. Subsequently, Mahan and Sloane[87] determined complete product velocity vector distributions for the reactive and non-reactive collisions of $C^+$ with $H_2$, HD and $D_2$. Even at the lowest initial relative energies investigated by them (2.0 eV), the $CH^+$ product distributions were not perfectly symmetric about the $\pm 90°$ barycentric axis. A slight forward peaking of the distribution was always evident, and became more prominent as the initial relative energy increased. The slight asymmetry is probably an indication that the collision complex persists for only approximately one rotational period even at the lowest energies investigated, rather than the several rotational periods which are necessary to produce a truly symmetric distribution. This finding seems more consistent with the rather short lifetime one would expect for a triatomic collision complex with a potential energy well depth ($\sim$4.3 eV) not much larger than the initial relative energy.

The distributions of the $CH^+$ and $CD^+$ products from $C^+$–HD collisions found by Mahan and Sloane[87] were quite similar, but not identical, at low relative energies. This is also consistent with the model of a collision complex in which the interaction between the atoms is strong, but which exists for only about one rotational period. As the initial relative energy was increased, the differences between the distributions of the isotopic products became more pronounced, and the general shapes of the distributions indicated an evolution into a direct interaction mode of reaction. The distribution of $C^+$ scattered non-reactively from $H_2$ showed both elastic and very strongly inelastic features. The latter could not be accounted for by a direct interaction model of inelastic scattering, and thus constitutes substantial evidence for the occurrence of a collision complex in which all the atoms are strongly interacting.

Mahan and Sloane analysed the $C^+$–$H_2$ reaction in terms of an electronic state correlation diagram which will be discussed in Section 2.3.2. Here we merely state the conclusion that the strongly bound $^2A_1$ ground state of $CH_2^+$ is accessible to the reactants at low energies if they move on a potential energy surface of $^2B_2$–$^2A'$–$^2A_1$ symmetry. That is, the reactants, initially in a $^2B_2$ state, must avoid the conical intersection of this state with the $^2A_1$

ground state of $CH_2^+$ by maintaining $C_s$ symmetry until the potential energy well of the $^2A'$–$^2A_1$ surface is reached. The resulting $CH_2^+(^2A_1)$ correlated adiabatically with the ground-state products $CH^+(^1\Sigma)$ and $H(^2S)$. A number of the features of the correlation diagram proposed by Mahan and Sloane[8] have been supported by the results of the *ab initio* calculations of Liskow *et al.*[88].

### 2.2.3.5 *The* $O^+$–$N_2$ *system*

The $O^+(N_2,N)NO^+$ reaction displays several intriguing features. It is an exoergic ion–molecule reaction whose cross-section at thermal energies ($2 \times 10^{-3}$ nm$^2$) is quite small[89]. As the relative translational energy of the reactants is increased, the cross-section at first falls slightly[90], then rises[90–93] to a broad maximum of *ca.* $5 \times 10^{-2}$ nm$^2$ at 8 eV, and then falls again. At low translational energies the cross-section is very sensitive[89] to vibrational excitation of $N_2$. Kaufman and Koski[94] and O'Malley[95] have proposed models intended to explain certain features of the reaction dynamics. One would hope that scattering experiments would be of assistance in elucidating the dynamics and in testing these models.

Recently, Cohen[93] has used the molecular beam technique to measure the total reaction cross-section as a function of the initial relative translational energy in the range 2–9 eV. The nitrogen issued from an oven whose temperature could be varied from 300 to 3000 K, and thus the effects of vibrational excitation of the $N_2$ could be explored. In this temperature range, Cohen found that the total reaction cross-section did not change by more than 15%. It was concluded that the reaction cross-section for first three *excited* vibrational states of $N_2$ could not be more than twice that of the ground vibrational state. Thus, in contrast to the behaviour observed at low translational energies, vibrational excitation has little effect on the cross-section when the translational energy is high.

Very recently, McFarland *et al.*[96] have used a flow-drift tube technique to measure the reaction rate constant from 0.2 to 2 eV relative energy, and have compared rate constants calculated from flow-drift tube, drift tube and molecular beam data. The agreement is spectacularly good in the range of data overlap, 0.2–2 eV.

Leventhal[97] has used an ion beam–scattering gas apparatus to measure the distribution of the translational energies of $NO^+$ from the $O^+(N_2,N)$-$NO^+$ reaction at a laboratory scattering angle of 0°. He found a prominent peak which appeared at the spectator stripping velocity in experiments in the 1–4 eV range of initial relative energies. In the one full intensity profile which was published, a second peak appears at low laboratory energy. This peak was dismissed as spurious. However, this interpretation is based on an erroneous analysis[75,76] of the problem of the laboratory to centre-of-mass transformation, and is probably not correct. The proper transformation[98,99] suggests that this low-energy peak is not spurious, but represents $NO^+$ formed by a small impact parameter rebound process, and is as important as the forward-scattered stripping peak. Some evidence that this is true comes from the merged-beam experiments of Neynaber and Magnuson[100].

Their product energy distributions show prominent back-scattering of $NO^+$, particularly at low initial relative energies (1–5 eV). They attribute much of the product at large scattering angles to reactions of $O^+$ with vibrationally excited $N_2$. However, if our interpretation of Leventhal's data[97] is correct, the ground vibrational state of $N_2$ must also give back-scattered $NO^+$.

Smith and Cross[101] have measured angular and energy distributions of $NO^+$ produced from $O^+$–$N_2$ collisions in the 1.5–15.8 eV range of initial relative energies. Their product velocity vector distributions show a prominent peak at the spectator stripping velocity, and essentially no back scattering. The absence of back scattering may be a consequence of the great difficulty of detecting very slow moving ions and the small total reaction cross-section. It is hard to believe that small impact parameter collisions do not lead to reaction. In summary, the energy dependence of the total reaction cross-section appears to be very well characterised, but more work needs to be done on the product velocity vector distributions before a definitive analysis of the reaction dynamics is possible.

### 2.2.4 Theoretical studies

#### *2.2.4.1 The* $H^+$–$H_2$ *system*

In the past there has been almost no attention given to ion–molecule reaction phenomena by scientists principally interested in the calculation of accurate potential energy surfaces or in investigating collision dynamics by classical trajectory calculations. Recently this situation has started to change. Portions of the first two $^1A_1$ potential energy surfaces for the $H_3^+$ system have been calculated by Bauschlicher *et al.*[102] at the *ab initio* SCF–CI level. Perspective plots of these surfaces for $H_3^+$ in the $C_{2v}$ conformation are shown in Figure 2.6. The coordinate $r$ is the separation of two hydrogen nuclei, and $R$ is the distance of the third nucleus from the centre-of-mass of the first two. At large $R$ both surfaces show a rather abrupt change in slope as $r$ reaches a critical value. This line of near discontinuity is the locus of the avoided intersection between the two surfaces. The perspective plots thus nicely demonstrate the original finding of Tully and Preston[68,69] that the transition from one surface to another occurs at large $R$ by what amounts to a vibrational motion of the two hydrogen nuclei which are in proximity.

The *ab initio* surfaces of Bauschlicher *et al.*[102] were compared with the diatomics-in-molecules surfaces of Tully and Preston. In general the agreement was found to be rather satisfactory except at small values of $R$. Since transitions from one surface to the other take place principally at relatively large values of $R$, the surface hopping probabilities calculated using the semi-empirical and *ab initio* surfaces are in rather good agreement.

#### *2.2.4.2 The* $HeH^+$–$H_2$ *system*

Benson and McLaughlin[103] have calculated an SCF potential energy surface for the $C_{2v}$ conformations of the atoms in the proton transfer reaction

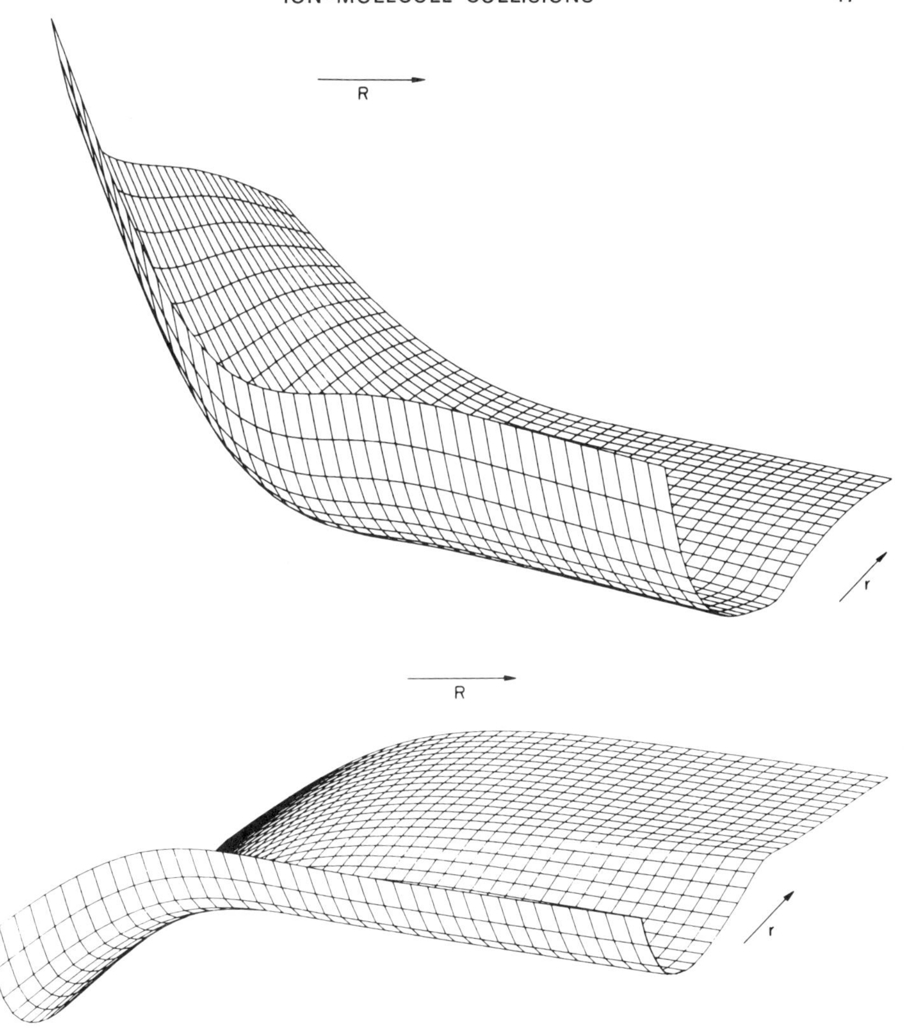


**Figure 2.6** Perspective plots of the ground and first excited $^1A_1$ surfaces of $H_3^+$ in $C_{2v}$ conformations. *R* is the altitude and *r* the base of the isosceles triangle of nuclei. The surfaces have been displaced vertically from one another for clarity. The locus of the intersection of the two surfaces is visible at large *R* as an abrupt change in slope in the *r* direction (From Bauschlicher *et al.*[102], by courtesy of the American Institute of Physics)

$HeH^+(H_2,He)H_3^+$. By performing a few configuration interaction calculations they were able to conclude that the SCF surface for this closed-shell system lies parallel to the SCF–CI surface to within a few percent of the total energy change for the reaction. Calculations of the energy of certain $C_s$ conformations indicate that the $C_{2v}$ conformation is the lowest potential energy path from reactant to products, as one might expect on intuitive grounds. The authors point out the formidable nature of the calculation of the potential energy surface for a four-atom system with unrestricted geometry. If along each of the six coordinates the energy is calculated at only 10 points, then $10^6$ energy calculations would be required to complete the potential energy grid. Thus the prospects for a completely *ab initio* treatment of a four atom system appear dim, unless powerful interpolation methods can be found.

McLaughlin and Thompson[104] have performed classical trajectory calculations using the Benson–McLaughlin[103] surface for $C_{2v}$ conformations of $HeH_3^+$. Two initial relative translational energies for $HeH^+$ and $H_2$ were examined, as well as the combinations resulting from vibrational quantum numbers of 0 or 2 in the diatomics. The reaction probability was found to be unity, and in most collisions greater than 90% of the energy available to products appears as vibration of $H_3^+$. The percentage of the total energy present as vibrational excitation of the products decreases slightly as the vibrational energy of the reactants is increased. So far, no experimental data are available to compare with these calculations.

### *2.2.4.3 The* $Ar^+$–$H_2$ *system*

An interesting study of the dynamics of the $Ar^+(H_2,H)ArH^+$ reaction has been carried out by Chapman and Preston[105], using the Kuntz–Roach[20] diatomics-in-molecules potential energy surfaces. Interaction between diabatic potential surfaces of the same symmetry proves to be very important in this system, as Kuntz and Roach[20] have emphasised. Of the four potential energy surfaces which can be generated from $Ar^+(^2P) + H_2$ and $Ar + H_2^+$-$(^2\Sigma)$, one $^2\Pi$–$^2A''$ surface does not interact with the other three and does not lead to ground-state products. The $^2\Pi$–$^2A'$ surface interacts only weakly with the lowest $^2\Sigma$–$^2A'$ surfaces which arise from $Ar^+$–$H_2$ and Ar–$H_2^+$ reactant configurations. In contrast, these lowest two surfaces interact strongly with each other, and it is therefore possible to reach the products $ArH^+ + H$ or $Ar + H_2^+$ starting with $Ar^+ + H_2$. Chapman and Preston computed total cross-sections for atom and charge exchange for $Ar^+$–$H_2$ collisions which are in reasonable agreement with experiment. The reaction and charge transfer probabilities were calculated as a function of impact parameter: for impact parameters less than 0.16 nm, reaction to $ArH^+ + H$ predominates, whereas at larger impact parameters, charge transfer is most important. The calculated energy and angular distribution of the chemical reaction products are in reasonable qualitative agreement with experimental results. Unfortunately only one energy and one isotopic combination of reactants was studied. The success of these calculations is sufficiently impressive that further comparisons with experimental data should be made.

Very recently, Kuntz and Roach[106] have carried out a classical trajectory study of exoergic ion–molecule reactions. The system was modelled after the $Ar^+(D_2,D)ArD^+$ reaction and its isotopic variants, but since no charge transfer channel was included the purpose of the work was more to study the effect of potential energy surface variation on the dynamics, than to match the experimental data for the real system. A number of quite interesting, if not particularly startling, points emerge from this work.

First, sharp forward-peaking of the reaction product with forward recoil at high translational energies is recovered from three exoergic surfaces having relatively weak angular dependence (but favouring isosceles geometry at small distances) and varying amounts of attractive energy release. A LEPS surface which favoured collinear geometry and was slightly endoergic produced a nearly isotropic angular distribution. The association between sharp forward-peaking and weak angular dependence of the potential can also be made from a straightforward geometric analysis of the collision problem (see Section 2.4).

Second, Kuntz and Roach note that the average (or most probable) value of $Q$, the translational exoergicity, is quite insensitive to changes in the potential energy surface. This seems quite reasonable in view of the accumulated experimental evidence that scattering at or very near the $Q$ values associated with spectator stripping is very prominent in exoergic hydrogen atom transfer reactions which go by a direct interaction mechanism. It would be hard to imagine that the potential surfaces for all these reactions are identical. Thus the similarity of the experimental results must be a consequence of the rather crude nature of even the best experiments, and the intrinsic insensitivity of low-resolution scattering data to subtle details of potential surfaces. Anyone unconvinced of the latter point should consider how much detail in the angular distribution is necessary in order to permit extraction of reliable interatomic potentials which are functions of one coordinate only.

Third, Kuntz and Roach carry out an interesting test of the ability of the direct interaction with product repulsion (DIPR) model to reproduce the angular and energy distributions found from the trajectory studies. The DIPR mode was found to be much less satisfactory for these hyperthermal ion–molecule reactions than it was for the thermal alkali atom–halogen reactions. The reason for this, in the opinion of this writer, is that the DIPR model attributes the dynamical features of the reaction entirely to the part of the reaction exoergicity released as product repulsion. It essentially neglects the momentum transfer effects in the A(BC,C)AB reaction of A hitting B, and B *then* hitting C. These effects should be particularly important at high initial relative energies and, indeed, Gillen, Mahan and Winn[81] reproduced the experimental results for the $O^+(D_2,D)OD^+$ reaction quite successfully with a model which involves only these momentum transfer effects.

Finally, the trajectories of Kuntz and Roach[106] revealed the tendency of the distant or 'wrong' atom C to appear in the product molecule when the initial relative energy was increased. The importance of 'wrong atom' reactions was also evident in the trajectory studies of Gillen *et al.*[81]. In fact, in the $O^+(HD,D)OH^+$ reaction, $OH^+$ is formed at large angles almost exclusively by 'wrong' atom reactions and at small angles principally by 'right' atom reactions at high energies.

## 2.3 CORRELATION DIAGRAMS

### 2.3.1 General considerations

The qualitative connection between the gross features of potential energy surfaces and product velocity vector distributions have been fairly well established as a result of experimental and calculational studies of reaction dynamics. Potential surfaces which are flat, or have barriers or only shallow wells along all possible reaction paths, lead to reaction by a short-lived or direct interaction of collision partners. This can usually be recognised experimentally by a product intensity distribution which is asymmetric about the $\pm 90°$ axis in the barycentric system, although there are examples[9,74] of reactions thought to be direct which give highly symmetric distributions. The dynamics of these direct reactions can be analysed and, it is to be hoped, predicted in terms of dynamic models which use a few important features of the potential surface, or by means of full trajectory calculations. On the other hand, if a potential surface has a well deep enough so that it constitutes an appreciable fraction of the total energy of the collision partners, and if this well is accessible to the reactants and products, the result may be that the collision complex persists for several rotational periods. This gives rise to a product velocity vector distribution which is symmetric about the $\pm 90°$ axis of the barycentric velocity system. For these reactions the product energy distributions, isotope effects and relative yields of the various channels may be calculable from statistical theories of reactions.

It is clear that it would be valuable to be able to anticipate the general form of a potential energy surface without doing any extensive numerical calculations. Molecular orbital and electronic state correlation diagrams offer a means of using known experimental information about the electronic properties of reactants, products and intermediates to deduce the major qualitative features of a potential energy surface. These methods have particular value in the study of ion–molecule reactions[64,80,87,94,109–111]. In these systems there are often several potential energy surfaces within a few electron volts of the ground state, and the reaction dynamics can be considerably influenced by interactions and intersections of these surfaces. In what follows we shall illustrate these points with two recent examples from the literature.

### 2.3.2 Specific examples: $C^+$–$H_2$, $O^+$–$H_2$

A number of examples of the use of correlation diagrams have appeared in the literature. Here we shall discuss in detail only the diagrams[80,87] for the $C^+(H_2,H)CH^+$ and $O^+(H_2,H)OH^+$ reactions, which illustrate some of the uses and limitations of correlation diagrams particularly clearly.

There are two types of correlation diagrams which can be of use. The primitive MO correlation diagram shows how the orbitals of the reactants evolve to those of the intermediate and products. These diagrams can be used to anticipate what the behaviour of a system with a particular electronic configuration might be, and this is sometimes sufficient. However, a single electronic configuration can produce a number of adiabatic electronic states or surfaces, and it is the correlation of these states with which we are most concerned. Although construction of an adiabatic electronic state correlation diagram can be made directly using the symmetry properties of the electronic states and the non-crossing rule, it is usually more edifying to construct the state correlation diagram after having assembled an orbital correlation diagram.

Construction of an orbital correlation diagram involves[112] ordering the atomic and molecular orbitals of the reactants according to their energies, then forming the molecular orbitals of the collision intermediate from linear combinations of the reactant orbitals. The process is repeated for the products. The correlation between reactant, intermediate and product molecular orbitals is completed with due regard for the interaction of orbitals of the same symmetry and substantial overlap. The process is not always unambiguous, mostly because the very idea of an orbital energy is not well defined, and the dependence of orbital energy on even the three coordinates appropriate to a triatomic system may not always be clear. Fortunately, data from photoelectron spectroscopy, appearance potentials and *ab initio* calculations can provide important calibration points. It is then usually possible to draw from the nodal properties of the orbitals a qualitative indication of their energy variation, and subsequently to deduce the behaviour of the states which arise from various electronic configurations.

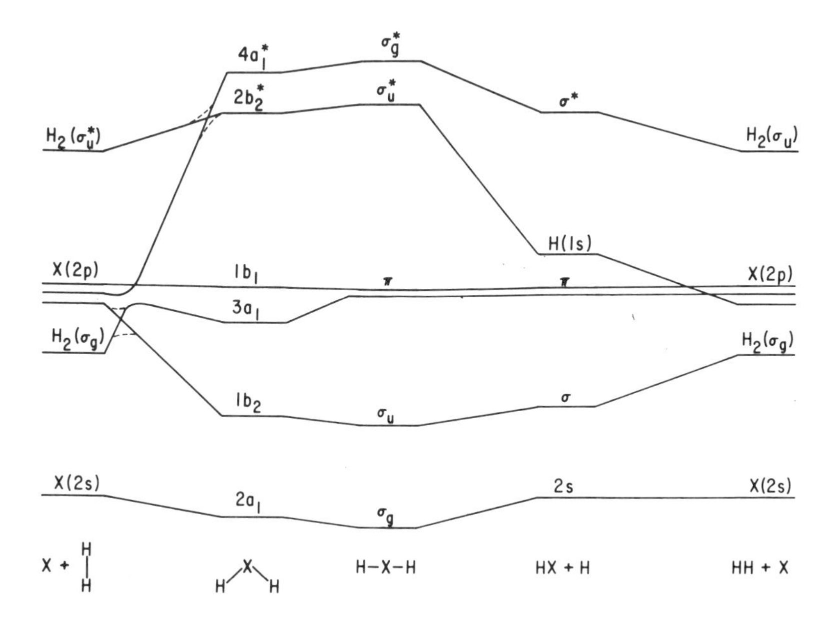


**Figure 2.7** A correlation diagram for the valence molecular orbitals of the X–$H_2$ system. On the left X approaches $H_2$ in $C_{2v}$ conformations, while on the right a collinear approach is assumed

In Figure 2.7 an example is given of an orbital correlation diagram for X + $H_2$ systems, where X is C, N or O. Correlations for conformations corresponding to collinear ($C_{\infty v}$) and perpendicular ($C_{2v}$) approaches are shown explicitly. It is important also to consider what happens in the less symmetric ($C_s$) conformations which are bound to predominate in most collisions. This is included in the diagram by indicating with dotted lines the mixing of the orbitals which occurs when the complex is distorted to lower symmetry.

As a specific example, the system $C^+$–$H_2$ is particularly attractive since there have been recent experimental investigations[86,87] of the $C^+(H_2,H)CH^+$ reaction (see Section 2.2.3.4) and certain limited regions of the potential surfaces have been calculated[88] using the SCF–CI approximation. Figure 2.8 is the electronic state correlation diagram presented by Mahan and Sloane[87], slightly modified to take account of the recent calculations of Schaefer and co-workers[88]. It is helpful to examine this state diagram while referring to the orbital correlation diagram, Figure 2.7.

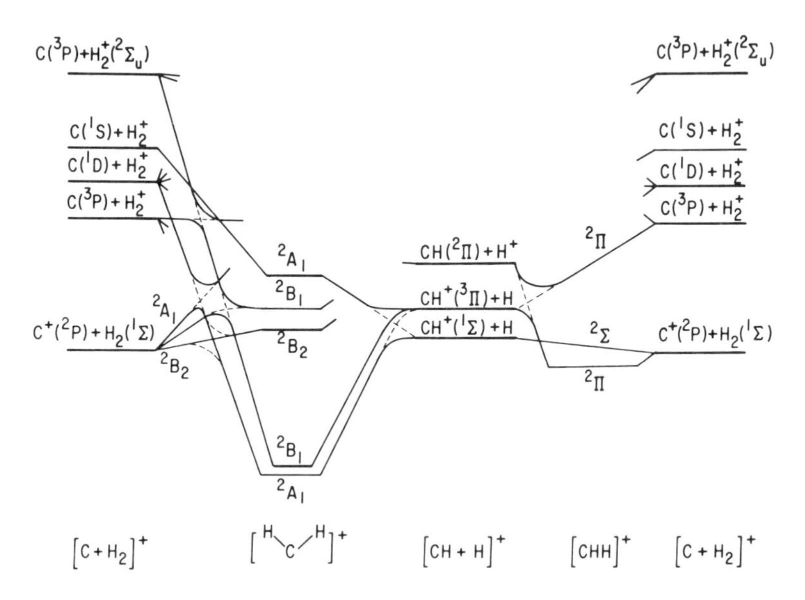


**Figure 2.8** An electronic state correlation diagram for the lower states of the $C^+$–$H_2$ system. On the left $C_{2v}$ conformations are assumed, while on the right $C^+$ approaches $H_2$ collinearly. Avoided surface intersections are indicated by crossed dotted lines, and conical intersections are represented by crossed solid lines accompanied by dotted, non-crossing lines

Starting with the collinear approach of reactants, we note that $C^+(^2P)$ and $H_2(^1\Sigma)$ give rise to three states, $^2\Sigma$ and (the doubly degenerate) $^2\Pi$, corresponding to the single p-electron of $C^+$ being in a p$\sigma$ or a p$\pi$ orbital. The orbital correlation diagram shows that $H_2(1\sigma_g)^2$ combined with the carbon valence electron configuration $(2s)^2(2p\sigma)^1$ evolves to the lowest energy configuration of the products $CH^+$ and H. Correspondingly, the state correlation diagram shows that the $^2\Sigma$ surface leads to the ground state products $CH^+(^1\Sigma)$ and $H(^2S)$. For the $^2\Pi$ surface the situation is different. The orbital correlation diagram shows that the configuration $C^+(2s)^2(2p\pi)^1$ and $H_2(1\sigma_g)^2$ leads to $H^+$ and $CH(2s)^2(2p\sigma)^2(2p\pi)^1$. However, the state correlation diagram shows that the energy of this set of products lies above that of $CH^+(^3\pi)$ and $H(^2S)$, which also can be formed via a $^2\Pi$ surface.

Thus there must be an avoided intersection of these $^2\Pi$ surfaces, and this is shown on the state correlation diagram. This example illustrates a major limitation of orbital correlation diagrams: different orbital configurations can give rise to states of the same symmetry which can interact strongly at certain conformations. Consequently, the orbital configuration is quite frequently not maintained as a reaction occurs.

A complementary point concerning the use of correlation diagrams can be made by examining the left sides of Figures 2.7 and 2.8. As $C^+(^2P)$ approaches $H_2(^1\Sigma^+_g)$ in the $C_{2v}$ conformation, one of the three surfaces generated is of $^2A_1$ symmetry. Since the ground state of symmetric $CH_2^+$ is $^2A_1$, there is an adiabatic correlation between the ground state reactants and the ground state of the symmetric intermediate $CH_2^+$. On the basis of this observation alone one might expect that $C^+$ can be easily inserted into $H_2$ on the $^2A_1$ surface. However, the valence electronic configuration of this $^2A_1$ surface at large $C^+$–$H_2$ distances is $(2a_1)^2(3a_1)^2(4a_1{}^*)^1$, while near the equilibrium conformation of $CH_2^+$, the ground state configuration is $(2a_1)^2(1b_2)^2(3a_1)^1$. Thus the orbital configuration at large distances implies that the $^2A_1$ surface is repulsive, and entry to the deep potential energy well can be gained only by crossing an energy barrier. This point, first anticipated from the orbital configurations alone[87], has since been substantiated by *ab initio* calculation[88].

Finally, the role of conformations which have lower than $C_{2v}$ symmetry is displayed in Figure 2.8 by the crossing of the $^2B_2$ and the lowest $^2A_1$ states. This crossing is allowed because the states belong to different species of the $C_{2v}$ point group. However, for $C_s$ conformations, both states are of $^2A'$ symmetry, and their intersection is avoided. This is indicated by the dotted lines in Figure 2.8. Consequently, it is possible to pass adiabatically from reactants to the ground state of the $CH_2^+$ intermediate if the collision complex departs from $C_{2v}$ symmetry. This is of course highly probable, and provides a satisfactory explanation for the fact that highly symmetric product velocity vector distributions characteristic of strongly interacting persistent collision complexes are found for this reaction[86,87].

The correlation diagram for the $O^+$–$H_2$ system shown in Figure 2.9 was used to anticipate and elucidate features of the experimental study of the $O^+(H_2,H)OH^+$ reaction[80,81]. As is well-known from appearance potential measurements and photoelectron spectroscopy, the ion $H_2O^+$ is strongly bound with its $^2B_1$ ground state lying 5.6 eV below the energy of any of its dissociation products. Thus the possibility exists that the $O^+(H_2,H)OH^+$ reaction might involve a persistent collision intermediate at low initial relative energies. However, consideration of the spin conservation rules and, more explicitly, examination of the state correlation diagram in Figure 2.9, shows that this possibility is tenuous at best. First, the reactants $O^+(^4S)$ and $H_2(^1\Sigma^+_g)$ have a different spin multiplicity and state symmetry ($A_2$) than the ground and lower excited states of $H_2O^+$. Second, the valence electron configuration of the reactants at great distances is $(2a_1)^2(1b_2)^1(3a_1)^2(1b_1)^2$-$(4a_1{}^*)^1$. The absence of one $1b_2$ bonding electron combined with the presence of a $4a_1{}^*$ antibonding electron should make the $^4A_2$ surface on which the reactants approach quite repulsive. Thus direct insertion of low-energy $O^+$ into $H_2$ is unlikely.

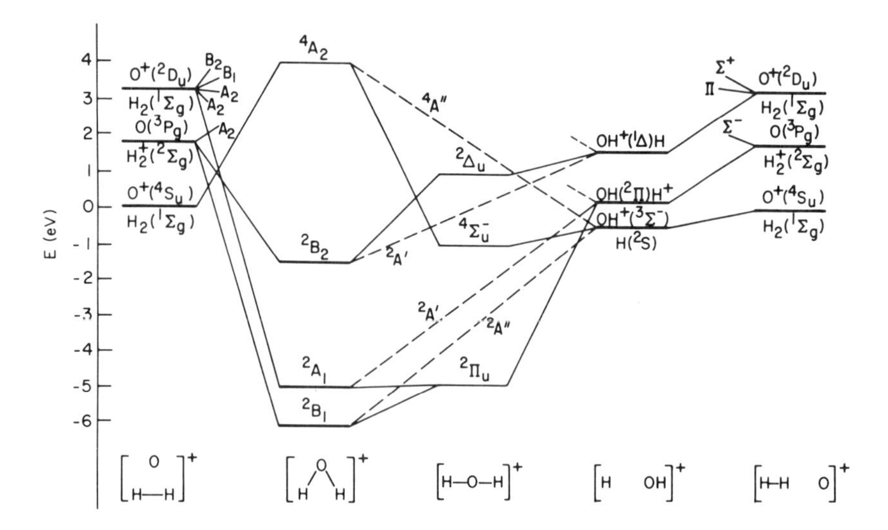

**Figure 2.9** An electronic state correlation diagram for the $H_2O^+$ system. Note that insertion of $O^+(^4S)$ into $H_2(^1\Sigma)$ is forbidden by both spin and orbital symmetry restrictions, whereas abstraction in collinear or quasicollinear conformations is a low-energy reaction path (From Gillen *et al.*[80], by courtesy of the American Institute of Physics)

For the collinear approach of $O^+$ to $H_2$, the orbital and state correlation diagrams show that ground state reactants can evolve adiabatically to ground state products along a $^4\Sigma^-$ surface which is not expected to have either barriers or wells of appreciable size. Thus $O^+(^4S)$ should react with $H_2(^1\Sigma^+{}_g)$ to produce $OH^+(^3\Sigma^-)$ and $H(^2S)$ by a direct interaction process. This has been observed experimentally to be the case at low, intermediate and high energies[79–81].

The correlation diagram of Figure 2.9 provided an explanation of another feature found in the experimental studies[80,81] of the $O^+(H_2,H)OH^+$ reaction. Product $OH^+$ was found in a region of velocity space where ground state $OH^+(^3\Sigma^-)$ would be unstable with respect to dissociation to $O(^3P)$ and $H^+$. In contrast, excited $OH^+(^1\Delta)$, which dissociates to the higher-energy products $O(^1D)$ and $H^+$, would be stable in the same velocity region. The correlation diagram shows that formation of $OH^+(^1\Delta)$ requires a transition from the $^4A_2$–$^4\Sigma^-$ surface of the reactants to the $^2B_2$–$^2\Delta$ surface of the intermediate. Examination of the transformation properties of the spin–orbit coupling operator shows that the $^4A_2$ and $^2B_2$ states can be mixed, and thus the crossing of these surfaces indicated in Figure 2.9 is in fact avoided. The minimum splitting between the states is expected to be of the order of the spin–orbit splitting in the oxygen atom, or *ca.* 0.01 eV. Thus in most collisions of several electron volts relative energy the system will behave diabatically and remain on the $^4A_2$–$^4\Sigma^-$ surface. In some cases, however, the transition to the $^2B_2$–$^2\Delta$ surface will occur, and lead to formation of $OH^+(^1\Delta)$, as the experiments[80] indicate.

The foregoing examples indicate the utility of MO and electronic state correlation diagrams in the study of ion–molecule reaction dynamics. As attention is more closely focused on non-adiabatic behaviour in chemical reactions, other applications of correlation diagrams will undoubtedly appear.

## 2.4 THE SEQUENTIAL IMPULSE MODEL OF DIRECT REACTIONS

### 2.4.1 Introduction

One of the major goals of the study of molecular dynamics is to develop models which allow the interpretation of the important features of reaction phenomena and provide the tools by which predictions of the behaviour of experimentally untried systems can be made. A number of simple models for the reaction process have been proposed[113–123] as alternatives to the expensive numerical calculation of exact trajectories. Even allowing for the necessity of using extremely simple approximations to potential energy surfaces and mechanical behaviour, most of these models are lacking in generality and rigour, and have not been particularly illuminating.

The sequential impulse model of Bates, Cook and Smith[113] is conceptually simple and has the capacity for considerable refinement. In brief, the reaction A(BC,C)AB is viewed as an event in which A hits B impulsively and elastically, B hits C in a like manner, and A combines with B (or C) if the appropriate energy of relative motion is less than the dissociation energy of the product molecule. Suplinskas[115], and George and Suplinskas[122], have elaborated this model and have shown that it can reproduce the major features of the $Ar^+$–$D_2$ reactive and non-reactive scattering. Gillen, Mahan and Winn[81] found that a version of this model in which atoms interact via hard-sphere potentials is consistent with the product angular distributions of the reaction of $O^+$ with $D_2$ and HD in the regime of high relative energies. These two sets of applications involved calculation of the final product velocities from sampled initial conditions using large digital computers. Although the simplicity of the potential energy surfaces makes these calculations relatively inexpensive, it would be valuable if the product distributions could be expressed analytically and evaluated with a small calculator. This proves to be possible, and in following sections we outline the basis of this analytical sequential impulse model and discuss certain experimental data in terms of it. The model enjoys a modicum of rigour, and is simple enough that its basis can be understood even by experts.

### 2.4.2 Velocity vector analysis

It is helpful to construct a velocity vector diagram which describes the sequential impulse process in which the projectile A (particle 1) hits B (particle 2), and then B hits C (particle 3). Figure 2.10 shows the projection of such a diagram in the plane of $V_1$, the initial velocity of A, and $V_3''$, the final velocity of C. We assume that BC is initially stationary in the laboratory

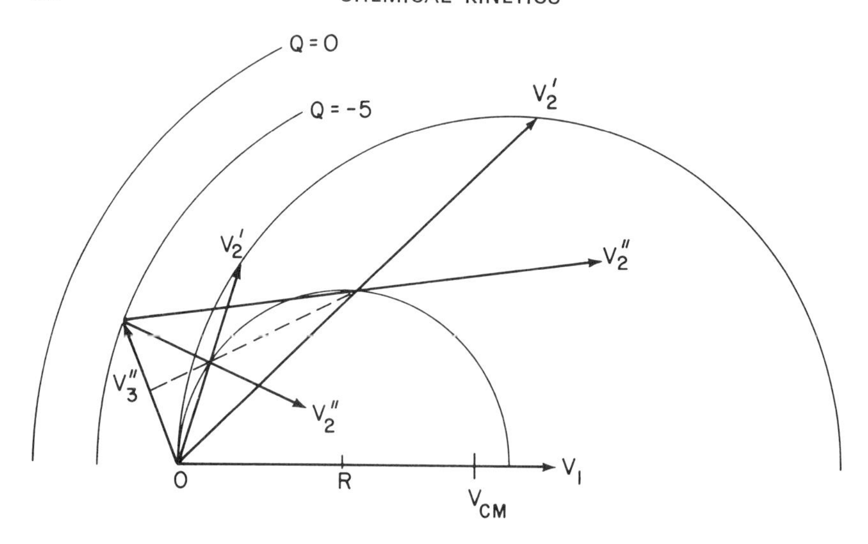


**Figure 2.10** A velocity vector diagram for the sequential impulse model of the reaction $O^+(D_2,D)OD^+$ at 20 eV initial relative energy, treated as a thermoneutral reaction. The impulsive collision sequence $1 \to 2$, $2 \to 3$ is assumed, and a prime is added to the velocities after each impulse has occurred. The limits of the stability zone for particle 3, the free product atom, are given by the circles marked $Q = 0$, $Q = -5$. In the plane of $V_1$ and $V_3''$, the perpendicular bisector of $V_3''$ (the dashed line) intersects the centroid circle (centred at $R$) at two points, which permit the two vectors $V_2'$ which can lead to $V_3''$ to be found. The velocity of the centre-of-mass of the total system is at $V_{CM}$

and that A moves with initial laboratory velocity $V_1$. As a result of an elastic collision with A, atom B is found with a laboratory velocity vector $V_2'$ which terminates on a sphere whose centre is at the velocity of the A–B centre-of-mass, and whose radius is $V_1A/(A+B)$. These various values of $V_2'$ are the possible initial relative velocities for the B–C collision, since C is still at rest in the laboratory. The centroids of the B–C pair are therefore located on a centroid sphere of radius

$$R = V_1 \frac{m_1}{m_{12}} \frac{m_2}{m_{23}}$$

whose centre is on $V_1$, at a distance $R$ from the origin. The result of the B–C elastic encounter is to rotate the vector $V_2'$ about the B–C centroid velocity and to produce a final laboratory velocity $V_3''$ for particle C. Once $V_3''$ is known, the final relative velocity and the scattering angle of the product AB can be found readily. In order for AB to exist as a bound molecule, $V_3''$ must fall into the stability zone[81] which corresponds to those final relative velocities consistent with AB internal excitation energies equal to or greater than zero, but less than the dissociation energy.

Because the A–B and B–C encounters are treated as independent events, there must be no long-range *forces* between atoms. The only types of potential energy surfaces that meet this criterion are those that result from AB and BC being pictured as square-well oscillators. For example, at large A–BC distances

$$\begin{aligned} \phi(\mathrm{BC}) &= \infty \quad r_{23} \leqslant d_{23} \\ &= 0 \quad d_{23} < r_{23} < \rho_{23} \\ &= D_{23} \quad r_{23} \geqslant \rho_{23} \end{aligned}$$

Here $\phi$ is the potential energy, $d_{23}$ is the mutual hard-sphere diameter of BC, $\rho_{23}$ is the BC separation at the outer wall of the square well, and $D_{23}$ is the BC dissociation energy. Similar relations hold for $\phi$(AB). Separation of BC and formation of AB can occur without any energy barriers being crossed if $\phi$(BC) is zero when $d_{23} < r_{23}$ and $d_{12} \leqslant r_{12} \leqslant \rho_{12}$. For this very simple surface there is an obvious reaction criterion: while $r_{12}$ is small, $r_{23}$ must increase sufficiently that the system enters the product channel, rather than returns to the reactant channel. For reactions at very high energy this criterion is almost equivalent to the requirement that the velocity of C be large enough so that AB remains bound. In applications of the sequential impulse model, this latter requirement is the only criterion for reaction which is used.

To find the intensity of product at a given value of $V_3''$, one must locate all impulse sequences that give scattering at $V_3''$, and sum their contributions. The first part of this proves to be rather simple. The B–C centroids which give scattering at $V_3''$ must satisfy two conditions. They must lie on the centroid sphere, and they must be on the perpendicular bisector of $V_3''$. In three dimensions, the perpendicular bisector of $V_3''$ (a plane) intersects the centroid sphere to generate a 'magic circle', the locus of all B–C centroids which can lead to scattering to the selected value of $V_3''$.

Figure 2.10 shows the part of the construction which occurs in the plane of $V_1$ and $V_3''$. The perpendicular bisector of $V_3''$ intersects the centroid sphere at two points. The corresponding B–C scattering angles represent limits of the range of values (and hence of B–C impact parameters) which can contribute to scattering at $V_3''$. To obtain the intensity at $V_3''$, one must integrate over the allowed range of BC impact parameters after properly weighting both the velocities $V_2'$ and the impact parameters.

The proper weighting of the final impact parameters can be found rather directly. For a given velocity $V_2'$ the probability that the BC axis will make a polar angle $\alpha$ with the $V_2'$ direction is simply

$$P(\alpha)\,\mathrm{d}\alpha = \tfrac{1}{2}\sin\alpha\,\mathrm{d}\alpha$$

To express this in terms of the B–C impact parameter $b$, we use for $\alpha < \pi/2$ the obvious relation

$$b = r_0 \sin\alpha$$

where $r_0$ is the BC bond distance, and obtain

$$P(b)\,\mathrm{d}b = \frac{b\,\mathrm{d}b}{2r_0^{\,2}\left(1-\dfrac{b^2}{r_0^{\,2}}\right)^{\frac{1}{2}}}; \quad \alpha < \frac{\pi}{2}$$

This proves to be the major factor which controls the intensity distribution of large-angle scattering.

For $\alpha > \pi/2$ there can be no BC collision if the atoms are hard spheres, so this range of orientations gives a zero value for $V_3''$. If the AB from such collisions is bound, it appears at the spectator stripping velocity. Another contribution to the stripping intensity comes from those impact parameters for which $\alpha < \pi/2$, but which are greater than $d_{23}$, the B–C collision diameter. If there were long-range attractive or repulsive forces between the separating atoms B and C, the stripping contribution would be scattered away from the zero-angle point, and the stripping peak would appear broadened or disappear entirely.

The argument just given indicates that a necessary criterion for the appearance of a sharp peak at the stripping velocity is a virtual absence of forces between the departing B and C, *and* an absence or weakness of angular dependent terms in the A–BC potential. For the $Ar^{+}$–$D_2$ system there is some evidence from diatomics-in-molecules calculations[20] that this weak angular dependence of the potential actually occurs. However, it is difficult to believe that the rigorous conditions for spectator stripping are met exactly in the large number of cases for which a stripping peak has been observed. It seems more likely that apparatus resolution effects make a very important contribution to the appearance of stripping peaks, and may even cause a minimum intensity at zero angle to appear as a maximum when there is considerable intensity at small angles. Re-examination of spectator stripping peaks at high resolution could be revealing.

To obtain the final expression for the specific intensity of atom C with velocity vector $\boldsymbol{V}_3''$, the rate at which B particles are formed in the volume element $(V_2')^2\, \mathrm{d}\Omega'\, \mathrm{d}V_2'$ is found, the ratio of this volume element to $(V_3'')^2\, \mathrm{d}\Omega''\, \mathrm{d}V_2''$ is evaluated, and the resulting expression is integrated over the allowed range of final impact parameters. The detailed derivations for reaction to form AB and AC and the final expressions for the intensity will be given elsewhere. Here we emphasise the interpretation of the experimental intensity distributions in terms of the vector diagram of Figure 2.10.

### 2.4.3 Discussion

The general qualitative form of the rebound scattering can be anticipated from the construction of Figure 2.10. Velocity vectors $\boldsymbol{V}_3''$ which are small and directed perpendicular to $\boldsymbol{V}_1$ will have bisectors which intercept the centroid sphere so as to give a large *range* of contributing B–C scattering angles. Thus there will be a high product intensity if $\boldsymbol{V}_3''$ lies in the zone of velocities which implies that AB is bound. In contrast, if $\boldsymbol{V}_3''$ is large enough, or makes a sufficiently obtuse angle with $\boldsymbol{V}_1$, its bisector may not make an intersection with the centroid sphere. Such velocities cannot be reached by sequential impulses, even though they may satisfy conservation of energy and product stability criteria. In particular, exactly backward recoil of C accompanied by forward recoil of AB is not possible if the overall reaction is thermoneutral, and the sequential impulse model applies. Very probably this is the reason why in the nearly thermoneutral reaction $O^{+}(H_2,H)OH^{+}$, no $OH^{+}$ is found to recoil forward when the initial relative energy is high, whereas such forward recoil is displayed by the exoergic reactions of $Ar^{+}$, $N_2^{+}$ and $CO^{+}$ with $H_2$.

The sequential impulse model indicates that large-angle scattering of AB will be of low intensity. For these processes $\boldsymbol{V}_3''$ lies nearly parallel to $\boldsymbol{V}_1$, and a construction similar to Figure 2.10 shows that only a small range of final scattering angles and impact parameters contributes to the intensity at $V_3''$. Apparently the sequential impulse model cannot give predominantly backward scattering of AB unless additional criteria for reaction, such as restricting $b$ to small values and $V_2'$ to large values, are imposed.

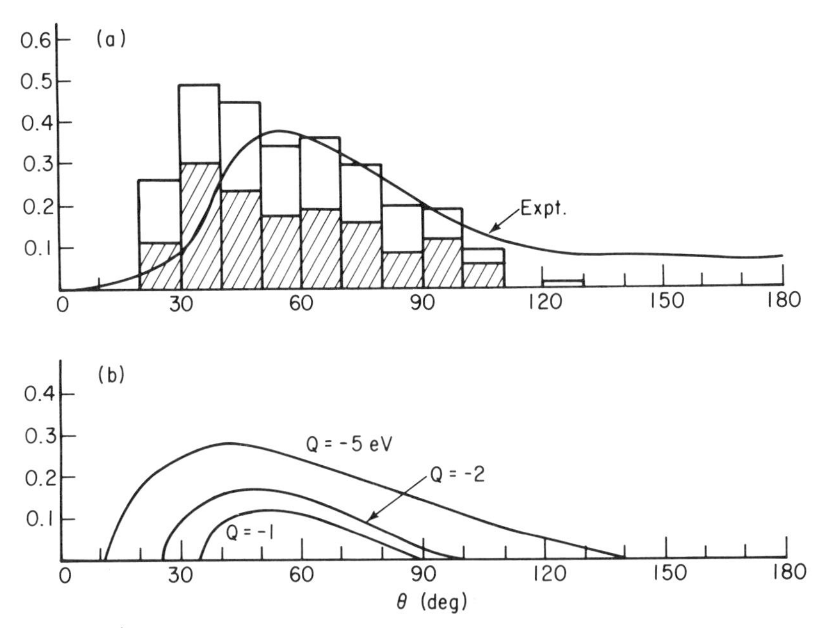


**Figure 2.11** A comparison of various calculated and experimental angular distributions of $OD^+$ from the $O^+(D_2,D)OD^+$ reaction at 20 eV initial relative energy. In the upper panel the histogram shows the results calculated numerically using exact hard-sphere trajectories. The shaded bars represent reaction with the atom hit first by $O^+$, and the open bars represent reaction with the trailing or 'wrong' atom. These calculations and the experimental data are from the work of Gillen *et al*[81]. The bottom panel shows the angular distributions of $OD^+$ at various values of $Q$ calculated with the analytical sequential impulse model. Only reactions with the initial atom struck are included, but the contribution of 'wrong' atom reactions is virtually identical in all respects. The vertical scale in the lower panel has been arbitrarily adjusted to facilitate comparison with the results in the upper panel

A comparison of the predictions of the analytical sequential impulse model with exact hard-sphere trajectory calculations and experimental data is made in Figure 2.11. The system chosen is the $O^+(D_2,D)OD^+$ reaction at 20 eV initial relative energy. At this high relative energy, the product formed by spectator stripping is unstable, so the only contributions to the product intensity come from events in which a fairly strong B–C interaction occurs. The minimum value of $Q$ which corresponds to the limit of product stability is $-5$ eV, while the maximum value of $Q$ is zero, since the reaction is treated as if it were thermoneutral.

In a strict sense, the three types of angular distributions presented in Figure 2.11 are not comparable. The experimental results are given as an angular distribution obtained by integrating the measured product intensity over the final relative velocity. Effects of the finite apparatus resolution have not been removed. The histogram gives the angular distribution obtained from the exact hard-sphere trajectory calculations[81], but has not been corrected for apparatus resolution effects. Finally, the curves from the analytical sequential impulse model are angular distributions at fixed values of the translational exoergicity $Q$. These would have to be weighted by the value of $u^2$, the square of the speed of $OD^+$ relative to the centre-of-mass, and integrated over $u$ to obtain a true differential cross-section. However, the value of $u^2$ varies by only 25% over the full range in which products are stable, so at most angles the major contributions to the intensity come from $Q$ values rather close to $-5$ eV. Thus the general shape of the differential cross-section for the analytical sequential impulse model can be inferred fairly accurately from the angular distribution at $Q = -5$ eV.

The shapes of the distributions from the analytical sequential impulse model and the results of the exact hard-sphere trajectory calculations are quite similar. This should be the case, for they are essentially the same model, the only difference being that shadowing effects and sequences of three impulses are included in the trajectory calculations. The form of the experimental distributions is reproduced fairly well, except at very small and very large scattering angles. Some of the discrepancy can be attributed to the finite resolution of the apparatus, which tends to produce extra intensity near 0° and 180°. In addition, some scattering at large angles may come from more complicated mechanical processes in which $O^+$ exerts large forces on both deuterium atoms simultaneously.

The analytical and numerical[81] versions of the impulse model reproduce the very dissimilar experimental forms of the angular distributions of $OH^+$ and $OD^+$ from $O^+$–HD collisions rather well. The deviations between the model and experiment can be attributed to the complete neglect of attractive forces in the model. The basic reason for the strong forward scattering of $OH^+$ observed experimentally can be deduced directly from the appropriate vector diagram. For formation of $OH^+$ by the process in which H is hit first by $O^+$, the centroid sphere is small ($R = 0.314V_1$), and only relatively small values of $V_3''$ are possible. When $OH^+$ is formed by events in which D is hit first by $O^+$, the centroid sphere is larger ($R = 0.592V_1$), so fairly large values of $V_3''$ are possible. Hence $OH^+$ can appear at larger angles by means of this 'wrong atom' process. However, in relatively few of these events is the translational energy imparted to the D atoms sufficient to stabilise $OH^+$, and consequently the $OH^+$ intensity is small at large angles.

The predominate scattering of $OD^+$ to large angles can also be rationalised. When $OD^+$ is formed by processes in which D is hit first, the centroid sphere is large, and the velocity vector $V_3''$ of the free H atom can be large enough to place appreciable $OD^+$ intensity in the 60–90° region, and to a lesser degree at larger angles. Formation of $OD^+$ at small angles occurs almost exclusively by processes in which $O^+$ hits H first. The sequential impulse model overestimates the small-angle scattering of $OD^+$, evidently because of the neglect of attractive forces between $O^+$ and the freed H atom.

The sequential impulse model can be extended to include certain forms of reactant and product attraction or repulsion. The simplest situation is the one which has been treated, at least partially, by the so-called modified stripping or polarisation-reflection models[120]. Attraction between the centres-of-mass of the reactants A and BC produces an expansion of the initial relative velocity vector $V_1$. Starting with the expanded value of $V_1$, elastic scattering of A from B and B from C then follows directly as outlined above to produce the angular distribution of reaction products. Any attraction between the centres-of-mass of the products AB and C is included by shrinking the final relative velocity vector. This procedure is, of course, approximate since it ignores the turning of the initial and final relative velocity vectors which long-range forces produce. However, at initial relative energies high enough so that the sequential impulse model might be expected to be a good approximation, the inclusion of reasonable attractive forces produces a very minor shift of the final relative translational energy distribution to larger values. At initial relative energies low enough so that the modifications due to the long-range attraction between reactants and products might produce small but significant shifts in both the energy and angular distributions of the products, the assumptions of the sequential impulse model are of questionable validity. The results in this case may be more illusory than illuminating. The same may be said of the primitive modified stripping or polarisation-reflection models[120] in which the effects of momentum transfer from A to B to C are either ignored or lumped into a single undetermined parameter.

The sequential impulse model also can include as a special case the DIPR model (direct interaction with product repulsion), which has been discussed and applied principally by Kuntz[118] and Marron[123]. In the DIPR model the scattering angle of atom C is computed by assuming that it is deflected only by a repulsive force which acts along the BC bond axis. While this force arises because of the presence of the incoming atom A, there is no explicit effect of the transfer of the original momentum of A to B and then to C. Thus, in terms of Figure 2.10, the only point on the centroid sphere which is considered is at the origin of the laboratory velocity system. The final relative velocity of B and C which appears is a parameter which may be adjusted to fit experimental data or be estimated from an appropriate potential energy surface. It is clear that the DIPR model may be an adequate description of the reaction process if the initial momentum of A is small, since this insures a small centroid sphere, and minimal effects of sequential momentum transfer. Thus the DIPR model or its close relatives should work well[124] for the thermal reaction of H with $Cl_2$. However, for the reaction of $Ar^+$ with $D_2$ at initial relative energies in the range of 2–5 eV, good results cannot be expected. This is consistent with the findings of Kuntz and Roach[106].

## 2.5 CONCLUSION

After a late start, research on ion-scattering phenomena has made rather impressive progress. A number of significant first accomplishments in chemical scattering have been made using ions: in the realm of elastic scattering, the first determination of the potential energy curves of chemically bonded ($D_0 \gtrsim 1$ eV) species[14,16], the first measurements of the angular distributions of inelastic scattering[36], the first resolution of discrete vibrational transitions induced by collision[39], the first measurements of reaction cross-sections for reactants in selected vibrational states[71], the first use of product velocity analysis to demonstrate spectator stripping[125], the first complete product velocity vector distributions for a chemical reaction[126], and the first comparisons of experimental velocity vector intensity maps with those calculated from an *ab initio* potential surface[66,67]. While experimental convenience has tended to restrict ion–molecule scattering experiments to initial relative energies above 1 eV, lower energies have been reached and will become more accessible as techniques are improved. At the same time, neutral–neutral scattering experiments are being carried out at energies extending well into the electron volt range. It is to be hoped that the rather superficial distinctions between ion–neutral and neutral–neutral scattering phenomena which have arisen or been imposed will soon disappear.

## Acknowledgement

This article was written during my tenure as Research Professor in the Miller Institute for Basic Research at the University of California, Berkeley. I would like to acknowledge the long-term support for my research in ion collision phenomena which has been provided by the U.S. Atomic Energy Commission through the Inorganic Materials Research Division of the Lawrence Berkeley Laboratory.

## Notes added in proof

In the time elapsed between the completion of this review and its publication, a substantial number of papers on ion–molecule collision phenomena have appeared. Time and space permit only a brief mention of those most closely connected with topics mentioned in this review.

A large book[128] which contains papers presented at the 1974 *NATO Advanced Study Institute on Ion–Molecule Interactions* has been published. Extended discussions of elastic scattering, direct reactive interactions, product angular distributions, classical trajectories, and potential energy surfaces for ion–molecule processes are included in a very broad collection of subjects.

The theory of the inelastic scattering of alkali ions by hydrogen molecules, particularly for the $Li^+$–$H_2$ system, has received considerable[129–134] attention. A number of edifying *ab initio* calculations of potential energy surfaces for molecule-ions have been done. Systems treated include[135–138] $N_2H^+$, $H_2O^+$, $HCO^+$ and $NeH_2^+$. Experimental work on reactive scattering which extends some of the ideas mentioned in this review includes an investigation of the $C^+$–$H_2$ system at very low kinetic energies[139], merged beam studies. of the $H_2^+(D,H)HD^+$ and $D^+(H_2,H)HD^+$ reactions[140,141], angular distribution measurements and correlation diagram analysis[142] of the $N^+(H_2,H)NH^+$ reaction, and product velocity vector distribution measurements[143] for the reactions of $NH_3^+$, $NH_2^+$, $NH^+$ and $N^+$ with HD. Total cross sections[144], isotope effects and product angular distributions[145] have been determined at several energies for the $F^+(H_2,H)HF^+$ reaction.

## References


1. Franklin, J. L., editor (1972). *Ion–Molecule Reactions* (New York: Plenum)
2. Dubrin, J. and Henchman, M. J. (1972). *MTP International Review of Science, Physical Chemistry, Series One*, Vol. 9, *Chemical Kinetics*, 213 (J.C. Polanyi, editor) (London: Butterworth)
3. Friedman, L. and Reuben, B. G. (1971). *Adv. Chem. Phys.*, **19,** 33
4. Bowers, M. T. and Su, T. (1973). *Adv. Electronics Electron Phys.*, **34,** 223
5. Gray, G. A. (1971). *Adv. Chem. Phys.*, **19,** 141
6. Tully, J. C. (1973). *Ber. Bunsenges. Phys. Chem.*, **77,** 557
7. Herman, Z. and Birkinshaw, K. (1973). *Ber. Bunsenges. Phys. Chem.*, **77,** 566
8. Durup, J. (1973). *Adv. Mass Spectrom.*, to be published
9. Henglein, A. (1972). *J. Phys. Chem.*, **76,** 3883
10. Dubrin, J. (1973). *Ann. Rev. Phys. Chem.*, **24,** 97
11. Parker, J. E. and Lehrle, R. S. (1971). *Int. J. Mass Spectrom. Ion Phys.*, **7,** 421
12. Weise, H. P. (1973). *Ber. Bunsenges. Phys. Chem.*, **77,** 578
13. Mittmann, H. U., Weise, H. P., Ding, A. and Henglein, A. (1971). *Z. Naturforsch.*, **26a,** 1112
14. Weise, H. P., Mittmann, H. U., Ding, A. and Henglein, A. (1971). *Z. Naturforsch.*, **26a,** 1122
15. Mittman, H. U., Weise, H. P., Ding, A. and Henglein, A. (1971). *Z. Naturforsch.*, **26a,** 1282
16. Bobbio, S. M., Rich, W. G., Doverspike, L. D. and Champion, R. L. (1971). *Phys. Rev.*, **A4,** 957
17. Rich, W. G., Bobbio, S. M., Champion, R. L. and Doverspike, L. D. (1971). *Phys. Rev.*, **A4,** 2253
18. Klingbeil, R. (1972). *J. Chem. Phys.*, **57,** 1066
19. Hyatt, D. and Stanton, L. (1970). *Proc. Roy. Soc.* (*London*), **A318,** 107
20. Kuntz, P. J. and Roach, A. C. (1972). *J. Chem. Soc. Faraday Trans. II*, **68,** 259
21. Inouye, H. and Kita, S. (1973). *J. Phys. Soc. Jap.*, **34,** 1588
22. Powers, T. R. and Cross, R. J. (1973). *J. Chem. Phys.*, **58,** 626
23. Inouye, H. and Kita, S. (1972). *J. Chem. Phys.*, **57,** 1301
24. Amdur, I., Jordan, J. E., Chien, K. R., Fung, L. W., Hance, R. L., Hulpke, E. and Johnson, S. E. (1972). *J. Chem. Phys.*, **57,** 2117
25. Inouye, H. and Kita, S. (1972). *J. Chem. Phys.*, **56,** 4877
26. Boerboom, A. J. H., van Dop, H. and Los, J. (1970). *Physica*, **46,** 458
27. Amdur, I., Inouye, H., Boerboom, A. J. H., Steege, A. N., Los, J. and Kistemaker, J. (1969). *Physica*, **41,** 566
28. Doering, J. P. (1973). *Ber. Bunsenges. Phys. Chem.*, **77,** 593
29. Moore, J. H., Jr. and Doering, J. P. (1970). *J. Chem. Phys.*, **52,** 1692
30. Moore, J. H., Jr. (1971). *J. Chem. Phys.*, **55,** 2760
31. Doering, J. P. and Moore, J. H. (1972). *J. Chem. Phys.*, **56,** 2176
32. Moore, J. H. (1972). *J. Geophys. Res.*, **77,** 5567
33. Moore, J. H. (1973). *Phys. Rev.*, **A8,** 2359
34. Augustin, S. D., Miller, W. H., Pearson, P. and Schaefer, H. F. (1973). *J. Chem. Phys.*, **58,** 2845
35. Gillen, K. T., Mahan, B. H. and Winn, J. S. (1972). Unpublished results
36. Gentry, W. R., Gislason, E. A., Mahan, B. H. and Tsao, C. W. (1967). *J. Chem. Phys.*, **47,** 1856
37. Cheng, M. H., Chiang, M. H., Gislason, E. A., Mahan, B. H., Tsao, C. W. and Werner, A. S. (1970). *J. Chem. Phys.*, **52,** 6150
38. Mahan, B. H. (1970). *J. Chem. Phys.*, **52,** 5221
39. Moore, J. H., Jr. and Doering, J. P. (1969). *Phys. Rev. Lett.*, **23,** 564
40. Herrero, F. A. and Doering, J. P. (1972). *Phys. Rev.*, **A5,** 702
41. Herrero, F. A. and Doering, J. P. (1972). *Phys. Rev. Lett.*, **29,** 609
42. Udseth, H., Giese, C. F. and Gentry, W. R. (1971). *J. Chem. Phys.*, **54,** 3642
43. Cosby, P. C. and Moran, T. F. (1970). *J. Chem. Phys.*, **52,** 6157
44. Petty, F. and Moran, T. F. (1972). *Phys. Rev.*, **A5,** 266
45. Shin, H. K. (1969). *J. Phys. Chem.*, **73,** 4321
46. Dittner, P. F. and Datz, S. (1968). *J. Chem. Phys.*, **49,** 1969

47. Dittner, P. F. and Datz, S. (1971). *J. Chem. Phys.*, **54,** 4228
48. van Dop, H., Boerboom, A. J. H. and Los, J. (1971). *Physica*, **54,** 223
49. Schöttler, J. and Toennies, J. P. (1968). *Z. Phys.*, **214,** 472
50. Held, W.-D., Schöttler, J. and Toennies, J. P. (1970). *Chem. Phys. Lett.*, **6,** 304
51. David, R., Faubel, M. and Toennies, J. P. (1973). *Chem. Phys. Lett.*, **18,** 87
52. Lester, W. A. (1971). *J. Chem. Phys.*, **54,** 3171
53. Schöttler, J. (1972). *Chem. Ber.*, **104,** 72; Max Planck Institute für Strömungforschung, Göttingen (unpublished)
54. Faubel, M. and Toennies, J. P. (1974). *Chem. Phys.*, **4,** 36
55. Inouye, H. and Kita, S. (1973). *J. Chem. Phys.*, **59,** 6656
56. Mahan, B. H. (1974). *J. Chem. Ed.*, **51,** 308, 377
57. Hirschfelder, J. O. (1969). *Int. J. Quantum Chem.*, **3S,** 17
58. Landau, L. and Teller, E. (1936). *Phys. Z. Sowjetunion*, **10,** 34
59. Kelley, J. D. and Wolfsberg, M. (1970). *J. Chem. Phys.*, **53,** 2967
60. Attermeyer, M. and Marcus, R. A. (1970). *J. Chem. Phys.*, **52,** 393
61. Gillen, K. T., Mahan, B. H. and Winn, J. S. (1973). *Chem. Phys. Lett.*, **22,** 344
62. Krenos, J. and Wolfgang, R. (1970). *J. Chem. Phys.*, **52,** 5961
63. Maier, W. B. (1971). *J. Chem. Phys.*, **54,** 2732
64. Holliday, M. G., Muckerman, J. T. and Friedman, L. (1971). *J. Chem. Phys.*, **54,** 1058
65. Holliday, M. G., Muckerman, J. T. and Friedman, L. (1971). *J. Chem. Phys.*, **54,** 3853
66. Krenos, J., Preston, R., Wolfgang, R. and Tully, J. (1971). *Chem. Phys. Lett.*, **10,** 17
67. Krenos, J. R., Preston, R. K., Wolfgang, R. and Tully, J. C. (1974). *J. Chem. Phys.*, **60,** 1634
68. Preston, R. K. and Tully, J. C. (1971). *J. Chem. Phys.*, **54,** 4297
69. Preston, R. K. and Tully, J. C. (1971). *J. Chem. Phys.*, **55,** 562
70. Tully, J. C. (1973). *Ber. Bunsenges. Phys. Chem.*, **77,** 557
71. Chupka, W. A. and Russell, M. E. (1968). *J. Chem. Phys.*, **49,** 5426
72. Chupka, W. A., Berkowitz, J. and Russell, M. E. (1969). *Proceedings of the Sixth International Conference on the Physics of Electronic and Atomic Collisions*, 71 (Cambridge, Mass.: M.I.T. Publications Office)
73. Rutherford, J. A. and Vroom, D. A. (1973). *J. Chem. Phys.*, **58,** 4076
74. Chiang, M. H., Gislason, E. A., Mahan, B. H., Tsao, C. W. and Werner, A. S. (1971). *J. Phys. Chem.*, **75,** 1426
75. Leventhal, J. J. (1971). *J. Chem. Phys.*, **54,** 3279
76. Leventhal, J. J. (1973). *J. Chem. Phys.*, **58,** 4710
77. Neynaber, R. H. and Magnuson, G. D. (1973). *J. Chem. Phys.*, **59,** 825
78. Pacák, V., Birkinshaw, K. and Herman, Z. (1973). *Proceedings of the Eighth International Conference on the Physics of Electronic and Atomic Collisions*, 106 (Beograd, Yugoslavia: Institute of Physics)
79. Harris, H. H. and Leventhal, J. J. (1973). *J. Chem. Phys.*, **58,** 2333
80. Gillen, K. T., Mahan, B. H. and Winn, J. S. (1973). *J. Chem. Phys.*, **58,** 5373
81. Gillen, K. T., Mahan, B. H. and Winn, J. S. (1973). *J. Chem. Phys.*, **59,** 6380
82. Henglein, A. (1966). *Adv. Chem. Ser.*, **58,** 63
83. Chiang, M., Gislason, E. A., Mahan, B. H., Tsao, C. W. and Werner, A. S. (1970). *J. Chem. Phys.*, **52,** 2698
84. Chiang, M. H., Mahan, B. H., Tsao, C. W. and Werner, A. S. (1970). *J. Chem. Phys.*, **53,** 3752
85. Iden, C. R., Liardon, R. and Koski, W. S. (1971). *J. Chem. Phys.*, **54,** 2757
86. Iden, C. R., Liardon, R. and Koski, W. S. (1972). *J. Chem. Phys.*, **56,** 851
87. Mahan, B. H. and Sloane, T. M. (1973). *J. Chem. Phys.*, **59,** 5661
88. Liskow, D. H., Bender, C. F. and Schaefer, H. F. (1974). *J. Chem. Phys.*, **61,** 2507
89. Schmeltekopf, A. L., Ferguson, E. E. and Fehsenfeld, F. C. (1968). *J. Chem. Phys.*, **48,** 2966
90. Johnsen, R. and Biondi, M. A. (1973). *J. Chem. Phys.*, **59,** 3504
91. Giese, C. F. (1966). *Adv. Chem. Ser.*, **58,** 20
92. Rutherford, J. A. and Vroom, D. A. (1971). *J. Chem. Phys.*, **55,** 5622
93. Cohen, R. B. (1972). *J. Chem. Phys.*, **57,** 676
94. Kaufman, J. J. and Koski, W. S. (1969). *J. Chem. Phys.*, **50,** 1942

95. O'Malley, T. F. (1970). *J. Chem. Phys.*, **52,** 3269
96. McFarland, M., Albritton, D. L., Fehsenfeld, F. C., Ferguson, E. E. and Schmeltekopf, A. L. (1973). *J. Chem. Phys.*, **59,** 6620
97. Leventhal, J. J. (1971). *J. Chem. Phys.*, **54,** 5102
98. Gentry, W. R., Gislason, E. A., Mahan, B. H. and Tsao, C. W. (1968). *J. Chem. Phys.*, **49,** 3058
99. Cross, R. J. and Wolfgang, R. (1969). *J. Phys. Chem.*, **73,** 743
100. Neynaber, R. H. and Magnuson, G. D. (1973). *J. Chem. Phys.*, **58,** 4586
101. Smith, G. P. K. and Cross, R. J. (1974). *J. Chem. Phys.*, **60,** 2125
102. Bauschlicher, C. W., O'Neil, S. V., Preston, R. K., Schaefer, H. F. and Bender, C. F. (1973). *J. Chem. Phys.*, **59,** 1286
103. Benson, M. J. and McLaughlin, D. R. (1972). *J. Chem. Phys.*, **56,** 1322
104. McLaughlin, D. R. and Thompson, D. L. (1973). *J. Chem. Phys.*, **59,** 4393
105. Chapman, S. and Preston, R. K. (1974). *J. Chem. Phys.*, **60,** 650
106. Kuntz, P. J. and Roach, A. C. (1973). *J. Chem. Phys.*, **59,** 6299
107. Mahan, B. H. (1968). *Accounts Chem. Res.*, **1,** 217
108. Dimpfl, W. L. and Mahan, B. H. (1974). *J. Chem. Phys.*, **60,** 3238
109. Tully, J. C., Herman, Z. and Wolfgang, R. (1971). *J. Chem. Phys.*, **54,** 1730
110. Mahan, B. H. (1971). *J. Chem. Phys.*, **55,** 1436
111. Mahan, B. H. and Winn, J. S. (1972). *J. Chem. Phys.*, **57,** 4321
112. Herzberg, G. (1966). *Electronic Spectra and Electronic Structure of Polyatomic Molecules* (Princeton, N.J.: Van Nostrand)
113. Bates, D. R., Cook, C. J. and Smith, F. J. (1964). *Proc. Phys Soc.*, **83,** 49
114. Light, J. C. and Horrocks, J. (1964). *Proc. Phys. Soc.*, **84,** 527
115. Suplinskas, R. J. (1968). *J. Chem. Phys.*, **49,** 5046
116. Kuntz, P. J. (1969). *Chem. Phys. Lett.*, **4,** 129
117. Light, J. C. and Chan, S. (1969). *J. Chem. Phys.*, **51,** 1008
118. Kuntz, P. J. (1970). *Trans. Faraday Soc.*, **66,** 2980
119. Chang, D. T. and Light, J. C. (1970). *J. Chem. Phys.*, **52,** 5687
120. Hierl, P. M., Herman, Z. and Wolfgang, R. (1970). *J. Chem. Phys.*, **53,** 660
121. Grice, R. and Hardin, D. R. (1971). *Mol. Phys.*, **21,** 805
122. George, T. F. and Suplinskas, R. J. (1971). *J. Chem. Phys.*, **54,** 1037
123. Marron, M. T. (1973). *J. Chem. Phys.*, **58,** 153
124. Herschbach, D. R. (1973). *Faraday Discuss. Chem. Soc.*, **55,** 233
125. Henglein, A., Lacmann, K. and Jacobs, G. (1965). *Ber. Bunsenges. Phys. Chem.*, **69,** 279
126. Gentry, W. R., Gislason, E. A., Lee, Y. T., Mahan, B. H. and Tsao, C. W. (1967). *Discuss. Faraday Soc.*, **44,** 137
127. Udseth, H., Giese, C. F. and Gentry, W. R, (1973). *Phys Rev.*, **A8,** 2483
128. (1975). *Interactions Between Ions and Molecules* (P. Ausloos, editor) (New York Plenum)
129. Brown, N. J. (1974). *J. Chem. Phys.*, **60,** 2958
130. DePusto, A. E. and Alexander, M. H. (1975). *J. Chem. Phys.*, **63,** 5327
131. Kouri, D. J. and Wells, C. A. (1974). *J. Chem. Phys.*, **60,** 2296
132. McGuire, P. (1974). *Chem. Phys.*, **4,** 249
133. Gentry, W. R. and Giese, C. F. (1975). *J. Chem. Phys.*, **62,** 1364
134. Schaefer, J. and Lester, W. A. (1975). *J. Chem. Phys.*, **62,** 1913
135. Leclec, J. C., Horsley, J. A. and Lorquet, J. C. (1974). *Chem. Phys.*, **4,** 337
136. Vasuden, K., Peyerimhoff, S. D. and Buenker, R. J. (1974). *Chem. Phys.*, **5,** 149
137. Bruna, P. J., Peyerimhoff, S. D. and Buenker, R. J. (1975). *Chem. Phys.*, **10,** 323
138. Vasudevan, K. (1975). *Mol. Phys.*, **30,** 437
139. Herbst, E., Champion, R. L. and Doverspike, L. D. (1975). *J. Chem. Phys.*, **63,** 3677
140. Wendel, K. L. and Rol, P. K. (1974). *J. Chem. Phys.*, **61,** 2059
141. Lees, A. B. and Rol, P. K. (1975). *J. Chem. Phys.*, **63,** 2461
142. Fair, J. A. and Mahan, B. H. (1975). *J. Chem. Phys.*, **62,** 515
143. Eisele, G., Henglein, A., Botschwina, P. and Meyer, W. (1974). *Ber. Bunsenges Phys. Chem.*, **78,** 1090
144. Lin, K. C., Cotter, R. J. and Koski, W. S. (1974). *J. Chem. Phys.*, **61,** 905
145. Wendell, K., Jones, C. A., Kaufman, J. J. and Koski, W. S. (1975). *J. Chem. Phys.*, **63,** 750

# 3
# Kinetic Investigation of Enzyme Catalysis

G. G. HAMMES
Cornell University



## 3.1 INTRODUCTION

Enzymes catalyse virtually all physiological reactions and are probably the most efficient known catalysts. Under ordinary conditions almost all biological reactions do not proceed at an appreciable rate in the absence of enzymes. Although many attempts have been made to synthesise enzyme-like catalysts, on a molar basis these synthetic catalysts fall short of enzymatic efficiencies by many orders of magnitude.

All enzymes are macromolecules, with molecular weights of approximately $10^4$–$10^6$. Many enzymes are entirely proteins, but others contain a protein, the apoenzyme, and a non-protein part, the prosthetic group. This prosthetic group is generally a complex organic molecule, but strongly bound metal ions are also quite prevalent.

Enzymes have been extensively studied by a variety of techniques for many years, and no attempt is made here to cover all of the many facets of the subject (*cf.* Ref. 1 for comprehensive reviews). In this chapter, selected examples of kinetic approaches to understanding enzyme catalysis are discussed. The overall goals are to illustrate the kinetic approaches available, to indicate the type of information that can be obtained, and finally to discuss some of the concepts underlying the present understanding of the mechanism of enzyme catalysis.

## 3.2 STEADY STATE KINETICS

Enzymes are such efficient catalysts that their concentrations must be kept very low if the catalysed reaction is to proceed at a rate convenient for study, i.e. if the overall reaction is to take place in seconds or longer. In addition, many enzymes are difficult to obtain in large quantities. Kinetic studies are often carried out at very low enzyme concentrations, typically $10^{-8}$–$10^{-10}$ M, and relatively higher substrate concentrations, typically greater than $10^{-6}$ M. Under these conditions the free enzyme and all of the reaction intermediates are present in sufficiently small concentrations that they can be assumed to be in a steady state (after a very short induction period, which is usually not studied). Experiments carried out under this steady state kinetic condition have been extensively used to characterise many enzymatic reactions. A large number of possible mechanisms exist for the great variety of reactions catalysed by enzymes, and only two reaction types are considered here: single substrate–single product reactions and two substrate–two product reactions. The general principles involved and the information which can be obtained are readily illustrated with these reaction types. More extensive discussions of steady state kinetics are available elsewhere[2,3].

### 3.2.1 Single substrate–single product reaction

A single substrate–single product reaction is the simplest reaction catalysed by an enzyme. An example of such a reaction is that catalysed by the enzyme fumarase:

$$\text{fumarate} + H_2O \rightleftharpoons \text{L-malate} \tag{3.1}$$

The simplest method for carrying out steady state kinetic studies, and the only method considered here, is to measure the initial velocity (rate) of the reaction. When this is done for the enzyme fumarase, the dependence of the initial velocity on the substrate concentration (at constant enzyme concentration) shown schematically in Figure 3.1 is obtained. The maximum velocity reached, $V_S$, is directly proportional to the total enzyme concentration $(E_0)$. The ratio $V_S/(E_0)$, the turnover number, is a direct measure of the catalytic efficiency of the enzyme. The substrate concentration at which the initial velocity is one-half the maximum velocity is called the Michaelis constant. A simple mechanism consistent with the experimental findings is

$$E + S \underset{k_{-1}}{\overset{k_1}{\rightleftharpoons}} X \underset{k_{-2}}{\overset{k_2}{\rightleftharpoons}} E + P \tag{3.2}$$

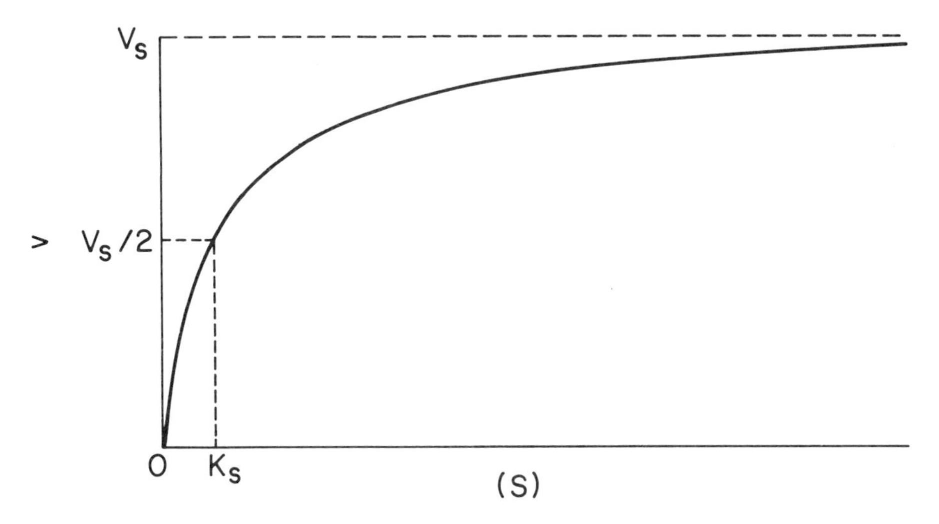


**Figure 3.1** Schematic plot of the initial velocity, $v$, *versus* the substrate concentration, (S), for a mechanism involving one substrate

where S and P represent the substrate and product, E is the free enzyme and X is a reaction intermediate. Since the total substrate concentration $(S_0)$ is much greater than the total enzyme concentration $(E_0)$, the intermediate X can be assumed to be in a steady state:

$$-d(X)/dt = (k_{-1} + k_2)(X) - k_1(E)(S) - k_{-2}(E)(P) = 0 \tag{3.3}$$

Conservation of mass and the assumption $(S_0) \gg (E_0)$ require that

$$(E_0) = (E) + (X) \tag{3.4}$$

$$(S_0) = (S) + (P) \tag{3.5}$$

The overall rate of the reaction is

$$-d(S)/dt = d(P)/dt = k_1(E)(S) - k_{-1}(X) \tag{3.6}$$

which can be combined with equations (3.3)–(3.5) to give

$$d(P)/dt = \frac{[k_1 k_2 (S) - k_{-1} k_{-2} (P)](E_0)}{k_1 (S) + k_{-2} (P) + k_{-1} + k_2} \tag{3.7}$$

This equation is usually written as

$$d(P)/dt = \frac{[V_S/K_S](S) - [V_P/K_P](P)}{1 + (S)/K_S + (P)/K_P} \tag{3.8}$$

where

$$V_S = k_2(E_0)$$
$$V_P = k_{-1}(E_0)$$
$$K_S = (k_{-1} + k_2)/k_1$$
$$K_P = (k_{-1} + k_2)/k_{-2}$$

The initial velocity of the reaction, $v$, can be obtained by setting (P) equal to zero, which gives

$$v = d(P)/dt = V_S/[1 + K_S/(S)] \tag{3.9}$$

This equation displays the dependence on substrate concentration illustrated in Figure 3.1: when $(S) = K_S$, $v = V_S/2$ and when $(S) \rightarrow \infty$, $v \rightarrow V_S$. The parameters $V_S$ and $K_S$ can be readily obtained from the data, either by a direct least-squares fit to equation (3.9) or by fitting other types of plots of the data; for example, a plot of $1/v$ *versus* $1/(S)$ gives a straight line with an intercept equal to $1/V_S$ and a slope equal to $K_S/V_S$ (*cf.* Ref. 3 for a discussion of statistical analyses of data). If the reaction is reversible, which is the case for fumarase, the same procedure can be carried out for the reverse reaction. Since four kinetic parameters are obtained, the four rate constants in equation (3.2) can be calculated. These four kinetic parameters are not independent, however, as they are related through the equilibrium constant, $K$. The relationship between the parameters can be derived from equation (3.8) by setting $d(P)/dt = 0$, which is the equilibrium condition. The equilibrium constant can then be written as

$$K = \frac{(P)}{(S)} = \frac{V_S K_P}{V_P K_S} = \frac{k_1 k_2}{k_{-1} k_{-2}} \tag{3.10}$$

While the mechanism of equation (3.2) is an adequate explanation of the experimental results, the uniqueness of this mechanism must be considered. In particular, the existence of only a single reaction intermediate seems unrealistic for such a complex mechanism. A careful analysis of this problem indicates that the form of the rate equation is independent of the number of intermediates in the mechanism[4]. Thus a general form of the mechanism for $n$ intermediates can be written as

$$E + S \underset{k_{-1}}{\overset{k_1}{\rightleftharpoons}} X_1 \rightleftharpoons \ldots \overset{k_i}{\rightleftharpoons} X_i \rightleftharpoons \ldots \underset{k_{-n}}{\overset{k_n}{\rightleftharpoons}} X_n \underset{k_{-(n+1)}}{\overset{k_{n+1}}{\rightleftharpoons}} E + P \tag{3.11}$$

The rate law for this mechanism is given by equation (3.8), except that the maximal velocities and Michaelis constants are now complex functions of the rate constants. Clearly it is no longer possible to determine all of the rate constants. However, lower bounds to the rate constants are given by the following inequalities[4]:

$$\begin{aligned} &k_{i+1} \geqslant V_S/(E_0) \quad i \neq 0 \\ &k_1 \geqslant [V_S + V_P]/K_S(E_0) \\ &k_{-i} \geqslant V_P/(E_0) \quad i \neq n+1 \\ &k_{-(n+1)} \geqslant [V_S + V_P]/K_P(E_0) \end{aligned} \tag{3.12}$$

This analysis indicates that steady state kinetics does not provide direct information about the reaction intermediates or about the detailed reaction mechanism, which is hardly surprising since the steady state assumption implies that the reaction intermediates are virtually impossible to detect experimentally. Information about the structural specificity of the enzyme can be obtained by varying the structure of the substrate and determining the associated maximum velocities and Michaelis constants. In some cases, further information can be obtained about the nature of the substrate binding site by studying the effect on the initial velocity of unreactive compounds that are structurally similar to the substrate. These compounds compete with the substrate for the catalytic site on the enzyme so that the mechanism of equation (3.2) must be augmented by the reaction

$$E + I \overset{K_I}{\rightleftharpoons} EI \tag{3.13}$$

where I is the inhibitor of the reaction and $K_I$ [= (EI)/(E) (I)] is the equilibrium constant for equation (8.13). The steady state initial velocity for this mechanism is

$$v = \frac{V_S}{1 + [K_S/(S)]\,[1 + K_I(I)]} \tag{3.14}$$

The inhibition constant, $K_I$, can be readily determined by studying the initial velocity as a function of substrate and inhibitor concentrations; by studying the differences in binding of various inhibitors, the structural specificity of the enzyme binding site can be determined.

### 3.2.2 King–Altman method

Before proceeding to a discussion of more complex mechanisms, a brief digression is made to present a very useful method developed by King and Altman for deriving the steady state rate equations for enzyme reactions[5]. This method permits the rate law to be written down by examination without having to solve simultaneous equations. The proof underlying the method involves a complex manipulation of determinants and need not be presented in order to use the method. The basic assumptions are that a catalytic process is involved and that the catalyst concentration is much smaller than that of the substrate.

As a vehicle for presenting the method of King and Altman, consider a simple Michaelis–Menten mechanism with two reaction intermediates:

$$E + S \rightleftharpoons X_1 \rightleftharpoons X_2 \rightleftharpoons E + P \tag{3.15}$$

The mechanism is first written in a manner which better indicates the catalytic nature of the reaction and the steady state condition that $(S_0) \gg (E_0)$:

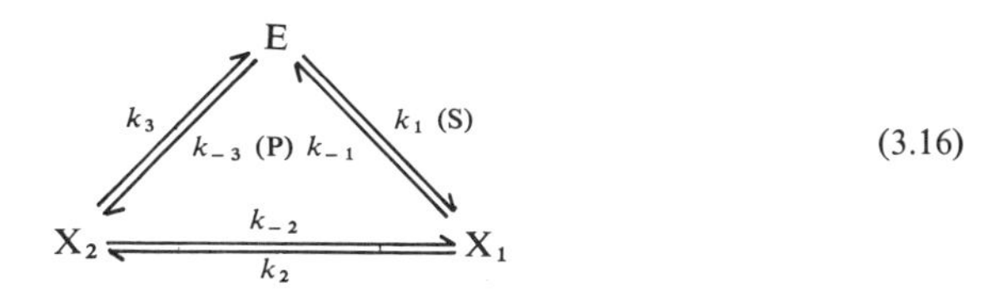


(3.16)

The ratio of any enzymic species to the total enzyme concentration can be written down by a schematic device. First, all possible paths leading directly to a given species within the mechanism are written down; then one individual step in the pathway is blocked out, and only those reaction arrows pointing to the species of interest are retained. This process is repeated for each set of reactions. Thus for the species E, the following three patterns are obtained:

$$X_2 \xrightarrow{k_3} E \xleftarrow{k_{-1}} X_1 \qquad X_2 \xrightarrow{k_3} E, \; X_2 \xleftarrow{k_2} X_1 \qquad X_2 \xrightarrow{k_{-2}} X_1 \xrightarrow{k_{-1}} E \tag{3.17}$$

The ratio (E)/($E_0$) is equal to the sum of the product of the rate constants involved in each of the above three diagrams divided by a denominator, $D$, to be defined shortly. Therefore

$$(E)/(E_0) = [k_{-1} k_3 + k_2 k_3 + k_{-1} k_{-2}]/D \tag{3.18}$$

In general each term of the numerator contains rate constants and concentrations associated with reaction steps leading to the species in question. If the number of enzyme species is $n$, the product of $n-1$ rate constants are found in each term in the numerator and are associated with $n-1$ different enzyme containing species. All of the possible combinations of $n-1$ rate constants which conform to this requirement are present as numerator terms. By constructing appropriate diagrams, the other enzyme species can be written as

$$(X_1)/(E_0) = [k_1 k_3 (S) + k_1 k_{-2}(S) + k_{-2} k_{-3}(P)]/D \tag{3.19}$$

$$(X_2)/(E_0) = [k_1 k_2(S) + k_2 k_{-3}(P) + k_{-1} k_{-3}(P)]/D \tag{3.20}$$

The denominator is simply the sum of all numerator terms:

$$\begin{aligned} D = {} & k_{-1}k_3 + k_2k_3 + k_{-1}k_{-2} + k_1k_3(S) + k_1k_{-2}(S) + k_1k_2(S) \\ & + k_{-2}k_{-3}(P) + k_2k_{-3}(P) + k_{-1}k_{-3}(P) \end{aligned} \tag{3.21}$$

The rate law for the mechanism is

$$d(P)/dt = k_3(X_2) - k_{-3}(E)(P) = (E_0)\left[\frac{k_3(X_2)}{(E_0)} - \frac{k_{-3}(P)(E)}{(E_0)}\right] \tag{3.22}$$

The initial velocity can now be obtained by substitution of equations (3.18)–(3.21) into equation (3.22) and setting (P) = 0. The result is equation (3.9) with

$$V_S = \frac{k_2 k_3 (E_0)}{k_3 + k_{-2} + k_2}$$

$$K_S = \frac{k_2 k_3 + k_{-1} k_3 + k_{-1} k_{-2}}{k_1 (k_{-2} + k_2 + k_3)}$$

This schematic method can be applied to all enzyme mechanisms and saves considerable time and effort for complex mechanisms.

### 3.2.3 Two substrates–two products reaction

Most enzymes utilise more than a single substrate and product. As an example of a more complex mechanism, a reaction with two substrates and two products is now considered. This reaction can be written schematically as

$$A + B \rightleftharpoons C + D \tag{3.23}$$

Examples of equation (3.23) are the reaction of glucose and adenosine 5′-triphosphate to give glucose 6-phosphate and adenosine 5′-triphosphate and the reaction of glutamate and oxalacetate to give ketoglutarate and aspartate. Two types of mechanisms are generally considered for this reaction: one assumes that A transfers a chemical moiety to the enzyme and then dissociates as C; the enzyme then binds B and transfers the chemical moiety to it to form D; the second mechanism supposes that both A and B must be simultaneously bound to the enzyme for reaction to occur. These mechanisms can be written as in equations (3.24) and (3.25), respectively:

$$A + E \underset{k_{-1}}{\overset{k_1}{\rightleftharpoons}} X_1 \underset{k_{-2}}{\overset{k_2}{\rightleftharpoons}} C + X_2 \qquad (3.24)$$

$$X_2 + B \underset{k_{-3}}{\overset{k_3}{\rightleftharpoons}} X_3 \underset{k_{-4}}{\overset{k_4}{\rightleftharpoons}} D + E$$

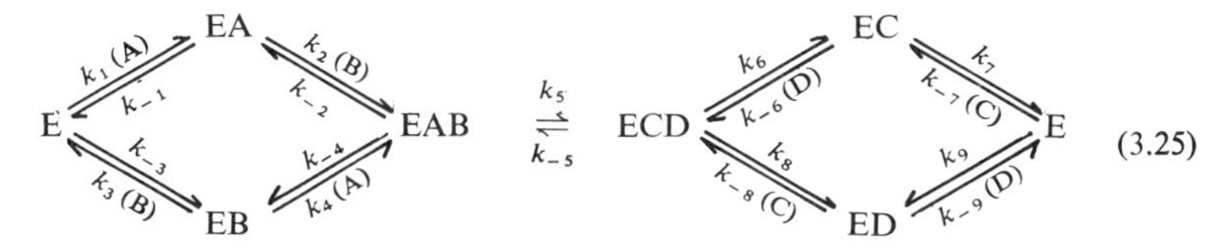

(3.25)

When (C) = (D) = 0, the initial steady state velocity for the mechanism of equation (3.24) can easily be obtained by use of the King–Altman method. The result can be written as

$$\frac{(E_0)}{v} = \phi_1' + \frac{\phi_2'}{(A)} + \frac{\phi_3'}{(B)} \qquad (3.26)$$

where

$$\phi_1' = \frac{k_2 k_4}{k_2 + k_4}$$

$$\phi_2' = \frac{k_{-1} + k_2}{k_1 k_2}$$

$$\phi_3' = \frac{k_{-3} + k_4}{k_3 k_4}$$

The steady state initial velocity for the mechanism of equation (3.25) can also be readily derived, but the resulting expression is so complex that it is rarely used. Instead, two limiting cases are usually considered. One assumes a compulsory addition of substrates to the enzyme, i.e. A must bind before B. This corresponds to deleting the four lowest sets of arrows in equation (3.25). The steady state initial velocity for this case is

$$\frac{(E_0)}{v} = \phi_1 + \frac{\phi_2}{(A)} + \frac{\phi_3}{(B)} + \frac{\phi_4}{(A)(B)} \qquad (3.27)$$

where

$$\phi_1 = \frac{k_6 k_7 + k_{-5} k_7 + k_5 k_7 + k_5 k_6}{k_5 k_6 k_7}$$

$$\phi_2 = \frac{1}{k_1}$$

$$\phi_3 = \frac{k_5 k_6 + k_{-2} k_6 + k_{-2} k_{-5}}{k_2 k_5 k_6}$$

$$\phi_4 = \frac{k_{-1}(k_5 k_6 + k_{-2} k_6 + k_{-2} k_{-5})}{k_1 k_2 k_5 k_6}$$

A second important limiting case of the mechanism of equation (3.25) is to assume that A and B bind to the enzyme independently of each other and furthermore that all of the binding equilibria are adjusted rapidly relative to the interconversion of EAB and ECD. The steady state initial velocity for this case is identical to equation (3.27) with

$$\begin{aligned}\phi_1 &= 1/k_5\\ \phi_2 &= 1/K_1k_5\\ \phi_3 &= 1/K_3k_5\\ \phi_4 &= 1/K_1K_3k_5\end{aligned}$$

where the $K_i$ represent equilibrium constants and the initial assumption requires that $K_1 = K_4$ and $K_2 = K_3$ (also $K_6 = K_9$ and $K_7 = K_8$).

The rate law for the first mechanism [equation (3.26)] can be easily distinguished from that for the two limiting cases of the second mechanism [equation (3.27)]. Both rate laws predict that a plot of $(E_0)/v$ *versus* $1/(A)$ should be linear at a constant concentration of B, but the lines obtained at different concentrations of B should be parallel for the first mechanism and intersecting for the second mechanism, as illustrated in Figure 3.2. The two

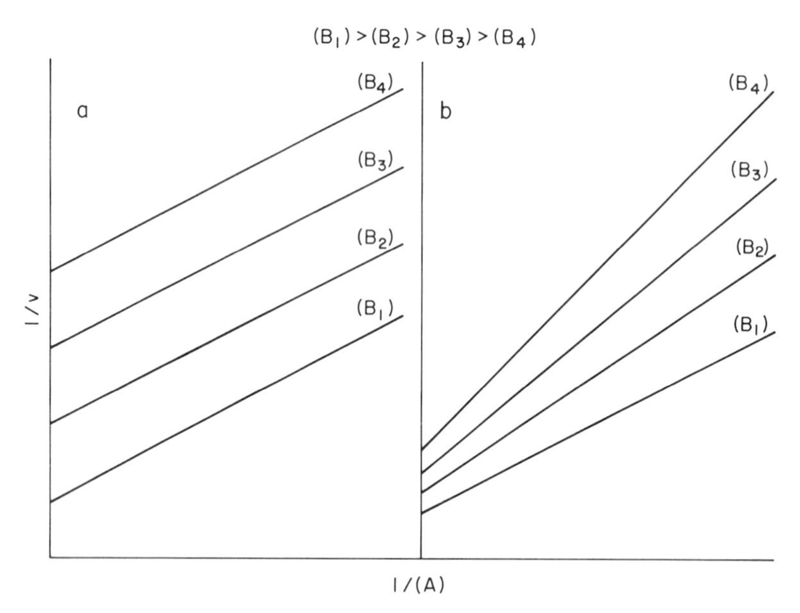


**Figure 3.2** Schematic plots of the reciprocal initial velocity, $1/v$, *versus* the reciprocal concentration of substrate A for a two-substrate reaction: (a) successive binary complex formation between enzyme and substrates [equations (3.24), (3.26)]; (b) ternary complex formation between both substrates and enzyme [equations (3.25), (3.27)]

limiting cases of the mechanism involving formation of a ternary complex between enzyme and the two substrates have identical rate laws. However, note that for random, rapid equilibration of substrates, $\phi_4 = \phi_2\phi_3/\phi_1$. Unfortunately the experimental precision is generally insufficient to utilise this relationship as a reliable mechanistic criterion. These two mechanisms can be distinguished by carrying out product inhibition studies of the steady state initial velocity. Thus experiments can be carried out with varying concentrations of A, B and D, but no C present, and with varying concentrations of A, B and C, but no D present. The steady state initial velocities for the compulsory order of addition of substrates for these two cases are

$$\frac{(E_0)}{v} = \phi_1 + \frac{\phi_2}{(A)}[1 + \theta_1(D)] + \frac{\phi_3}{(B)} + \frac{\phi_4}{(A)(B)}[1 + \theta_1(D)] \tag{3.28}$$

$$\frac{(E_0)}{v} = \phi_1[1 + \theta_2(C)] + \frac{\phi_2}{(A)} + \frac{\phi_3}{(B)}[1 + \theta_3(C)] + \frac{\phi_4}{(A)(B)}[1 + \theta_3(C)] \tag{3.29}$$

In these equations the $\theta_i$ are constants which can be expressed in terms of the rate constants.

For the random addition of rapidly equilibrating substrate binding, the corresponding equations are

$$\frac{(E_0)}{v} = \phi_1 + \frac{\phi_2}{(A)} + \frac{\phi_3}{(B)} + \frac{\phi_4}{(A)(B)}[1 + K_6(D)] \tag{3.30}$$

$$\frac{(E_0)}{v} = \phi_1 + \frac{\phi_2}{(A)} + \frac{\phi_3}{(B)} + \frac{\phi_4}{(A)(B)}[1 + K_7(C)] \tag{3.31}$$

Since the forms of the rate laws are different for these two limiting mechanisms, the mechanisms can be distinguished experimentally in principle. In practice this distinction is not always clear-cut because the rate laws are sufficiently complex and the experimental error sufficiently large so as to make this difficult and because additional mechanistic steps must often be added to completely account for all of the data. Consideration of equations (3.28) and (3.29) reveals that A and B (and C and D) do not appear symmetrically in the rate law so that it is also possible to determine which substrate binds first to the enzyme and which binds second for a compulsory pathway of substrate addition. (Similar information can be obtained for the products.) Obviously many other mechanistic possibilities can be considered within equation (3.25). However, the analysis presented should be sufficient to indicate how steady state kinetic studies can be utilised to distinguish between complex reaction pathways.

As might be anticipated, from the previous discussion of the single substrate–single product reaction, a detailed analysis of the two substrate–two product mechanisms indicates that the form of the rate law is independent of the number of reaction intermediates and that lower bounds for the rate constants can be obtained from the steady state parameters[6].

### 3.2.4 pH dependence of steady state parameters

The rates of enzymatic reactions are generally quite dependent on pH, and mechanistic information can be obtained from the variation of steady state kinetic parameters with pH. At extreme values of the pH, substantial structural changes can be induced in the structure of the enzyme which often lead to inactivation of the enzyme. However, such structural changes are not usually responsible for the pH dependence of steady state kinetic parameters under milder conditions. Instead, ionisations are assumed to occur at the catalytic site, with only particular ionisation states producing catalytic activity. For example, the enzyme ribonuclease A catalyses the hydrolysis of pyrimidine 2′,3′-cyclic phosphates to pyrimidine 3′-monophosphates. The dependence of the steady state initial velocity on the enzyme and substrate concentrations conforms to equation (3.9), and the dependence of the maximum velocity and Michaelis constant on pH for the hydrolysis of uridine 2′,3′-cyclic phosphate is shown in Figure 3.3[7]. The observed pH dependence can be explained by the following expanded Michaelis–Menten mechanism:

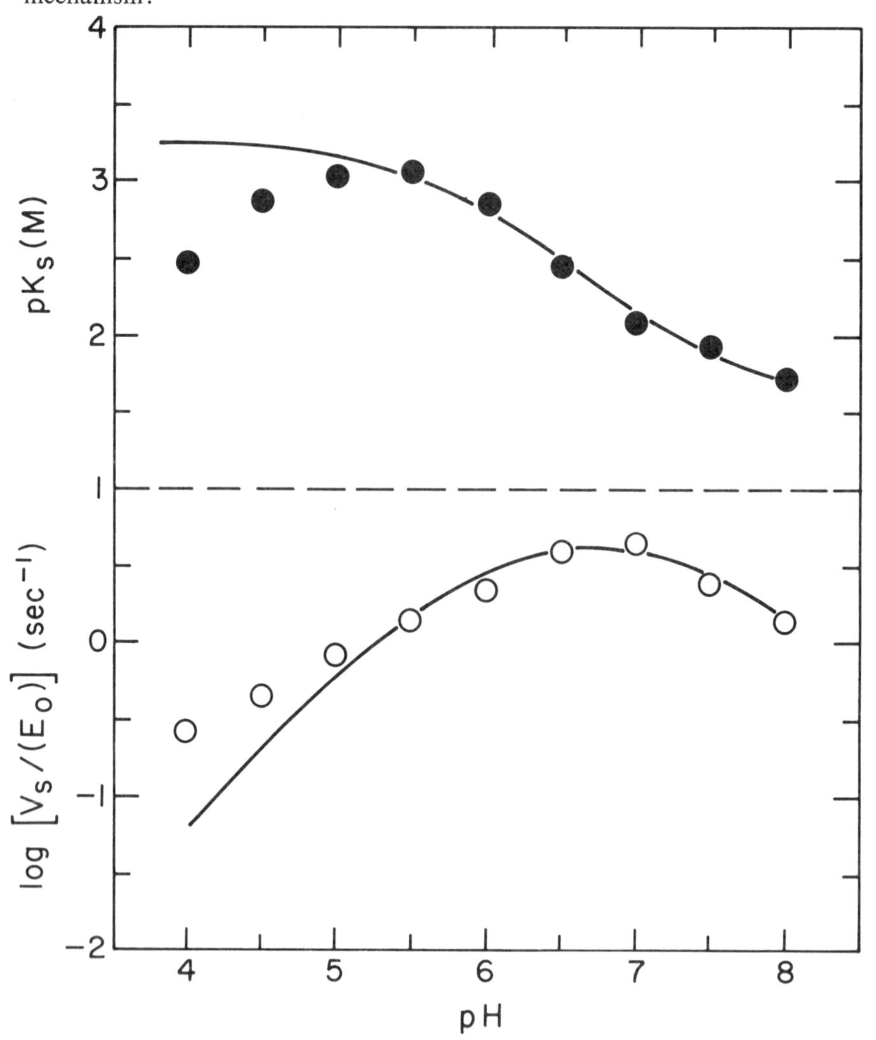


**Figure 3.3** Variation with pH of the logarithm of the turnover number, $V_S/(E_o)$ (○), and the negative logarithm of the Michaelis constant, $K_S$ (●), for the ribonuclease catalysed hydrolysis of uridine 2′,3′-cyclic phosphate. The solid line was calculated[7] according to equations (3.35) with $pK_a = 5.4$, $pK_b = 6.4$, $pK_{Xa} = 5.9$, $pK_{Xb} = 7.6$, $k_2 = 5.11\ s^{-1}$ and $(k_{-1} + k_2)/k_1 = 1.74 \times 10^{-3}$ mol $l^{-1}$

$$\begin{array}{ccccc} EH_2 & & XH_2 & & EH_2 \\ K_a \upharpoonleft\downharpoonright & & K_{Xa} \upharpoonleft\downharpoonright & & \upharpoonleft\downharpoonright \\ EH + S & \underset{k_{-1}}{\overset{k_1}{\rightleftharpoons}} & XH & \xrightarrow{k_2} & EH + P \\ K_b \upharpoonleft\downharpoonright & & K_{Xb} \upharpoonleft\downharpoonright & & \upharpoonleft\downharpoonright \\ E & & X & & E \end{array} \quad (3.32)$$

Hydrogen ions and formal charges have been omitted from equation (3.32) for the sake of simplicity, and the $K_i$'s define ionisation constants, e.g. $K_a = (EH)(H^+)/(EH_2)$. This mechanism postulates that the catalytic site can exist in three ionisation states, but only the singly protonated state (EH) is catalytically active.

The steady state initial velocity for equation (3.32) can be readily derived if the assumption is made that the protolytic equilibria are adjusted rapidly relative to all other reactions; this assumption is generally valid, especially in buffered systems where proton transfer reactions occur very rapidly.

The total enzyme concentration can be written as

$$\begin{aligned} (E_0) &= (E) + (EH) + (EH_2) + (X) + (XH) + (XH_2) \\ &= (EH)\,[1 + H^+)/K_a + K_b/(H^+)] + (XH)[1 + (H^+)/K_{Xa} \\ &\quad + K_{Xb}/(H^+)] \end{aligned} \quad (3.33)$$

and the steady state assumption can be written as

$$d[(X) + (XH) + (XH_2)]/dt = 0 = -(k_{-1} + k_2)\,(XH) + k_1(EH)\,(S) \quad (3.34)$$

The final rate equation has the same form as equation (3.9) except that

$$K_S = \left(\frac{k_{-1} + k_2}{k_1}\right) \frac{1 + (H^+)/K_a + K_b/(H^+)}{1 + (H^+)/K_{Xa} + K_{Xb}/(H^+)} \quad (3.35)$$

$$V_S = [k_2(E_0)] \frac{1}{1 + (H^+)/K_{Xa} + K_{Xb}/(H^+)}$$

The ratio $V_S/K_S(E_0)$, which is plotted *versus* pH in Figure 3.4, is dependent only on the ionisation constants of the free enzyme, and the bell-shaped curve can be analysed to give $pK_a$ and $pK_b$, which are 5.4 and 6.4, respectively. for the data shown. The maximum velocity–pH curve is also bell-shaped; the values of $pK_{Xa}$ and $pK_{Xb}$ are 5.9 and 7.6, respectively[7]. The curves shown in Figures 3.3 and 3.4 have been calculated according to equation (3.35) with the above p$K$ values.

The type of mechanism in equation (3.32) is commonly used to explain the pH dependence of steady state kinetic parameters. However, this mechanism is undoubtedly an oversimplification. As indicated earlier, the form of the rate law is independent of the number of reaction intermediates. If the pH dependence of a general $n$ intermediate mechanism [equation (3.11)] is considered, the predicted pH dependence of the steady state kinetic parameters is considerably more complex than equation (3.35)[4]. However, the ratio $V_S/K_S(E_0)$ remains a function only of the ionisation constants of the free enzyme. On the other hand, the pH dependence of the maximum velocity cannot be interpreted in terms of a single set of p$K$ values. A further complication can arise if all of the ionisation states of the enzyme (E, EH and $EH_2$)

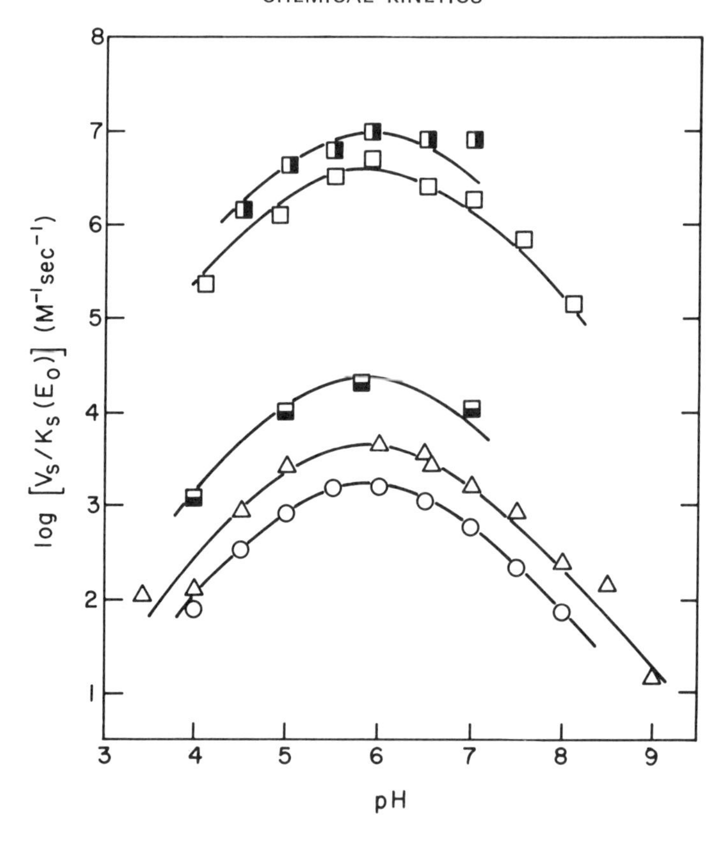


**Figure 3.4** Logarithm of the ratio of the turnover number to the Michaelis constant as a function of pH for the reaction of ribonuclease A with several substrates; uridine 2′,3′-cyclic phosphate, ○[7]; cytidine 2′,3′-cyclic phosphate, △[35]; UpU, ⬓[36]; UpA, □[36]; and CpA, ◨[36]. The solid curves were calculated using equations (3.35) with $pK_A = 5.4$, $pK_b = 6.4$ and values for $k_1k_2/(k_{-1} + k_2)$ equal to 2940, $3.88 \times 10^4$, $6.8 \times 10^6$ and $1.59 \times 10^7$ l $mol^{-1}$ $s^{-1}$, respectively

possess different catalytic activities. (The deviations of the data in Figure 3.3 from the theoretical curves at low pH are probably due to a small catalytic activity of $EH_2$.) In this case no simple interpretation of the pH dependence of the steady state kinetic parameters is possible. Whether this situation prevails can be determined if the enzyme can utilise more than a single substrate. If only EH has appreciable activity, then plots of $V_S/K_S(E_0)$ *versus* pH should be characterised by the same set of p$K$ values for all substrates. This is actually the case for many enzymes, including ribonuclease A, as shown in Figure 3.4. This situation can become somewhat more complex if the substrate also has ionisable groups; however, since the p$K$ values of these groups can be determined independently, the effect of substrate ionisable groups can usually be taken into account. Thus the pH dependence of the steady state kinetic parameters can often give information about the p$K$ values of ionisable groups on the free enzyme critical for catalytic activity;

moreover, the identity of the ionisable groups can sometimes be inferred from the p$K$ values. For reactions involving more than a single substrate, the analysis becomes more complex, but essentially the same type of information can be obtained.

Reaction rates in enzyme systems are also often sensitive to other experimental conditions such as temperature, ionic strength, solvent, etc. However, the dependence of steady state kinetic parameters on these variables is usually very difficult to interpret in mechanistic terms and has not proved to be a generally useful mechanistic tool.

In summary, steady state kinetic studies can provide information about complex mechanistic pathways, about the structural specificity of the catalytic site, about lower bounds of the rate constants, and about the nature of ionisable groups on the enzyme critical for catalysis.

## 3.3 TRANSIENT KINETICS

In order to obtain information about the intermediates in enzyme catalysis, experiments must be done at sufficiently high enzyme and substrate concentrations that the intermediates can be detected with currently available experimental methods. However, because of the unusually great catalytic efficiency of enzymes, the rates of enzymatic reactions at sufficiently high concentrations for detection of reaction intermediates are usually too fast for measurement by conventional methods. In fact, most of the elementary steps in enzyme catalysis take place in times of much less than one second. A variety of experimental techniques are available for the study of very fast reactions (*cf.* Ref. 8), and reactions with characteristic time constants as small as $10^{-11}$ s can be studied. Since molecular vibration periods are $10^{-12}$–$10^{-13}$ s, virtually all chemical reactions are accessible to study in principle. Two techniques primarily have been used for enzyme systems: rapid mixing methods[9] and the temperature jump method[10]. With rapid mixing methods, the reactants are mixed together as rapidly as possible, and the time progress of the reaction occurring after mixing is followed, usually by spectrophotometric techniques. The time resolution of rapid mixing methods is limited by how rapidly the two reactant solutions can be mixed: a few tenths to one ms typically can be achieved. With the temperature jump method, the reactants are equilibrated, and the temperature is rapidly raised, usually 5–8 °C, by a high voltage discharge through the solution. Since equilibrium constants are temperature dependent, the equilibrium concentrations will be altered by the temperature change. By measuring the rate of change of the equilibrium concentrations after the temperature jump, the kinetics of the chemical reactions can be studied. The present resolution time of the temperature jump method is *ca.* $10^{-7}$–$10^{-8}$ s. In order for the temperature jump method to be applicable, equilibration of the reactants must lead to an appreciable concentration of reaction intermediates. If a reaction proceeds essentially to completion, this is not the case. Such 'irreversible' reactions can be studied by using a combined stopped flow–temperature jump method; with this method a steady state of reaction intermediates is established in a few ms, and a temperature jump perturbation is applied to the reacting mixture. This technique is applicable if the overall reaction has a half time of greater than 10–20 ms.

### 3.3.1 Elementary steps in enzyme catalysis

A variety of different elementary processes or reactions can occur during the course of enzyme catalysis. Three types of reactions, which appear to be of general importance in enzyme catalysis, are now considered in detail. These reactions are the initial combination of enzyme and substrate, macromolecular conformational changes and covalent bond breaking and forming involving acid–base catalysis.

#### *3.3.1.1 Enzyme–substrate reactions*

All enzymatic reactions are initiated by the combination of enzyme and substrate, and many such reactions have been studied. In many cases, a two-step mechanism occurs: the bimolecular reaction of enzyme and substrate followed by an isomerisation or conformational change of the enzyme–substrate complex[11]. This mechanism can be written as

$$E + S \underset{k_{-1}}{\overset{k_1}{\rightleftharpoons}} X_1 \underset{k_{-2}}{\overset{k_2}{\rightleftharpoons}} X_2 \tag{3.36}$$

Some typical rate constants which have been determined for several different enzyme systems are presented in Table 3.1.

The maximum rate of combination of enzyme and substrate is determined by how fast the reactants can diffuse together. Using simple diffusion theory, the upper limit for the second-order rate constant can be calculated[19] to be *ca.* $10^9$ l mol$^{-1}$ s$^{-1}$. Although the enzyme displays a great deal of stereospecificity in its interaction with substrates, the measured second-order rate constants are very close to the diffusion controlled limit. The dissociation rate constant for enzyme–substrate complexes, on the other hand, varies greatly and is a measure of the stability of the complex since the association rate constants are all quite similar.

#### *3.3.1.2 Conformational changes*

The rate constants associated with the isomerisation (conformational change) vary considerably, but are approximately in the range $10^2$–$10^4$ s$^{-1}$. Obviously these rate constants must be greater than the turnover number $[V_s/(E_0)]$ of the enzyme if the conformational transition is involved in the catalytic process; this is the case for the examples cited in Table 3.1. The conformational changes involve only non-covalent changes in the enzyme–substrate complex. The most important non-covalent interactions in these transitions are hydrogen bonding, solvation and hydrophobic interactions. The dynamics of these processes have been extensively studied in model

**Table 3.1 Some typical rate constants associated with enzyme–substrate complex formation**

$$E + S \underset{k_{-1}}{\overset{k_1}{\rightleftharpoons}} X_1 \underset{k_{-2}}{\overset{k_2}{\rightleftharpoons}} X_2$$

| *Enzyme* | *Substrate* | $10^{-7} k_1$ /l $mol^{-1}$ $s^{-1}$ | $10^{-3} k_{-1}$ /$s^{-1}$ | $10^{-3} (k_2 + k_{-2})$ /$s^{-1}$ | *Ref.* |
|---|---|---|---|---|---|
| Ribonuclease A | Cytidine 3′-phosphate | 4.6 | 4.2 | 1.8 | 12 |
| | Uridine 3′-phosphate | 6.1 | 11 | 1.8 | 12 |
| | Cytidine 2′-3′-cyclic phosphate | 2–5 | 10–20 | 13 | 13 |
| | Uridine 2′,3′-cyclic phosphate | 1.1 | 21 | 35 | 14 |
| | Cytidylyl 3′,5′-cytidine | 1.4 | 7 | 8.6 | 15 |
| Creatine kinase | Adenosine 5′-diphosphate (ADP) | 2.2 | 18 | 19.4 | 16 |
| | MgADP | 0.53 | 5.1 | 19.4 | 16 |
| | CaADP | 0.17 | 1.2 | 19.4 | 16 |
| | MnADP | 0.74 | 4.1 | 19.4 | 16 |
| Chymotrypsin | Furylacryloloyl-L-tryptophanamide | 0.62 | 2.7 | ~100 | 17 |
| Aspartate aminotransferase | *erythro*-β-Hydroxyaspartate | 0.31 | 11 | 3.7 | 18 |

systems, and the results obtained provide some insight into the molecular details of the corresponding processes in proteins.

The dynamics of hydrogen bonding are difficult to study in water because water itself is a potent hydrogen bond donor and acceptor. However, many kinetic studies of hydrogen bonding have been made in non-aqueous solvents. As an example, consider the dimerisation of 2-pyridone:

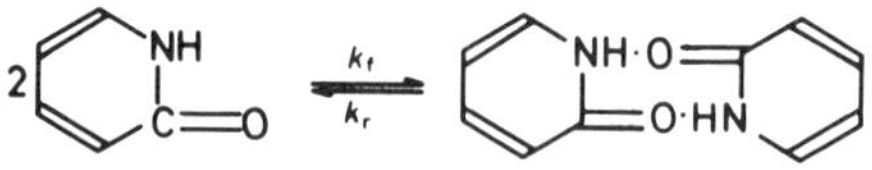

2 (2-pyridone: NH, C=O) $\underset{k_r}{\overset{k_f}{\rightleftharpoons}}$ (dimer: NH·O=, =O·HN) (3.37)

The rate constants for this reaction, as determined by ultrasonic attenuation measurements, in a variety of solvents are given in Table 3.2. In weakly

**Table 3.2 Rate constants for the dimerisation of 2-pyridone**

$$2A \underset{k_r}{\overset{k_f}{\rightleftharpoons}} A_2$$

| *Solvent* | $10^{-9}\,k_f$ $/\mathrm{l\,mol^{-1}\,s^{-1}}$ | $10^{-7}\,k_r$ $/\mathrm{s^{-1}}$ | *Ref.* |
|---|---|---|---|
| $CHCl_3$ | 3.3 | 2.2 | 20 |
| Dioxane | 2.1 | 13 | 20 |
| 1% $H_2O$–dioxane | 1.7 | 17 | 21 |
| $CCl_4$–dimethyl sulphoxide (1.1 M) | 0.26 | 14.8 | 20 |
| $CCl_4$–dimethyl sulphoxide (1.1 M) | 0.069 | 27 | 20 |

hydrogen bonding solvents (the first three entries in Table 3.2), the association rate constants are *ca.* $10^9$ l mol$^{-1}$ s$^{-1}$, which is essentially the value expected for a diffusion controlled reaction. Many other similar types of reactions have been studied in inert solvents with similar results. The mechanism of this reaction can be written schematically as

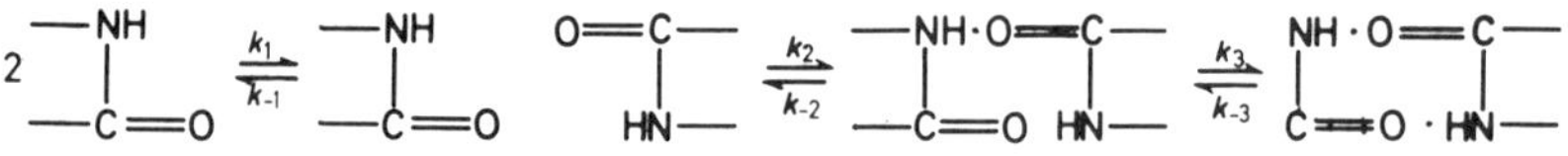

2 (—NH, —C=O) $\underset{k_{-1}}{\overset{k_1}{\rightleftharpoons}}$ (—NH, —C=O  O=C—, HN—) $\underset{k_{-2}}{\overset{k_2}{\rightleftharpoons}}$ (—NH·O=C—, —C=O HN—) $\underset{k_{-3}}{\overset{k_3}{\rightleftharpoons}}$ (NH·O=C—, C=O·HN—) (3.38)

This sequence of reactions represents diffusion together of the reactants, the formation of the first hydrogen bond and the formation of the second hydrogen bond. Since only a single rate process is observed, the intermediates can be assumed to be in a steady state. The observed forward and reverse rate constants then can be written as

$$k_f = \frac{k_1}{1 + (k_{-1}/k_2)(1 + k_{-2}/k_3)} \tag{3.39}$$

$$k_r = \frac{k_{-3}}{1 + (k_3/k_{-2})(1 + k_2/k_{-1})} \tag{3.40}$$

The experimental finding that $k_f$ is equal to $k_1$ requires that $k_2$ must be much greater than $k_{-1}$, i.e. desolvation of the solutes and formation of the first hydrogen bond must be faster than diffusion apart of the reactants. The value of $k_{-1}$ can be readily calculated and is approximately $10^{10}$ $s^{-1}$, so that $k_2$ must be $10^{11}$–$10^{12}$ $s^{-1}$. This extremely rapid rate is comparable with molecular vibration frequencies. Under these conditions, the observed dissociation rate constant is $k_r = k_{-1}(k_{-2}/k_2)(k_{-3}/k_3)$. Since $k_{-1}$ is essentially the same for all cases, the reverse rate constant is a direct measure of the relative thermodynamic stability of solute–solute hydrogen bonds compared with solute–solvent hydrogen bonds. For the last two entries in Table 3.2, high concentrations of dimethyl sulphoxide, which can form strong hydrogen bonds, are present, and the association rate is no longer diffusion controlled. Instead, desolvation of the solute, with a specific rate constant of *ca.* $10^8$ $s^{-1}$, is rate determining[20].

Direct measurements of desolvation rates have been made with ultrasonic and n.m.r. techniques; some typical rate constants are presented in Table 3.3.

**Table 3.3 Representative desolvation rate constants**

| *Molecular species* | $k/s^{-1}$ | *Ref.* |
|---|---|---|
| $NH_3 \cdot H_2O$ | $2.2 \times 10^{11}$ | 22 |
| $(PhCH_3)_2NCH_3 \cdot H_2O$ | $2.7 \times 10^9$ | 22 |
| Dioxane$(H_2O)_2$ | $2.8 \times 10^8$ | 23 |
| (Dioxane)$_2(H_2O)_2$ | $1.0 \times 10^8$ | 23 |
| Glycine, di-, tri-glycine* | $4 \times 10^8$ | 24 |

*Only the sum of the solvation and desolvation rate constants, i.e. the reciprocal relaxation time, could be determined

These results indicate that the rate of water dissociation is decreased as hydrophobic groups are placed around the hydrogen bond acceptor. This can be viewed as being due to the fact that a sheath of strongly interacting molecules form around the hydrophobic groups, making dissociation more difficult than for normally structured water.

Although the elementary step of hydrogen bond formation has a specific rate constant of $10^{11}$–$10^{12}$ $s^{-1}$, desolvation of hydrophobic groups, which very likely is often rate limiting for hydrogen bond formation, has a specific rate constant of about $10^8$ $s^{-1}$. These rate constants imply that the rate of conformational transitions should be considerably faster than is observed, so that a crucial factor must be missing in the model systems considered thus far. A possible explanation can be found in studies of simple polymers. For example, polyglutamic acid can exist in either an $\alpha$-helical or random coil configuration[25]. A transition between the two states in aqueous solution can be triggered by very small changes in pH. This is a cooperative process, and the transition is very sharp. This cooperative transition is found only with polymers containing more than six residues[26]. Although only hydrogen bonding and solvation process are involved, the rate[27] of this process at the midpoint of the transition is only *ca.* $10^6$ $s^{-1}$. The rate constant for the elementary step in helix formation in this system can be estimated from theory and experiment to be *ca.* $10^8$ $s^{-1}$, suggesting desolvation is rate determining. Thus the overall rate is considerably slower than the elementary steps because a cooperative transition occurs. This suggests that the conformational transitions in proteins are also slow relative to the elementary steps because the conformational transitions are highly cooperative. Furthermore, highly cooperative transitions require a large number of interactions, such as found in a macromolecule, and all enzymes are macromolecules.

### *3.3.1.3 Acid–base catalysis*

Covalent bond breaking and forming is often, but not always, the rate determining step in enzyme catalysis. Some typical turnover numbers for enzymes are presented in Table 3.4. A turnover number of *ca.* $10^3\ s^{-1}$ seems to be

**Table 3.4 Approximate maximum turnover numbers of typical enzymes**

| *Enzyme* | $V_S/(E_0)$ $/s^{-1}$ | *Ref.* |
|---|---|---|
| Chymotrypsin | $10^2$–$10^3$ | 28 |
| Carboxypeptidase | $10^2$ | 28 |
| Urease | $10^4$ | 28 |
| Fumarase | $10^3$ | 23 |
| Ribonuclease A | $10^4$ | 30 |
| Transaminases | $10^3$ | 31 |
| Carbonic anhydrase | $10^6$ | 32 |
| Acetylcholinesterase | $10^4$ | 33 |

about the norm, but appreciably higher values are found in several instances. Because of the marked pH dependence of most enzymatic reactions, acid–base catalysis is presumed to be involved. The elementary steps in acid–base catalysis are proton transfer reactions, and the rates of such reactions have been extensively studied so that the governing principles are quite well understood[34]. In catalytic reactions, the acid or base involved in catalysis must always return to its initial state of protonation. Thus for solvent mediated acid–base catalysis the catalytic cycle must contain the two reactions

$$B + H^+ \rightleftharpoons BH^+ \tag{3.41}$$

$$BH^+ + OH^- \rightleftharpoons B + H_2O \tag{3.42}$$

For 'normal' acids and bases, these two forward reactions are diffusion controlled with typical rate constants of $10^{10}\ l\ mol^{-1}\ s^{-1}$. Since these rates are diffusion controlled, the actual intramolecular proton transfer, after the molecules have diffused together, must have a specific rate constant of *ca.* $10^{12}\ s^{-1}$ (in analogy with the earlier discussion of hydrogen bonding). The explanation of this rapid proton transfer in molecular terms is that rapid proton conduction can occur through structured water. As might be expected, deviations from diffusion control occur when the water structure is perturbed, e.g. by internal hydrogen bonding or by an unusually high charge density. Since the forward rate constants are known to be *ca.* $10^{10}\ l\ mol^{-1}\ s^{-1}$, the reverse rate constants of equations (3.41) and (3.42) can be written as $10^{10}\ K_A\ s^{-1}$ and $10^{10}\ K_W/K_A\ s^{-1}$, respectively, where $K_A$ is the acid dissociation constant of $BH^+$ and $K_W$ is the ionisation constant of water. The catalytic rate is maximal when both of the rate constants are maximised, and this occurs with a p$K_A$ of *ca*. 7. Many enzymes have maximal activity around pH 7, and imidazole residues, which have a p$K_A$ of about 7, have been implicated in many enzyme mechanisms. According to these considerations, the maximum turnover number for solvent mediated acid–base catalysis is *ca*. $10^3$ $s^{-1}$, which is similar to that observed for many enzymes (Table 3.4).

Direct proton transfer, not mediated by water, can also occur. The overall reaction is

$$DH + A \rightleftharpoons HA + D \tag{3.43}$$

where D and A are the proton donor and acceptor, respectively. Many reactions of this type have been studied. For 'normal' acids and bases, the forward reaction is diffusion controlled if the p$K$ of the acceptor is much higher than that of the donor. Again this implies the specific rate constant for intramolecular proton transfer after the reactants have diffused together is *ca*. $10^{12}$ $s^{-1}$. The rate constant for the reverse reaction then is proportional to the equilibrium constant for the reaction, $K_A/K_D$ (the ratio of the ionisation constants of acceptor and donor), and the rate constant for intramolecular proton transfer in the reverse direction is approximately $10^{12}$ $K_A/K_D$ $s^{-1}$ ($K_D > K_A$). Since the catalytic cycle must involve both protonation and deprotonation, the maximum theoretical rate of $10^{12}$ $s^{-1}$ cannot be expected. For example, if the p$K$ difference between enzyme and substrate is unfavourable by seven p$K$ units, the maximum rate constant is $10^5$ $s^{-1}$. Furthermore, the concentration of the intermediate formed would be exceedingly low so that its rate of appearance and disappearance could not be studied directly. Also the rate constant for further reactions of the intermediate must be greater than $10^{12}$ $s^{-1}$ if a turnover number of $10^5$ $s^{-1}$ is to be achieved (since the maximum intermediate concentration is less than $10^{-7}$ of the total enzyme concentration). Actually this is an optimistic example since the p$K_A$ difference between protein ionisable groups and the substrate are usually greater than 7, and the rates of proton transfer for very poor acids and bases (e.g. carbon acids and bases), which most substrates are, are considerably slower than normal because changes in electronic structure often accompany protonation and deprotonation. Thus for this mechanism to achieve the observed turnover numbers, the enzyme must enhance the acidity or basicity of the substrate through interactions with protein groups.

Another possible mechanism for acid–base catalysis involves concerted proton transfers, that is simultaneous proton acceptance and donation by the substrate. This mechanism eliminates the necessity for forming an unstable reaction intermediate in very low concentrations. The rate of such a process cannot be reliably estimated. An upper bound is the direct proton transfer rate between the two ionisable groups on the enzyme. If the p$K$ difference between these groups is about two units, the maximum turnover rate is $10^{10}$ $s^{-1}$. The poor acid–base properties of the substrate can be expected to lower this rate another three to four orders of magnitude so that a maximum turnover number of *ca*. $10^6$ $s^{-1}$ for a mechanism involving concerted proton transfers is not unreasonable.

For all of the proton transfer mechanisms considered, the observed turnover numbers for most enzymes are surprisingly close to the estimated upper bounds of the rate constants for proton transfer reactions. For many enzymes, a large number of reaction intermediates can be detected, but the intermediates involved in bond breaking and forming are usually present in concentrations that are too low to be detected.

An overall picture of enzyme catalysis consistent with the available information is that the enzyme appears to break down the catalysis into a number of steps, with the enzyme optimising its configuration for each step through conformational changes. These conformational changes are cooperative, and a macromolecular structure is necessary for both the cooperativity and the multiple configurations. The actual chemical events occur at close to their maximum possible rates through the entire catalytic cycle.

## 3.4 AN EXAMPLE: RIBONUCLEASE A

Bovine pancreatic ribonuclease A is an enzyme which has been extensively studied with many different techniques: the amino acid sequence is known, the three-dimensional structure is known, steady state and transient kinetic studies have been made, chemically modified enzymes have been prepared, and many other chemical and physical studies have been carried out (*cf.* Ref. 30). Ribonuclease A catalyses the breakdown of ribonucleic acid in a two-step reaction as shown in Figure 3.5. First an oxygen phosphorus bond is cleaved, if the 3′-linked nucleoside has a pyrimidine base, and a 2′,3′-cyclic phosphate is formed; the cyclic phosphate is then hydrolysed to give a

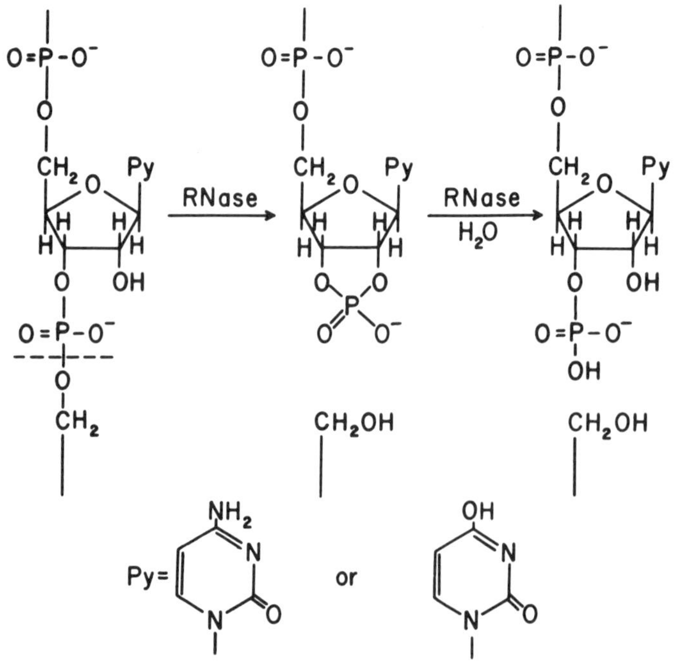

Figure 3.5 The two step hydrolysis of ribonucleic acid catalysed by the enzyme bovine pancreatic ribonuclease A

terminal pyrimidine 3′-phosphate. Kinetic studies have rarely employed ribonucleic acid itself as a substrate because the system becomes inhomogeneous as ribonucleic acid is degraded, making a kinetic analysis very difficult. However, convenient model substrates of known structure are available: dinucleosides, pyrimidine 2′,3′-cyclic phosphates and pyrimidine 3′-phosphates. At equilibrium essentially only the 3′-monophosphates are present, and the catalytic breakdown of the dinucleosides is faster than hydrolysis of the corresponding cyclic phosphates. Therefore, the interaction of ribonuclease with all three types of substrates can be readily studied.

Steady state kinetic studies have been made with a large variety of substrates, and some typical steady state parameters are presented in Table 3.5.

**Table 3.5 Some typical steady state parameters for ribonuclease A***

| *Substrate* | $K_m$/mmol $l^{-1}$ | $V_S/(E_0)$ /$s^{-1}$ | *Ref.* |
|---|---|---|---|
| Cytidine 2′,3′-cyclic phosphate | 11.1 | 18 | 35 |
| Uridine 2′,3′-cyclic phosphate | 7.3 | 4.2 | 7 |
| CpA | 1.0 | 3000 | 36 |
| UpA | 1.9 | 1200 | 36 |
| CpC | 4.0 | 240 | 36 |
| UpC | 3.0 | 40 | 36 |
| CpU | 3.7 | 27 | 36 |
| UpU | 3.7 | 11 | 36 |

*pH 7, ~ 25 °C; C, U and A designate cytosine, uridine and adenosine and p a 3′5′-phosphate linkage between the nucleosides.

The detailed structural requirements for both the base and sugar have been determined from steady state investigations[30]. In general, substrates with uracil undergo catalysis slower than those with cytidine as the pyrimidine base. With dinucleosides the nature of the neighbouring base also markedly influences the catalytic rates. These results can be interpreted in terms of the molecular structure of the enzyme–substrate complex deduced from x-ray crystallographic studies[30]. The pyrimidine bases are found to be specifically hydrogen bonded with protein groups. For both dinucleosides and cyclic phosphate, the pH dependence of $V_S/K_S(E_0)$ is similar (*cf.* Figure 3.4). The bell-shaped plots suggest that ionisable groups with p$K$ values of 5.4 and 6.4 are critical for catalysis for all substrates. Thus the mechanism in equation (3.32) appears to be valid, and parallel catalytic paths through multiple protonation states of the enzyme are negligible at pH values close to neutrality. Transient kinetic studies of the interaction of all three types of substrates with ribonuclease have been carried out using temperature jump and stopped flow–temperature jump techniques[12–15,37].

When a temperature jump is applied to an enzyme solution in the absence of substrates, a single relaxation process is observed[37]. The associated relaxation time is independent of enzyme concentration, but varies with pH. This relaxation process is due to an isomerisation of the enzyme; the simplest mechanism, which is quantitatively consistent with the data, is

$$\mathrm{E'H} \underset{k_{-1}}{\overset{k_1}{\rightleftharpoons}} \mathrm{EH} \overset{K_A}{\rightleftharpoons} \mathrm{E} + \mathrm{H}^+ \tag{3.44}$$

The protolytic equilibrium is adjusted rapidly relative to the first step. At 25 °C, $k_1 = 780\ s^{-1}$, $k_{-1} = 2470\ s^{-1}$ and $pK_A = 6.1$. The rate constants are considerably smaller in $D_2O$ than in $H_2O$ and the relaxation process is modified when molecules bind at the active site, suggesting a conformational change involving hydrogen bonding and associated with the active site occurs. This isomerisation also can be interpreted in terms of the three-dimensional structure of the enzyme[38,39]. Ribonuclease is a compact kidney-shaped molecule with the active site located along a groove. Two histidine residues (numbers 12 and 119 of the amino acid sequence) are located at the active site and a third histidine residue (48) is at the top of the 'hinge' of the groove. The observed relaxation process could be associated with an opening and closing of the groove such that the imidazole residue of histidine 48 is buried in E′H and has a p$K$ of 6.1 when exposed in the EH isomer.

The interaction of dinucleosides, pyrimidine 2′,3′-cyclic phosphates and pyrimidine 3′-phosphates with the enzyme is characterised by two relaxation processes, in addition to the process associated with the unliganded enzyme[12–15]. In all cases the results obtained can be described by the two-step mechanism of equation (3.36) and some of the rate constants obtained are included in Table 3.1. The role of the two isomers of the unliganded enzyme in this mechanism is uncertain; the assumption that the substrates bind equally well to both isomers is consistent with the data, although a preference for one of the isomers cannot be excluded. However, this uncertainty does not alter the basic nature of the overall mechanism.

Many of the rate constants have been determined as a function of pH and temperature. The pH dependence of the second-order rate constant is similar to that of $V_S/K_S(E_0)$ observed in steady state studies[7,35,36]. The two ionisable groups inferred to be present at the active site from these data, one in its basic form, the other in its acid form, are the imidazole rings of histidines 12 and 119. These residues also have been directly implicated in the catalytic mechanism by structural and chemical studies. The pH dependence of the relaxation time associated with the conformational change following the binding of substrates is similar to that of the relaxation time associated with the conformational change of the unliganded enzyme, and these two conformational changes are very probably similar in nature. Direct evidence supporting the role of histidine 48 in this conformational transition is found in n.m.r. studies, which indicate the environment of the histidine 48 imidazole residue is perturbed by the binding of a pyrimidine 3′-nucleotide[40].

An overall mechanism for the enzymatic reaction can be envisaged as follows. The enzyme exists in dynamic equilibrium between two forms differing in the structure of the active site groove. When the enzyme binds a substrate, almost as fast as the two can diffuse together, the groove shape is altered and lysine 41 swings over to the enzyme to assist in the binding process. The substrate is oriented very precisely so that the imidazole residues (histidines 12 and 119) can catalyse the chemical reaction. The elementary steps associated with the proton transfer reactions cannot be studied directly, and only the overall rate of the conversion of substrate to product (and the reverse reaction) can be studied. Apparently the concentrations of the intermediates are very small. Although the details of the proton transfer processes remain to be determined, detailed stereochemical studies and consideration of the three-dimensional structure indicate that the mechanism probably involves an in-line concerted proton transfer between the two imidazole residues and the substrate[41,42]. The conformational change is then reversed, and the product dissociates. This mechanism is shown schematically in Figure 3.6 for the transesterification reaction. The hydrolysis step proceeds in a similar manner.

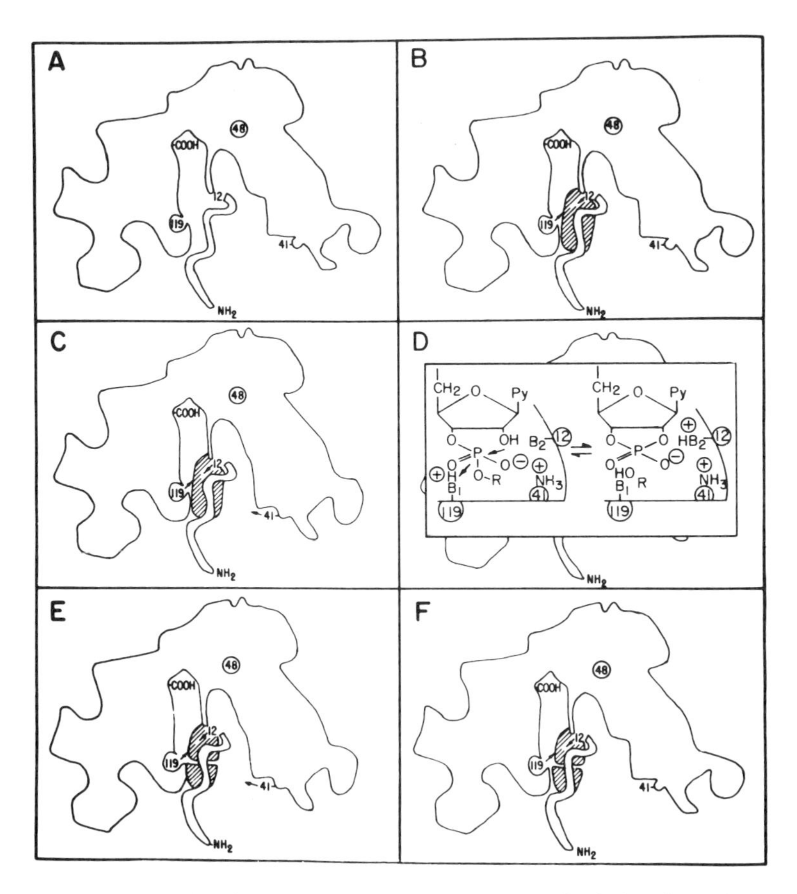


**Figure 3.6** A pictorial representation of the first half of the ribonuclease reaction. The free enzyme (A) can exist in two conformations differing by small movements about the hinge region joining the two halves of the molecule; the substrate is bound (B) and a conformational change occurs bringing lysine 41 close to the active site (C); acid–base catalysis occurs (D); products are formed (E); a conformational change occurs (F); and the product dissociates to give the free enzyme (A)

Thus a combination of detailed kinetic, chemical and structural studies has led to a rather complete picture of the catalytic process for ribonuclease. Moreover, almost the entire time course of the reaction has been studied, and the reaction has been resolved into discrete elementary steps.

## 3.5 CONCLUSIONS

This chapter has presented some of the basic concepts in kinetic analyses of enzymatic reactions. Only the catalytic aspects of enzyme function have been considered; a second important function of enzymes is the regulation of metabolic fluxes. The kinetic behaviour of regulatory enzymes is generally more complex than has been discussed; in particular, the initial velocities are often not hyperbolic functions of substrate concentrations. Space limitations do not permit a detailed discussion of regulatory enzymes; however, comprehensive reviews are available[43–46].

Both steady state and transient kinetic studies can provide mechanistic information about enzyme catalysis. Steady state kinetics are useful for establishing the specificity of enzymes, the mechanistic pathways for multisubstrate enzymes and the nature of ionisable groups at the active site; transient state kinetics provide information about the reaction intermediates and their rates of interconversion and about the nature of ionisable groups on the protein involved in the individual reaction steps. Only by exploring the entire time course of the catalytic process and by determining the elementary steps in the reaction mechanism can a detailed molecular picture of the reaction mechanism be developed.

### References

1. (1970). *The Enzymes*, 3rd ed. (P. D. Boyer, editor) (New York: Academic Press)
2. Cleland, W. W. (1970). *The Enzymes*, **2,** 1
3. Plowman, K. M. (1972). *Enzyme Kinetics* (New York: McGraw-Hill)
4. Peller, L. and Alberty, R. A. (1959). *J. Amer. Chem. Soc.*, **81,** 5907
5. King, E. L. and Altman, C. (1956). *J. Phys. Chem.*, **60,** 1375
6. Bloomfield, V., Peller, L. and Alberty, R. A. (1962). *J. Amer. Chem. Soc.*, **84,** 4367
7. del Rosario, E. J. and Hammes, G. G. (1969). *Biochemistry*, **8,** 1884
8. (1974). *Investigation of Rates and Mechanisms of Reactions. Part II. Investigations of Elementary Reaction Steps in Solution and Very Fast Reactions* (G. G. Hammes, editor); *Techniques of Chemistry*, Vol. VI (A. Weissberger, editor) (New York: Wiley)
9. Chance, B., Ref. 8, p. 5
10. Hammes, G. G., Ref. 8, p. 147
11. Hammes, G. G. and Schimmel, P. R. (1970). *The Enzymes*, **2,** 67
12. Hammes, G. G. and Walz, F. G. Jr. (1969). *J. Amer. Chem. Soc.*, **91,** 7179
13. Erman, J. E. and Hammes, G. G. (1966). *J. Amer. Chem. Soc.*, **88,** 5607
14. del Rosario, E. J. and Hammes, G. G. (1970). *J. Amer. Chem. Soc.*, **92,** 1750
15. Erman, J. E. and Hammes, G. G. (1966). *J. Amer. Chem. Soc.*, **88,** 5614
16. Hammes, G. G. and Hurst, J. K. (1969). *Biochemistry*, **8,** 1083
17. Hess, G. P., McConn, J., Ku, E. and McConkey, G. (1970). *Phil. Trans. Royal Soc. B*, **257,** 89
18. Hammes, G. G. and Haslam, J. L. (1969). *Biochemistry*, **7,** 1591
19. Alberty, R. A. and Hammes, G. G. (1958). *J. Phys. Chem.*, **62,** 154
20. Hammes, G. G. and Lillford, P. J. (1970). *J. Amer. Chem. Soc.*, **92,** 7578
21. Hammes, G. G. and Spivey, H. O. (1966). *J. Amer. Chem. Soc.*, **88,** 1621
22. Grunwald, E. and Ralph, E. K., III (1967). *J. Amer. Chem. Soc.*, **89,** 4405
23. Hammes, G. G. and Knoche, W. (1966). *J. Chem. Phys.*, **45,** 4041
24. Hammes, G. G. and Pace, N. C. (1968). *J. Phys. Chem.*, **72,** 2227
25. Doty, P., Wade, A., Yang, J. T. and Blout, E. R. (1957). *J. Polymer Sci.*, **23,** 851
26. Applequist, J. and Doty, P. (1959). *Abstracts*, 135th Meeting, American Chemical Society, Boston

27. Barksdale, A. F. and Steuhr, J. E. (1972). *J. Amer. Chem. Soc.*, **94,** 3334
28. Laidler, K. J. (1955). *Discuss. Faraday Soc.*, **20,** 83
29. Alberty, R. A. and Pierce, W. H. (1957). *J. Amer. Chem. Soc.*, **79,** 1523
30. Richards, F. M. and Wyckoff, H. W. (1971). *The Enzymes*, **4,** 647
31. Velick, S. F. and Vavra, J. (1962). *J. Biol. Chem.*, **237,** 2109
32. Lindskog, S., Henderson, L. E., Kannon, K. K., Liljas, A., Nyman, P. O. and Strandberg, B. (1971). *The Enzymes*, **5,** 587
33. Froede, H. C. and Wilson, I. B. (1971). *The Enzymes*, **5,** 87
34. Eigen, M. (1963). *Angew. Chem.*, **75,** 489
35. Herries, D. G., Mathias, A. P. and Rabin, B. R. (1962). *Biochem. J.*, **85,** 127
36. Witzel, H. (1963). *Prog. Nucl. Acid*, **2,** 221
37. French, T. C. and Hammes, G. G. (1965). *J. Amer. Chem. Soc.*, **87,** 4669
38. Kartha, G., Bello, J. and Harker, D. (1967). *Nature*, **213,** 862
39. Wyckoff, H. W., Hardman, K. D., Allewell, N. M., Ingami, T., Johnson, L. N. and Richards, F. M. (1967). *J. Biol. Chem.*, **242,** 3984
40. Meadows, D. H. and Jardetzky, O. (1968). *Proc. Nat. Acad. Sci. U.S.A.*, **61,** 406
41. Usher, D. A., Richardson, D. I., Jr. and Eckstein, F. (1970). *Nature*, **228,** 663
42. Usher, D. A., Erenrich, E. S. and Eckstein, F. (1972). *Proc. Nat. Acad. Sci. USA*, **69,** 115
43. Koshland, D. E., Jr. (1970). *The Enzymes*, **1,** 342
44. Stadtman, E. R. (1970). *The Enzymes*, **1,** 398
45. Atkinson, D. E. (1970). *The Enzymes*, **1,** 461
46. Hammes, G. G. and Wu, C.-W. (1974). *Ann. Rev. Biophys. Eng.*, **3,** 1

# 4
# Ion Chemistry of Planetary Atmospheres

E. E. FERGUSON
NOAA Environmental Research Laboratories, Boulder, Colorado



## 4.1 INTRODUCTION

The modern space programme has stimulated a large-scale programme of gas-phase ion chemistry research in a number of laboratories, leading to a large increase in data and knowledge on ion–molecule reactions. This research has been carried on largely outside the conventional boundaries of chemical kinetics. The motivation has been to supply basic data necessary for the understanding of the observed ion densities and ion compositions of planetary atmospheres as these observations became available from scientifically instrumented rockets and satellites.

The understanding of the ion composition of a planet requires a great deal more information than just a knowledge of ion–neutral reaction rate constants. The ionisation sources must be known, e.g. the solar u.v. and x-ray spectrum and the photoionisation cross-sections of atmospheric constituents.

The composition and density of the neutral atmosphere of the planet must be known, both because it provides the initial ion source and because it provides the neutral reactants for the primary ions. Laboratory data on electron recombination, attachment and other processes are also required. The complexity of the overall atmospheric ion chemistry (coupled as it is to the neutral chemistry) is such that, when all available knowledge, i.e. observational, laboratory and theoretical, are applied to the problem, uncertainties still remain. The advance in knowledge since about 1965 has been particularly great, yet there remain serious gaps in our knowledge in all these areas, particularly in certain atmospheric ion and neutral composition determinations and in critical laboratory measurements.

The only planet for which direct ion composition measurements have been made is the Earth. However, a measured electron density profile alone ($n_e$ *vs.* altitude) is a sufficient constraint to significantly test planetary ion chemistry models. Radiowave occultation measurements from spacecraft transmitters to Earth can be used to determine the variation of refractive index as a function of radius of a planet and the refractive index measures the electron density in the planetary ionosphere where the neutral density is low. Electron density profiles have been obtained in this way for Mars and Venus by Mariner spacecraft and, together with airglow emission observations, these have stimulated fairly detailed ion reaction schemes and a substantial amount of laboratory investigation. It seems quite probable that the ion chemistry of Mars and Venus (i.e. the ion chemistry of largely $CO_2$ planets) is simpler than that of Earth. The Pioneer 10 encounter with Jupiter on December 4, 1973 provided an electron density profile of that planet which can be expected to stimulate considerations of the Jovian ionosphere. These data were augmented in December 1974 by data from the Pioneer 11 encounter. Quite remarkably, Pioneer 10 also discovered an ionosphere on Jupiter's satellite Io, proving for one thing that Io has an atmosphere[1].

When the first terrestrial ion composition measurements became available (in the early 1960s), the specific ion chemistry that is involved in atmospheric ion production and loss had not been studied by workers in the field of ion–molecule reactions, a subject which had only become an active field of chemical kinetics in the early 1950s. Chemists were concerned with quite different problems for the most part, problems mainly relating to the structure and reactions of hydrocarbons, radiation chemistry, etc. The techniques in use in ion–molecule reaction studies up to that time, almost entirely conventional mass spectrometers with special ion sources, were not suited for making the kinds of measurements needed in aeronomy (aeronomy is a general name for upper atmospheric science). New techniques were therefore required and developed in response to the problems posed by ionospheric chemistry. These developments have been made in large part by physicists (who were more closely tied into the space programme), particularly physicists engaged in using gas discharge devices to study atomic and molecular properties of ions and neutrals such as electron attachment, excitation, quenching, diffusion, electric mobility, etc. This is the field generally referred to as gaseous electronics. This research field has its own national and international forums and it is not too surprising that the work in this field has stayed somewhat isolated from conventional chemical kinetics. This separation is of course to the disadvantage of both gaseous electronics and chemical kinetics and happily is being removed. Presently, techniques developed specifically for ionospheric studies are being vigorously utilised for a broad range of other chemical problems and new techniques developed by chemists (such as ion cyclotron resonance spectroscopy) are being exploited for ionospheric problems.

The present chapter will be presented from the point of view of an aeronomer involved in laboratory reaction rate studies and is directed towards an audience of kineticists (and chemists more generally) who are interested in the application of reaction kinetics to geophysics and in some of the insights into kinetics that have arisen from geophysically oriented studies.

A system of nomenclature has developed historically for the Earth's ionosphere based on electron density profiles. Electron density profiles can be obtained from ground-based radio reflection and these measurements were available long before ion composition measurements were made. The nomenclature consists of a set of capital letters for different altitude regions. In the early observations, the ionosphere appeared to form distinct 'layers' of electron density, but modern measurements show the layering to be fuzzy at best and the altitude demarcations are somewhat arbitrary. Nevertheless, this terminology is so ingrained in the subject that it is difficult to dispense with it entirely even for a discussion of ionospheric chemistry. Some division into altitude regions is actually useful even in a chemical discussion, since the chemistry does change markedly with altitude.

The ionosphere is quite variable; it varies markedly with time of day, with latitude, with season, with the 11-year solar cycle and with various changes in solar activity (e.g. solar flares) which are somewhat erratic. This means that one cannot simply describe '*the* ionosphere'. The study of ionospheric chemistry is complicated by these geophysical variations in ion composition and also by the sporadic, uncertain and evolving nature of ion composition measurements. Each ionospheric experiment selects a given altitude range for measurement so that one often has different altitude segments of the ion composition, obtained at different times and under different conditions, and the results do not lend themselves to simple presentation. In addition, some experiments report only ion current profiles as a function of altitude because of the difficulty of reducing the ion currents sampled into a supersonic vehicle to ambient ion densities. Indeed, some available data are still in the form of telemetry records. Such data are of great value in the qualitative and exploratory stage of ionospheric chemistry. The mere existence of certain ions often poses a challenging chemical problem.

The altitude region above 140 km is referred to as the F-region, and this is divided into an F1 region below 200 km and an F2 region above 200 km. The F2 region contains the electron density maximum. The F region is dominated by atomic ions and this fact, coupled with the low neutral atmospheric density, leads to the simplest ion chemistry of the Earth's atmosphere. To a large degree the ion chemistry of this region can be considered to be in a state of satisfactory, quantitative understanding. A very definitive test and detailed refinement of F-region ion chemistry understanding is expected to result in the next few years from the Atmospheric Explorer satellite series, the first of which was launched in December 1973. These are very comprehensively instrumented satellites which will provide detailed simultaneous data for analysis. The April 1973 issue of *Radio Science* is devoted to a discussion of this programme.

The altitude region from 140 km down to 90 km is described as the E region. Here the ions are largely molecular $O_2^+$ and $NO^+$ and their chemistry is relatively simple and well understood. Atomic metal ions, arising from meteors, also exist in the E region. Their chemical behaviour has not as yet been studied in great detail.

A fascinating account of the historical development of knowledge of the normal E and F layers was presented by Bates as the Sydney Chapman Memorial Lecture at the University of Colorado in April 1973. This account has recently been published[2]. The British school of theoretical atomic physicist-aeronomers, notably Massey, Bates, Dalgarno and their students, have played a dominant role in the development of aeronomy. The late Sydney Chapman who pioneered so much of ionospheric and atmospheric science coined the term 'aeronomy' to describe this discipline.

Below 90 km is the so-called D region, which is chemically very complex. Negative ions exist as well as positive ions; three-body reactions become important as well as binary reactions; and gas phase solvation effects become prominent. This is the most active ionospheric region so far as current ion chemistry research is concerned.

The order of this chapter will be a discussion of certain aspects of the Earth's ion chemistry proceeding from high altitude to low altitude, parallel-

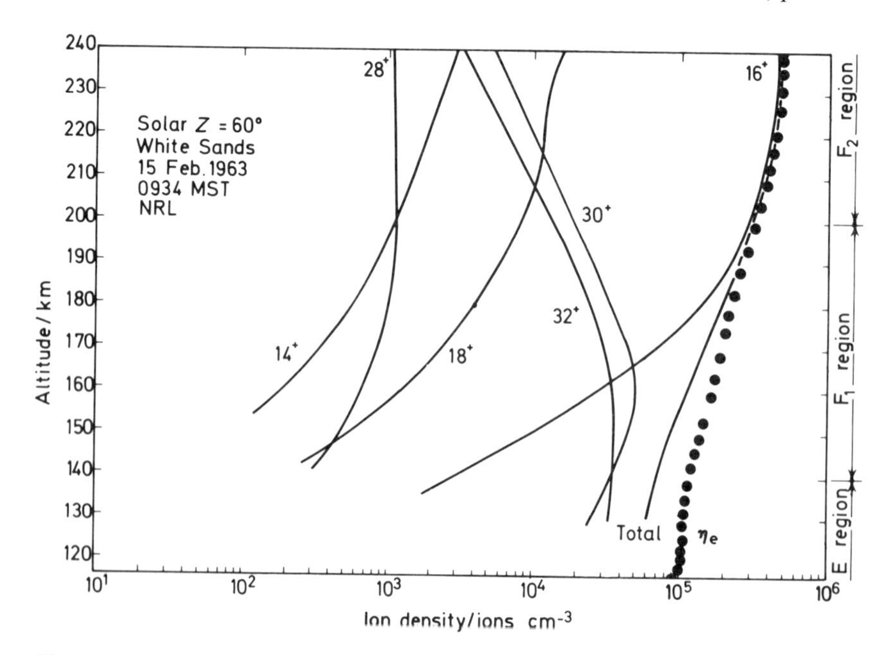


**Figure 4.1** E and F region positive ion composition $16^+ = O^+$, $30^+ = NO^+$, $32^+ = O_2^+$, $18^+ = H_2O^+$ (a contaminant), $28^+ = N_2^+$, $14^+ = N^+$, $\eta_e$ = electron density (From Holmes *et al.*[3], by courtesy of North Holland Publishing Co.)

ing the historical development and the increasing chemical complexity. Only a brief discussion of ion chemistry in other planets will be presented.

In order to give the reader some feeling for the ion chemistry to be considered, Figures 4.1 and 4.2 show some typical ionospheric measurements. Figure 4.1 shows an early E and F region ion composition measurement obtained by Holmes, Johnson and Young of the Naval Research Laboratory[3]. This particular ion composition profile played a very important role in ionospheric development. For a number of years, laboratory measurements and theory were tested for agreement with this observed profile. The ions observed were $O^+$ ($16^+$), $NO^+$ ($30^+$), $O_2{}^+$ ($32^+$), $N_2{}^+$ ($28^+$) and $N^+$ ($14^+$). The $H_2O^+$ ($18^+$) signal was recognised as being due to rocket outgassing contamination. The electron density is also shown in Figure 4.1.

For most of the present discussion the ions will be in very dilute concentration in the neutral atmosphere. This is nearly a prerequisite for chemistry to be important in a planetary atmosphere. At higher altitudes, where the ion concentration approaches the neutral concentration, the density is so low that collisions (and hence reactions) rarely occur. Figure 4.1 does not include all of the ions that have been observed in the E and F regions. It does not include the many metal ion species of the E region nor the $He^+$, $H^+$, $O^{2+}$ and $N^{2+}$ observed at higher altitudes, for example.

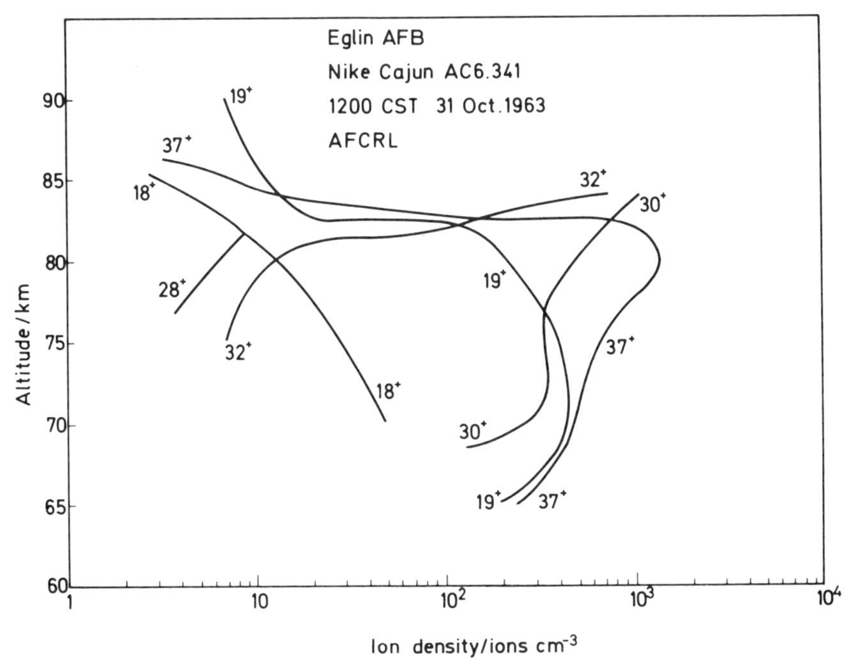


**Figure 4.2** D region positive ion composition $18^+ = H_2O^+$ (a contaminant) $19^+ = H_3O^+$, $37^+ = H_5O_2{}^+$, $28^+ = Si^+$ (or $N_2{}^+$), $30^+ = NO^+$, $32^+ = O_2{}^+$ (From Narcisi and Bailey[4], by courtesy of the American Geophysical Union)

Figure 4.2 shows another historically very important result, the first observation of D region positive ions by Narcisi and Bailey[4] of the Air Force

Cambridge Research Laboratories. This result can fairly be said to have opened up D region ion chemistry research.

The early knowledge of the upper atmosphere was obtained primarily by the interaction of radio waves with the ionosphere. It was not until 1946 that the first direct observations of the *in situ* neutral atmosphere and solar ultraviolet radiation were made possible through early high-altitude rocket flights. Since the necessary laboratory data on many of the relevant atomic and chemical processes have only been obtained concurrently with the advent of rocket technology, it is not surprising that many of the earlier inferences on atmospheric properties were in error.

A very comprehensive treatise on the physics and chemistry of the Earth's upper atmosphere and ionosphere has been compiled by Banks and Kockarts[5]. They discuss in detail the composition and physical conditions of the

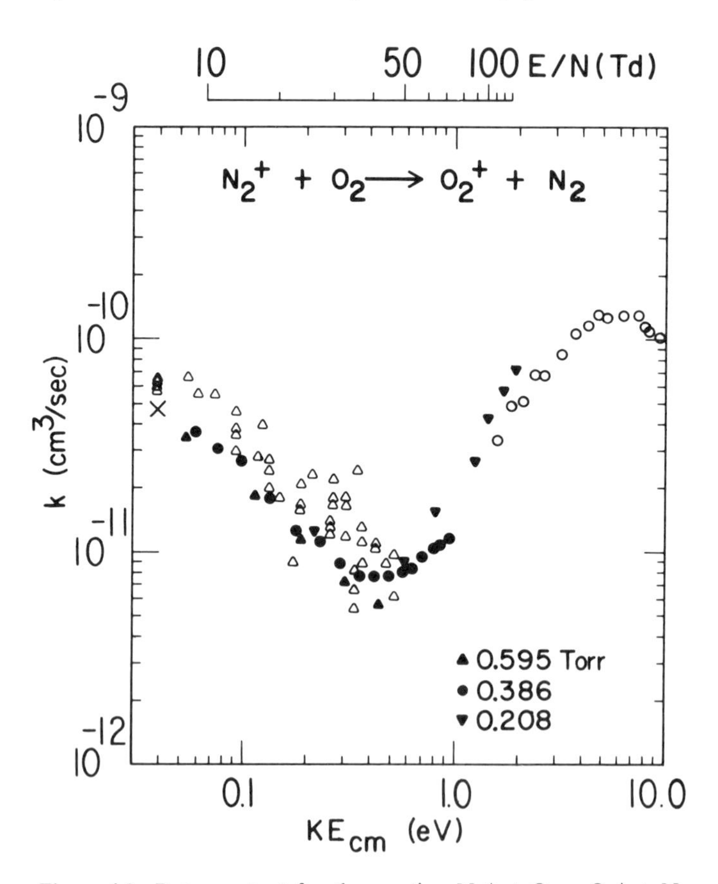


**Figure 4.3** Rate constant for the reaction $N_2^+ + O_2 \rightarrow O_2^+ + N_2$ as a function of $N_2^+$ kinetic energy in the centre of mass[18]. Solid symbols are flow-drift tube data at various pressures; × is earlier flowing afterglow data (Dunkin *et al.* (1968). *J. Chem. Phys.*, **49**, 1365); Δ, drift tube data (Johnsen *et al.* (1970). *J. Chem. Phys.*, **52**, 5080); ○, ion beam data (Neynaber *et al.*, Gulf Radiation Technology Report, A12 209)

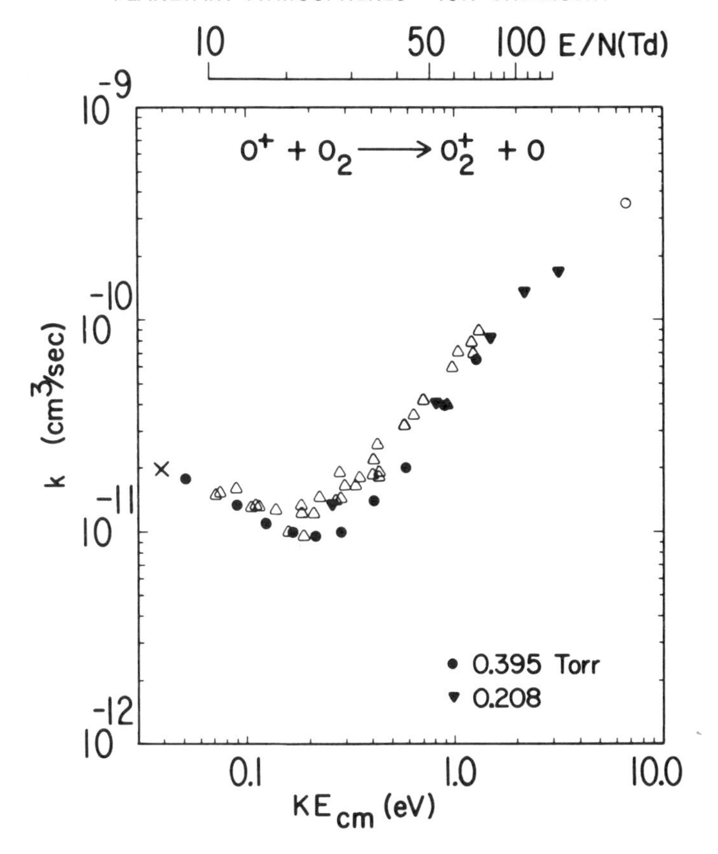


**Figure 4.4** Rate constant for the reaction $O^+ + O_2^+ \rightarrow O_2^+ + O$ as a function of $O^+$ kinetic energy in the centre of mass[18]. Solid symbols are flow-drift tube data; ×, flowing afterglow data (Dunkin *et al.* (1968). *J. Chem. Phys.*, **49,** 1365); Δ, drift tube data (Johnsen and Biondi (1973). *J. Chem. Phys.*, **59,** 3504); ○, ion beam data (Stebbings *et al.* (1966). *J. Geophys. Res.*, **71,** 771)

planet, solar radiation and photoabsorption, collision processes and reactions, and transport processes.

A number of detailed discussions of the experimental methods of ion chemistry have recently been published[6–15]. The ionospherically relevant rate constants now available number in the hundreds[16]. The ions studied include $He_2^+$, $NO^+(H_2O)_3$, $Fe^+$, $MgO^+$, $SiO^+$, $Ca^{2+}$, $Mg^{2+}$, $O_4^+$, $H_3O^+ \cdot OH$, $O_2^-$, $O_3^-$, $O_4^-$, $NO^-$, $O_2^-(H_2O)_2$, $CO_3^-(H_2O)_2$, two different forms of $NO_3^-$, $AlO^-$ and many others, as well as more conventional ions like $O^+$, $N_2^+$ and $CO_2^+$. The neutral reactants so far studied include O, N and H atoms, electronically excited $O_2(^1\Delta_g)$, $O_3$, $NO_2$, $H_2O$, Na and vibrationally excited $N_2$, as well as the experimentally easier $O_2$, $N_2$, $CO_2$, etc. This does *not* include all of the neutral reactants of interest for atmospheric problems, unfortunately.

A flowing afterglow system is now being operated[17] over the temperature range 80–900 K, obtaining temperature dependences of rate constants that are useful for atmospheric applications as well as providing interesting kinetic data. Another combined flowing afterglow–drift tube has recently been constructed[18] that allows the determination of rate constants as a function of ion kinetic energy from thermal to several electron volts, as do the static drift tube studies of Johnsen and Biondi[19].

One of the more serious deficiencies in ionospheric data in the past has been the lack of knowledge of reaction rate constant energy dependences. That problem is now being rapidly solved. Two examples of data just recently available on the energy dependences of rate constants are shown in Figures 4.3 and 4.4 from Ref. 18. The agreement of current data from static drift tubes and the new flow-drift tube is very good, as is the agreement with the earlier 300 K flowing afterglow results. The agreement at higher energies with the overlapping beam data is excellent. Many more such recent results are available[18,19].

## 4.2 HIGH-ALTITUDE (F-REGION) ION CHEMISTRY

At very high altitudes in a planetary atmosphere the neutral composition is less complex than at lower altitudes. Low mass species are favoured by the gravitational separation (following the barometric equation $p = \exp(-mgh/kT)$ or a Boltzmann distribution $n \approx \exp(-U/kT)$ where $U$ is the potential energy). In addition, solar u.v. light dissociates molecules and as the pressure becomes lower and consequently association less frequent, simpler molecules exist. In the case of the Earth, molecular oxygen is strongly photodissociated above ~100 km. Also above ~100 km (coincidentally) the various gases start to separate gravitationally according to their individual mass, rather than mixing uniformly by turbulence as they do below this, so that above 200 km atomic oxygen is the dominant neutral species. At still higher altitudes (the precise altitude varies greatly with the solar cycle which induces large temperature variations), helium, which is present only in 5 p.p.m. at the Earth's surface, becomes the dominant neutral. Ultimately atomic hydrogen, arising from the dissociation of water, methane and ammonia, becomes the dominant neutral.

At high altitudes (above ~500 km, the so-called exosphere) collisions are infrequent and particles escape so that dynamical processes must be taken into account and a hydrostatic model cannot be assumed for the neutral atmosphere. Photochemical steady-state treatments of ion composition are not valid in much of the ionosphere where the ion composition is influenced and even controlled by dynamic factors such as diffusion and mass transport. In all ionospheric analyses one must be very careful to isolate the chemical from the non-chemical factors determining ion concentration.

The main body of the ionosphere (the E and F layers) arises from ionisation by solar u.v. radiation of $\lambda < 1026$ Å. This radiation ionises the major neutral constituents O, $O_2$ and $N_2$.

### 4.2.1 Formation of planetary F-2 layers

At high altitudes in planetary atmospheres, atomic ion production dominates molecular ion production because atomic neutrals are dominant. This leads to an interesting situation. In addition to simply determining the ion composition in a planetary atmosphere, the reactions of atomic ions with molecular gases can control the electron density! This situation provides the coupling between laboratory ion chemistry and the ion chemistry of Mars and Venus for which only electron densities have been measured. This occurs in the following way in the case of the planet Earth.

Above about 200 km in the Earth's atmosphere, atomic oxygen becomes the dominant neutral species. Therefore the major positive ion produced by the solar u.v. incident on the atmosphere is $O^+$. (Photoionisation cross sections for u.v. wavelengths well beyond threshold are rather insensitive to the neutral species being ionised.) Above about 130 km, the atmosphere is optically thin for most of the solar u.v. Therefore the rate of ionisation is proportional to the concentration of ionisable species in this altitude range. The ion production rate then decreases with altitude as

$$P\ (\text{ions cm}^{-3}\ \text{s}^{-1}) \propto [\text{O}] \propto \exp(-m_O gh/kT) \qquad (4.1)$$

where $m_O$ is the mass of atomic oxygen.

The loss of atomic positive ions in planetary atmospheres invariably involves a chemical reaction. The processes of electron recombination with atomic ions, principally radiative recombination, have very low rate constants. For example, radiative recombination of atomic oxygen ions

$$\text{O}^+ + \text{e}^- \rightarrow \text{O}^+ + h\nu \qquad (4.2)$$

has a rate constant $\sim 10^{-12}$ $cm^3$ $s^{-1}$ and since the electron density hardly ever exceeds $\sim 10^6$ $cm^{-3}$, this ion loss rate does not exceed $10^{-6}$ $s^{-1}$. Stated differently, the lifetime against radiative recombination loss of an ion (atomic or molecular) usually exceeds $10^6$ s, or many days. Both chemical and dynamical (diffusion and transport) processes dominate such a loss. In particular, in the ~200–500 km altitude region of the earth, $O^+$ ions are lost mainly by the reactions:

$$\text{O}^+(^4\text{S}) + \text{N}_2(^1\Sigma) \rightarrow \text{NO}^+(^1\Sigma) + \text{N}(^4\text{S}) \qquad (4.3)$$

and

$$\text{O}^+(^4\text{S}) + \text{O}_2(^3\Sigma) \rightarrow \text{O}_2^+(^2\Pi) + \text{O}(^3\text{P}) \qquad (4.4)$$

Since $[N_2] > [O_2]$, because of the much greater $O_2$ photodissociation, (4.3) dominates (4.4) even though the rate constant for (4.4) is *ca.* 20 times that for (4.3). The formation of molecular ions removes the bottleneck to electron recombination since molecular ion–electron dissociative recombination

$$\text{NO}^+ + \text{e}^- \rightarrow \text{N} + \text{O} \qquad (4.5)$$

$$\text{O}_2^+ + \text{e}^- \rightarrow \text{O} + \text{O} \qquad (4.6)$$

is very fast, over $10^5$ times faster than radiative recombination. Thus, reaction (4.3) [and to some extent (4.4)] becomes the rate-limiting step for electron loss.

Reaction (4.6) produces excited O atoms which then radiate airglow emission which can be observed. The measured emission intensity then depends on the rate constant for reaction (4.4) and careful airglow measurement and analysis have provided a critical test of the ion chemistry[20]. Information on ion chemistry derived from airglow observations is particularly valuable in the case of the planets where ion composition measurements have not yet been made.

The rate of (4.3) will decrease with altitude in the same way as the $N_2$ concentration decreases, neglecting a slight effect due to the temperature dependence of (4.3). (Both the variation of atmospheric temperature in the relevant altitude and the variation of rate constant with temperature are slight.) Thus the loss rate will be given by

$$L \propto [N_2] \propto \exp(-m_{N_2}gh/kT) \qquad (4.7)$$

Since the mass of $N_2$ exceeds that of O, it is clear $L$ decreases faster with altitude than $P$ and therefore the electron density, $n$, which is proportional to $P/L$, will *increase* with altitude,

$$n \propto [O]/[N_2] \propto \exp\{(m_{N_2} - m_O)gh/kT\} \qquad (4.8)$$

This leads to the situation in which the electron density maximum occurs at a higher altitude than does the maximum electron production rate. In the case of the Earth, the electron density maximum (the so-called F2 peak) occurs around 250–300 km, depending on latitude, period of solar cycle, etc., whereas the electron production rate peaks broadly between 100 and 150 km, which is the altitude of unit optical depth for the dominant solar ionising u.v. radiation (suitably averaged over wavelength). Different wavelengths, of course, are absorbed at somewhat different altitudes so that the ion production peak is broad.

The reason why the electron density does not increase indefinitely [as would be predicted by (4.8)] is that chemical loss of $O^+$ does not remain the dominant loss process at very low pressures but rather diffusion, which increases with decreasing pressure, eventually dominates. It turns out that the F region peak occurs at about the altitude at which diffusive loss and chemical loss become equal. This explains why observation of electron density profiles alone for Mars and Venus have supplied sufficient constraints on their ion chemistry to provide information on their atmospheres.

It is a curious fact that reactions (4.3) and (4.4) are anomalously slow reactions for their type, with the consequence that the Earth has an anomalously dense ionosphere, a fact which has profound implications for radio propagation. For Mars and Venus, on the other hand, the reaction analogous to (4.3) or (4.4) is

$$O^+(^4S) + CO_2(^1\Sigma) \rightarrow O_2^+(^2\Pi) + CO(^1\Sigma) \qquad (4.9)$$

Reaction (4.9) converts the main atomic ion produced in these atmospheres to a molecular ion by reaction with the major neutral constituent. This reaction is 1000 times faster than (4.3), which has the implication that $CO_2$ planets cannot have such dense ionospheres as has the Earth. This has been found to be the case for both Mars and Venus (making the trivial and small scaling correction for solar–planetary distance) and neither of these planets exhibits the dominant F-2 layer occurring on Earth. Chemically this situation was quite surprising and rather interesting. The slow reactions (4.3) and (4.4) are spin-allowed while the fast reaction (4.9) does not conserve spin!

Reaction (4.9) and other reactions cause $O_2^+$ to be the dominant ion in the Venus ionosphere in the region of the ionosphere where the electron density was determined by Mariner 5 occultation measurements. This is deduced by Kumar and Hunten[21] from model studies in which they fit the electron density profile. The altitude dependence of the electron and positive ion density is coupled to the temperature and mass of the ions. Kumar and Hunten[21] find that a minimum atomic oxygen ratio of $\sim$0.5% is required to provide a reasonable fit to Mariner 5 data while more than 1.5% atomic oxygen would produce a dense F-2 layer, contrary to observation. This case illustrates very well the value of ion chemistry analysis in deducing important planetary atmospheric properties.

### 4.2.2 The terrestrial $O^+ + H \rightleftarrows H^+ + O$ process

A very interesting problem in the ion composition of the high terrestrial ionosphere relates to the charge-transfer process

$$O^+ + H \underset{k_{-10}}{\overset{k_{10}}{\rightleftarrows}} H^+ + O \qquad (4.10)$$

The near equality of the ionisation potentials of O and H leads to an accidental near-resonance, which led to the conjecture[22] that the rate constants $k_{10}$ and $k_{-10}$ would be very large and that as a consequence the $O^+/H^+$ ratio would be in equilibrium up to very high altitudes

$$[H^+]/[O^+] = (k_{10}/k_{-10})[H]/[O] \qquad (4.11)$$

where the equilibrium constant $K = k_{10}/k_{-10}$ can be calculated readily from statistical mechanics. At temperatures greater than $\sim$2000 K, $\Delta E$ for (4.10) $\ll kT$ and therefore $K$ is given to a good approximation by the statistical weight ratio 9/8, and this has been used in most ionospheric analyses. If the equilibrium assumption is correct, reaction (4.10) is the major $H^+$ source in the ionosphere, outweighing the contribution of direct photoionisation of atomic hydrogen.

There has been much interesting controversy on the validity of the assumption of large values of $k_{10}$ and $k_{-10}$ at atmospheric temperatures, both from the point of view of aeronomy and from the point of basic quantum chemistry. Aeronomically, the equilibrium (4.11) has been used to deduce atmospheric atomic hydrogen concentrations. Both $O^+$ and $H^+$ ion concentrations have been measured directly by rocket-borne mass spectrometers and the neutral atomic oxygen has been measured directly several ways at different altitudes and its altitude dependence extended theoretically. There has been some question as to whether atomic hydrogen concentrations thus deduced from (4.11) agree satisfactorily with concentrations determined by solar Lyman α-scattering from the terrestrial hydrogen geocorona[23]. The applicability of (4.11) has thus been ambiguous due to uncertainties in the measurements and their analysis.

The theoretical question has been concerned with the behaviour of accidental charge-transfer rate-constants (or cross-sections) at low energy. Bates and Lynn[24] deduced that the cross section for an accidentally resonant process would be very small at low energy, as a result of what is usually referred to as the adiabatic principle. Simply stated, this principle is that an electronic transition, due to a perturbation, decreases in probability the more slowly the perturbation is applied. A thermal energy collision is viewed as a slow perturbation, but this of course depends on the energy change of the electronic transition. Rapp[25], on the other hand, predicted large cross sections for (4.10). Until recently, the only experimental data on (4.10) were beam measurements of $\sigma_{-10}$ from 50 to 10 000 electron volts $H^+$ kinetic energy[26] and of $\sigma_{10}$ down to 0.6 eV centre of mass kinetic energy[27]. Recently $k_{-10}$ was measured at 300 K in a flowing afterglow experiment[28] and $k_{10}$ was then calculated from the equilibrium constant. The measured value of $k_{-10}$ is $3.8 \times 10^{-10}$ cm$^3$ s$^{-1}$ and this yields $k_{10} = 6.8 \times 10^{-10}$ cm$^3$ s$^{-1}$ or a velocity averaged cross section $\sigma_{10} = 27$ Å$^2$. This is consistent with extrapolation of the beam data using a formula derived only for symmetric resonance

$$\sigma^{\frac{1}{2}} = a - b \log E \tag{4.12}$$

Symmetric resonance is charge transfer in which reactants and products (i.e. $A^+ + A \rightarrow A + A^+$) are identical. Alternatively, the theoretical question could have been posed as to the validity of the extension of (4.12) to asymmetric resonances.

The experimental problem hindering measurement of simple processes such as (4.10) at thermal energy has been the difficulty in reacting ions with chemically unstable neutrals such as O and H atoms. Only the flowing afterglow technique described above has so far been successful in this.

### 4.2.3 The loss of terrestrial helium ions

The chemical loss of terrestrial helium ions has been a problem of considerable aeronomical interest since 1962 when Bates and Patterson[29] proposed that the chemical loss of helium ions might lead to the ejection of helium from the Earth's atmosphere. That helium becomes a dominant neutral species at sufficiently high altitudes was first deduced from the orbital decay of the Echo I satellite and Nicolet[30] pointed out that the rate of photoionisation of the He in the Earth's atmosphere is $\sim 10^6$ ionisations cm$^{-3}$ s$^{-1}$, which is close to the rate of production of neutral helium into the Earth's atmosphere by radioactive decay of thorium and uranium in the Earth's crust. Therefore if the deionisation process caused ejection of the helium atoms from the Earth's atmosphere, a terrestrial steady-state helium concentration could be explained. It had been shown earlier that the evaporative loss of neutral helium from the Earth's atmosphere was insufficient to balance the helium production. Kockarts[31] very recently has reviewed this problem.

The proposal of Bates and Patterson was that the reaction

$$He^+ + O_2 \rightarrow He + O + O^+ + 5.9 \text{ eV} \tag{4.13}$$

might proceed through an $HeO^+$ complex, which upon dissociation would impart more than the required 2.4 eV for planetary escape to the light helium atom. This proposal had to be abandoned when it was shown[32,33] that $He^+$ reacts more rapidly with $N_2$ than with $O_2$, and since the $N_2/O_2$ ratio is very large at the high altitudes where $He^+$ is destroyed (because of $O_2$

photodissociation). The $He^+$ charge-transfer with $N_2$ is resonant into $N_2^+$ excited states and does not lead to energetic helium atoms. The Bates and Patterson assumption that $He^+$ would not react (charge-transfer) with $N_2$ was an implicit assumption of the validity of the adiabatic principle in fore-stalling low-energy electronic transitions. Just as in the case of reaction (4.10), the thermal charge-transfer, however, was very fast with $N_2$. In the case of molecules, an accidental resonance can almost always be established because of the multitude of vibration–rotation levels that can be populated in the resulting molecular ion. There apparently are few Franck–Condon or other restrictions on such charge-transfers, since by far the largest fraction of exo-thermic ion–molecule reactions studied at thermal energy have been found to be very fast.

### 4.2.4 Comparison of laboratory and ionospheric $O^+$ reaction rate constants

It is of interest to inquire as to how compatible the laboratory measurements of ionospheric reaction rate constants are with ionospheric observation and analysis. In the case of the F-2 region, the most important reactions are the loss of $O^+$ ions in reaction with $N_2$ (4.3) and $O_2$ (4.4). These reactions, which convert the dominant atomic ions to molecular ions, control the electron density as has been described. There are many parameters involved in the ionospheric analysis of the $O^+$ concentration of course: the neutral atomic oxygen concentration, the solar ionising flux and ionisation cross-sections, and the neutral $N_2$ and $O_2$ concentration, as well as the reaction rate constants. Since all these parameters have uncertainties, the ionospheric analysis cannot be a precise test of the laboratory rate constant measure-ments; nevertheless the test is quite a good one. In this high-altitude regime, the $O^+$ concentration almost equals the electron density which is well measured, and neutral composition measurements are also quite good.

The earliest ionospheric analyses preceded the measurement of the energy dependence of reactions (4.3) and (4.4). In fact it was shown by Donahue[34] in 1966 that the rate constants for all the important E and F region reactions must decrease with temperature to be compatible with the ionospheric data, which all refer to temperatures in excess of 300 K. Ionospheric analysis[23,35] established the $O^+$ rate constants to be equal to the laboratory values within an uncertainty of no more than a factor of two. A rather unusual experiment was carried out by Danilov and Semenov[36] in which a small (half litre) air sample was released in front of the sampling port of a rocket-borne mass spectrometer at high altitude where the ionosphere was dominated by $O^+$. The ratio of the $O_2^+$ sampled (produced by $O^+ + O_2$) to the $NO^+$ sampled (produced by $O^+ + N_2$) was then interpreted as a measure of the relative rate constants, leading to a ratio of $23 \pm 3$ which agrees quite well with the ratio obtained from laboratory studies in the $\sim$0.1 eV energy range appro-priate to the F region.

The general conclusion is that the ionospheric ion composition at high altitudes is consistent with the available information on atmospheric para-meters and laboratory rate constants. The simultaneous measurements of a wide range of atmospheric parameters (neutral composition, solar flux, temperature and ion densities) on the Atmospheric Explorer satellites in the coming years should provide a much more critical and precise test in this regard.

There are still substantial uncertainties to be resolved in the laboratory. Improved branching ratios are needed for several reactions including:

$$CO_2^+ + H \rightarrow COH^+ + O \quad (4.14a)$$
$$\rightarrow H^+ + CO_2 \quad (4.14b)$$

$$He^+ + N_2 \rightarrow He + N + N^+ \quad (4.15a)$$
$$\rightarrow He + N_2^+ \quad (4.15b)$$

$$N_2^+ + O \rightarrow NO^+ + N \quad (4.16a)$$
$$\rightarrow O^+ + N_2 \quad (4.16b)$$

$$N^+ + O_2 \rightarrow NO^+ + O \quad (4.17a)$$
$$\rightarrow O_2^+ + N \quad (4.17b)$$

The general agreement which now exists between essentially all current laboratory measurements of overall reaction rate constants does not extend to the branching ratios where wide discrepancies occur. The branching ratios are difficult to measure and can have sharp energy dependences.

Rate constants for excited state ion reactions have not been measured at thermal energies. Important ionospheric reactions include

$$O^+(^2D) + N_2 \rightarrow N_2^+ + O \quad (4.18)$$

and

$$O_2^+(a^4\Pi_u) + N_2 \rightarrow N_2^+ + O_2 \quad (4.19)$$

These are extremely difficult to measure for obvious reasons, the difficulty in producing and quantitatively measuring excited state ion concentrations most particularly.

## 4.3 MIDDLE ALTITUDE (E-REGION) ION CHEMISTRY

### 4.3.1 Molecular ion chemistry

The chemical distinction between what we refer to as high-altitude ion chemistry and middle-altitude ion chemistry is to some extent a distinction between the chemistry of atomic and molecular ions. Since, as we have mentioned, atomic ions have very low electron recombination rates whereas molecular ions almost invariably have very large ones, it is a matter of considerable importance as to whether atomic or molecular ions predominate. The middle altitude ionospheric molecular ions (principally $O_2^+$ and $NO^+$, Figure 4.1) have comparable recombination coefficients so that the role of ion chemistry there is largely that of determining the relative concentrations of different ions. If the ions in an ionosphere all had equal recombination coefficients and these were independent of altitude (i.e. temperature), then the electron density maximum would of course occur at nearly the same altitude as the electron production maximum, i.e. near unit optical depth for the ionising radiation, or below 150 km in the case of the Earth. This would nearly be the situation if all the dominant ions were molecular.

In spite of the great abundance of $N_2$ in the Earth's atmosphere and the consequent high production rate of $N_2^+$, $N_2^+$ is not found to be a major ion at any altitude (Figure 4.1). This is a result of the fast reactions

$$N_2^+ + O \rightarrow NO^+ + N \tag{4.16a}$$

and

$$N_2^+ + O_2 \rightarrow O_2^+ + N_2 \tag{4.20}$$

the first dominating at higher altitudes where atomic oxygen predominates, the second at lower altitudes where molecular oxygen dominates. Reaction (4.16a) was inferred to have a large rate constant by Norton, Van Zandt and Denison[37] from ionospheric analysis well in advance of laboratory measurement.

The reactions (4.16a) and (4.3) lead to $NO^+$ being an important and often dominant ion in much of the Earth's atmosphere, in spite of the fact that the parent neutral NO is only a very minor constituent. The observed $NO^+$ is almost entirely a secondary ion in the middle and the high altitude ionosphere.

On the other hand, $O_2^+$ is observed to be a major ion in the ionosphere, owing to the fact that the only chemical losses of $O_2^+$ are with trace neutral components of the atmosphere, specifically

$$O_2^+ + NO \rightarrow NO^+ + O_2 \tag{4.21}$$

$$O_2^+ + N \rightarrow NO^+ + O \tag{4.22}$$

and

$$O_2^+ + M \rightarrow M^+ + O_2 \tag{4.23}$$

where M is one of the metals deposited in the atmosphere by the ablation of meteors. Reaction (4.21) is fast, as exothermic charge-transfer reactions with molecules generally are, but of course there is very little NO in the atmosphere, particularly above ~100 km. Nevertheless this reaction competes in the E region with electron recombination (there is also a very low concentration of electrons). Reaction (4.22) is also fast but there is also very little atomic nitrogen in the atmosphere. The precise role of (4.22) is uncertain because the atomic nitrogen concentration in the atmosphere has so far defied measurement, but theory suggests that $[N] < [NO]$. The metals are present only in very small concentrations so that while (4.23) may be an important source of metal ions it will not be an important loss of $O_2^+$.

It will be noted that many reactions produce $NO^+$. Energetic considerations dictate that every combination of an oxygen ion and neutral nitrogen or a nitrogen ion and neutral oxygen can exothermically yield $NO^+$. In addition to reactions (4.3), (4.16a) and (4.22), the reactions

$$O_2^+ + N_2 \rightarrow NO^+ + NO \tag{4.24}$$

and

$$N^+ + O_2 \rightarrow NO^+ + O \tag{4.17a}$$

$$\rightarrow O_2^+ + N \tag{4.17b}$$

are exothermic. Reaction (4.24) is observed not to occur efficiently, however, $k_{24} < 10^{-15}$ cm$^3$ s$^{-1}$, which is characteristic of reactions in which two bonds need to be broken. However, even a very small rate constant for (4.24) would lead to a significant source of the minor constituent NO, whose atmospheric origin at high altitude has been a major mystery. This demonstrates the need for a wide dynamic range for rate constant measurements. This upper limit of $10^{-15}$ cm$^3$ s$^{-1}$ is as small an ion–neutral rate constant as has yet been determined. It would be desirable to push this and some other reaction rate constants to still lower values.

The currently accepted belief is that NO in the high atmosphere is produced by the neutral reaction[38]

$$N(^2D) + O_2 \rightarrow NO + O \tag{4.25}$$

The chain of evidence has not as yet been satisfactorily closed, however, as the atmospheric source of the metastable $N(^2D)$ atoms remains to be proven. The dissociative recombination of $NO^+$, reaction (4.5), is presumed to be the main $N(^2D)$ source, but as yet the product states of (4.5) have not been experimentally determined.

An important point can be made by reference to reaction (4.16a), namely that the product states of ion–molecule reactions have almost never been measured. In the case of (4.16a) this is of considerable importance since energetically it is possible that the reaction produces the metastable $N(^2D)$, which has the important role in neutral chemistry described above. The radiative lifetime of this metastable atom is 26 h so that the difficulty of detecting metastable production by its radiation (at 5200 Å) in any laboratory apparatus is obvious since the atom will almost invariably be quenched before radiating. In the case of (4.21) there is sufficient energy to produce the metastables $O_2(^1\Delta_g)$ and $O_2(^1\Sigma)$ as well as ground-state molecular oxygen. In many cases it would be of interest to know product states, but their determination has so far defied the ingenuity of experimentalists.

Often, of course, only ground electronic state products are energetically possible, e.g. reactions (4.3) and (4.9). Even here, however, it would be of interest to know whether the reaction exothermicity goes into vibrational and rotational energy or into translational energy of the products.

While it is generally true that exothermic charge-transfer of either atomic or molecular ions with molecular neutrals is efficient, there are exceptions. An important ionospheric one is

$$O^+(^4S) + NO(^2\Pi) \rightarrow NO^+(^1\Sigma) + O(^3P) \tag{4.26}$$

which is so far immeasurably slow ($k < 8 \times 10^{-13}$ cm$^3$ s$^{-1}$)[39]. A typically fast reaction here would have been of importance under certain extreme atmospheric conditions, namely some peculiar auroras in which remarkably large (and as yet unexplained) enhancements of NO have been detected, such that the NO concentration becomes comparable with the oxygen concentration. The few cases like (4.26) disturb our complacency about generalities and argue the need for laboratory measurements of *all* important processes. Reaction (4.26) is not easy to understand on simple theoretical models. For example, arguments have been made that Franck–Condon factors are important in charge-transfer. However, the reaction

$$H^+ + NO \rightarrow NO^+ + H \tag{4.27}$$

has the same exothermicity as (4.26), owing to the accidental resonance

(4.10) and therefore would involve the same $NO \rightarrow NO^+$ Franck–Condon overlaps. But reaction (4.27) is extremely fast! Clearly a more detailed theoretical model is required. If the potential surfaces for a reaction were known, then it would be possible in principle to predict the course of a reaction, but this is almost never the case. The reactants $O^+$ and NO apparently do not come together at thermal energy to form a 'well-mixed' $NO_2^+$ complex since an immeasurably slow isotopic interchange has been found[40].

$$^{18}O^+ + N^{16}O \rightarrow {}^{16}O^+ + N^{18}O \quad k < 5 \times 10^{-12}\ cm^3\ s^{-1} \tag{4.28}$$

This contrasts to many fast isotopic exchanges, e.g.[40]

$$^{18}O^+ + C^{16}O \rightarrow {}^{16}O^+ + C^{18}O \quad k = 4.4 \times 10^{-10}\ cm^3\ s^{-1} \tag{4.29}$$

and

$$D^+ + H_2 \rightarrow H^+ + HD \quad k = 1 \times 10^{-9}\ cm^3\ s^{-1} \tag{4.30}$$

### 4.3.2 Metal ion chemistry

There is a constant influx of small meteors into the Earth's atmosphere. These meteors ablate upon entry into the ~100 km altitude range, depositing a variety of metals, i.e. Mg, Fe, Na, K, Al, Cr, Co, Ca and Ni (and silicon) in roughly their cosmic abundance. The presence of sodium in the atmosphere has long been known because it emits the well known D-line doublet at 5890 Å in the twilight and night sky. Since the metals have low ionisation potentials, the atmospheric ions $NO^+$ and $O_2^+$ readily charge-transfer with them. This has been well established by the extensive laboratory beam studies of Rutherford and his colleagues[41–45]. The resulting ions are atomic and one then has the same situation as arises in the F-region, namely ions that do not readily recombine with electrons. This has contributed to a very striking phenomenon, called sporadic-E by radiophysicists, which is the occurrence of very thin layers of atomic metal ions (and electrons) in the ~100 km altitude range. The layers may be only a few kilometres thick and have an electron density an order of magnitude larger than the ambient electron density (above and below the layer). The sporadic existence of these layers has been known for a long time by virtue of their profound effect on the propagation and reflection of radio waves. It was only with the advent of rocket-borne mass spectrometers that the ions were found to be metallic.

A question arose rather early as to the nature of the loss prccesses for ionospheric sporadic-E. By analogy with the F-region problem it was natural to look for chemical reactions that would convert the atomic ions into molecular ions, which could then rapidly recombine with electrons. Because of the relatively low ionisation potentials of the metals they cannot exothermically react with the major atmospheric molecules $O_2$ and $N_2$ and one was forced to look for a molecule with a weak bond. Ozone was the obvious candidate and so reactions such as

$$Mg^+ + O_3 \rightarrow MgO^+ + O_2 \tag{4.31}$$

followed by

$$MgO^+ + e^- \rightarrow Mg + O \tag{4.32}$$

were proposed as possible loss processes for the metallic sporadic-E layers. Reaction (4.31) was indeed found to be fast[46] and presumably (4.32) is fast as well. However, an unexpected situation arose; the reaction

$$MgO^+ + O \rightarrow Mg^+ + O_2 \quad (4.33)$$

was also found to be fast, and, indeed, the rate constants of (4.32) and (4.33) are about the same[46]. Since the atomic oxygen concentration far exceeds that of ozone in the E region of the ionosphere, the effect of (4.31) is mitigated by (4.33), i.e. the $Mg^+/MgO^+$ ratio will remain very large so that the effective electron recombination coefficient will not be significantly increased. Reactions like (4.31) were also found to occur rapidly for $Fe^+$ and $Ca^+$ ions but not for $Na^+$ and $K^+$ ions. Presumably the reaction

$$Na^+ + O_3 \rightarrow NaO^+ + O_2 \quad (4.34)$$

and the analogous reaction for $K^+$ are endothermic, although the available thermochemistry is not definitive on this point.

Reactions with atomic oxygen have not been carried out for $CaO^+$ and $FeO^+$ but these are probably fast. Metal oxide ions have not been observed in the ionosphere except[47] for a trace of $FeO^+$. There is a peculiar difficulty in making such observations, namely the coincidence in masses of $K^+$ and $NaO^+$, $Fe^+$ and $CaO^+$, $Ca^+$ and $MgO^+$. Rocket-borne mass spectrometers are necessarily of fairly low resolution because of the sensitivity required.

The ion $SiO^+$ apparently has been observed in the ionosphere, associated with the meteor ion layers. Since this ion is known to be rapidly destroyed by the reaction[48]

$$SiO^+ + O \rightarrow Si^+ + O_2 \quad (4.35)$$

it is extremely difficult to account for the presence of a detectable concentration of $SiO^+$. It has been proposed[49] that perhaps the molecular oxygen in the E layer is vibrationally excited so that the reverse of (4.35) occurs often enough to maintain the observed $SiO^+/Si^+$ ratio. No other source of $SiO^+$ has yet been proposed but the idea of vibrationally excited $O_2$ being responsible has not received support. Interestingly, the laboratory observation of (4.35), which implied its exothermicity, showed that the accepted dissociation energy of $SiO^+$ was $\sim$1 eV in error. This has subsequently been traced to an error in the reported ionisation potential of SiO [50].

In a recent[51] and very remarkable experiment in which ion mass spectra were obtained during the Geminid meteor shower, the following ions were identified in the E region, with enhanced layers at 95 and 119 km: $^{23}Na$, $^{24}Mg$, $^{25}Mg$, $^{26}Mg$, $^{27}Al$, $^{28}Si$, $^{29}Si$, $^{39}K$, $^{40}Ca$, $^{41}K$, $^{52}Cr$, $^{54}Fe$, $^{56}Fe$, $^{57}Fe$, $^{58}Fe$, $^{58}Ni$, $^{59}Co$, $^{60}Ni$. Ions at mass 43 and 44 were also observed and tentatively identified as $AlO^+$ and $SiO^+$. Atmospheric metal ion chemistry is thus a very full subject, and one which has scarcely been touched. When reliable metal ion concentrations are known and when all of the relevant chemistry is known, a detailed analysis should lead to information on the meteor source strength and atmospheric dynamics.

## 4.4 LOW ALTITUDE (D-REGION) ION CHEMISTRY

### 4.4.1 Terrestrial positive ion chemistry

The serious beginning of D-region (below $\sim$90 km) ion chemistry was the pioneering first observation of the ion composition by Narcisi and Bailey[4], who observed the positive ions above about 65 km in late 1963 (Figure 4.2). This was a very difficult task. In contrast to the E and F regions where earlier ion composition measurements had been made with rocket-borne mass spectrometers, D-region pressures are relatively high so that a vacuum pump must also be flown on the rocket, complicating the experiment a great deal. There is also a serious problem in reducing observed ion currents to ambient ion concentrations, for two reasons. One is the difficulty in determining the proportionality factor between ion current and ion concentration for a supersonic probe (vehicle) sampling in a relatively high pressure region. This is essentially an unsolved problem which is generally treated by normalising to other measurements of total ion density. An even more serious problem is that the sampling procedure can alter the nature of some of the ions. This is because the ions are mainly weakly bound cluster ions that can be broken up both by the hydrodynamic shock wave ahead of the supersonic vehicle and by the electric field used to draw the ions into the sampling orifice. Both effects have been directly demonstrated. The upshot of this is that all D-region ion composition measurements are still rather qualitative. In view of the expense and difficulty involved in improving this situation, the current rate of progress is not very brisk. There are proposals to drop mass spectrometers into the D region on parachutes which will go subsonic, thereby greatly simplifying the sampling measurements.

The observations of Narcisi and Bailey provided a very unexpected discovery, namely that the dominant ions below ~82 km were water cluster ions $H^+(H_2O)_n$, with the dominant ion being $H_5O_2^+$ (Figure 4.2). It was several years before even a qualitative explanation was provided for the production of water cluster ions in the atmosphere and the details are not yet quantitatively understood.

The main ionisation source in the D region was shown by Nicolet[52] in 1945 to be photoionisation of the trace constituent nitric oxide by solar Lyman-alpha u.v.

$$NO + \text{Ly-}\alpha\ (1215\ \text{Å}) \rightarrow NO^+ + e^- \tag{4.36}$$

This is a result of several somewhat chance circumstances, the low ionisation potential of NO (9.25 eV) relative to the major atmospheric constituents, the large intensity of Lyman-α (the strongest solar line), and the coincidence of a narrow $O_2$ optical transmission window at the Lyman-α wavelength. Thus the expectation was for $NO^+$ to be the major D-region ion, just as it is in the lower E region.

Hunten and McElroy[53] also showed that some D-region $O_2^+$ will arise from photoionisation of metastable oxygen $O_2(^1\Delta_g)$, which can be photoionised by a band of wavelengths that are not absorbed by ground-state $O_2$ and which therefore penetrate down into the D region. The $O_2(^1\Delta_g)$ is produced by the photolysis of ozone

$$O_3 + h\nu \rightarrow O_2(^1\Delta_g) + O \tag{4.37}$$

and its concentration is known from measurements of the infrared atmospheric airglow, using rocket-borne photometers

$$O_2(^1\Delta_g) \rightarrow O_2(^3\Sigma) + h\nu\ (1.2\ \mu m) \tag{4.38}$$

The origin of these water cluster ions was a major mystery in ionospheric chemistry for some time. It was discovered in 1969 that both the $NO^+$ and the $O_2{}^+$ produced in the D region (or lower) in the atmosphere will lead to water cluster ions[54,55]. Subsequently a number of groups have worked out the rather involved kinetics in some detail[56-60].

It should be pointed out that an obvious way of producing water cluster ions in the D region that was well-known in ion chemistry, namely

$$H_2O^+ + H_2O \rightarrow H_3O^+ + OH \quad k \approx 10^{-9}\ \mathrm{cm^3\ s^{-1}} \tag{4.39}$$

followed by

$$H_3O^+ + H_2O + M \rightarrow H_5O_2{}^+ + M, \text{ etc.} \tag{4.40}$$

is not effective in the D region. The water concentration in the D region is of the order of a few p.p.m. and the reaction

$$H_2O^+ + O_2 \rightarrow O_2{}^+ + H_2O \tag{4.41}$$

was found to be very fast[61], $k \approx 2 \times 10^{-10}$ cm$^3$ s$^{-1}$ so that any $H_2O^+$ produced, which could only be a very small amount, would lead to $O_2{}^+$. The currently established D-region positive ion chemistry scheme is illustrated in Figure 4.5. The $O_2{}^+$ conversion largely proceeds as follows:

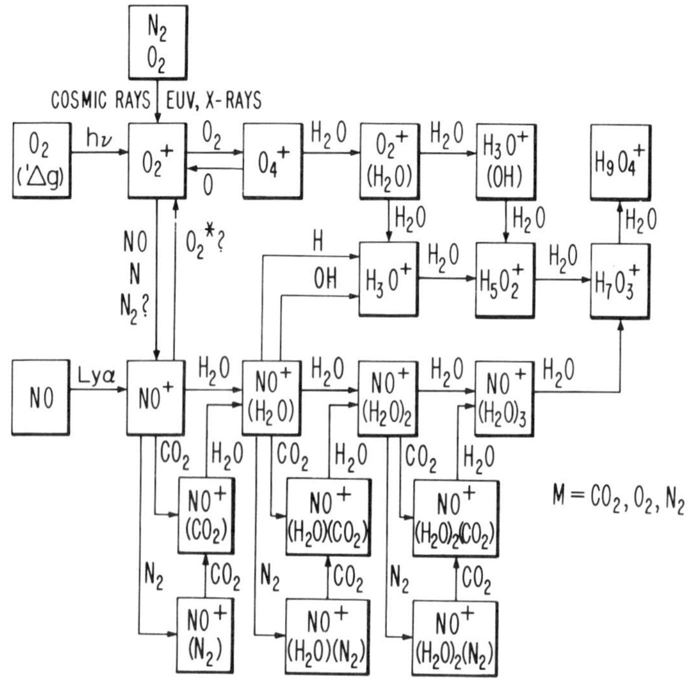


**Figure 4.5** Schematic diagram of D region positive ion chemistry, particularly the conversions of $O_2{}^+$ and $NO^+$ to $H^+(H_2O)_n$.

$$O_2{}^+ + O_2 + M \rightarrow O_4{}^+ + M \tag{4.42}$$

$$O_4{}^+ + O \rightarrow O_2{}^+ + O_3 \tag{4.43}$$

$$O_4{}^+ + H_2O \rightarrow O_2{}^+ \cdot H_2O + O_2 \tag{4.44}$$

$$O_2^+ \cdot H_2O + H_2O \rightarrow H_3O^+ \cdot OH + O_2 \quad (4.45a)$$

$$\rightarrow H_3O^+ + OH + O_2 \quad (4.45b)$$

$$H_3O^+ \cdot OH + H_2O \rightarrow H_3O^+ \cdot H_2O + OH \quad (4.46)$$

Reaction (4.45) 'deionises' the $O_2$, effecting the conversion from $O_2^+$ to $H_3O^+$. Reaction (4.45) branches, going mostly into channel *a* but also producing a sizeable channel *b*. The branching ratio is much less certain than the total rate constant. We believe that channel *b* is endothermic and probably occurs as a result of vibrational excitation in the $O_2^+ \cdot H_2O$ produced in (4.44). These reactions are difficult to study, being highly coupled, so that one cannot for example isolate individual steps but must unravel the complex scheme. In view of the difficulty in studying such systems, the agreement between different investigators shown in Table 4.1 is remarkable.

**Table 4.1** $O_2^+$ **reaction sequence in moist air at** 296 K

| *Reaction* | *Rate constant* | | | |
|---|---|---|---|---|
| | *Ref.* 60 | *Ref.* 56 | *Ref.* 58 | |
| $O_2^+ + O_2 + O_2 \rightarrow O_4^+ + O_2$ | 2.5 | 2.4 | — | $10^{-30}$ cm$^6$ s$^{-1}$ |
| $O_4^+ + H_2O \rightarrow O_2^+ \cdot H_2O$ | 1.5 | 1.3 | 2.2 | $10^{-9}$ cm$^3$ s$^{-1}$ |
| $O_2^+ \cdot H_2O + H_2O \rightarrow H_3O^+ \cdot OH + O_2$ | 1.0 | 0.9 | 1.9 | $10^{-9}$ cm$^3$ s$^{-1}$ |
| $\rightarrow H_3O^+ + OH + O_2$ | 0.2 | 0.3 | $\leqslant 0.3$ | $10^{-9}$ cm$^3$ s$^{-1}$ |
| $H_3O^+ \cdot OH + H_2O \rightarrow H_3O^+ \cdot H_2O + OH$ | 1.4 | $\geqslant 1$ | 3.2 | $10^{-9}$ cm$^3$ s$^{-1}$ |

The $NO^+$ conversion scheme can proceed as follows:

$$NO^+ + H_2O + M \rightarrow NO^+ \cdot H_2O + M \quad (4.47)$$

$$NO^+ \cdot H_2O + H_2O + M \rightarrow NO^+(H_2O)_2 + M \quad (4.48)$$

$$NO^+(H_2O)_2 + H_2O + M \rightarrow NO^+(H_2O)_3 + M \quad (4.49)$$

$$NO^+(H_2O)_3 + H_2O \rightarrow H_7O_3^+ + HNO_2 \quad (4.50)$$

Since $NO^+$ is a product of so many reactions in air (it can be considered as a sort of ion 'sink' in dry air), it follows that the above scheme will inevitably occur in moist air. An end product of air ionisation is therefore nitrous acid.

Reaction (4.42), in spite of being a three-body reaction, is fast at D-region and higher pressures because only major constituent neutrals are involved. Reaction (4.43) was discovered somewhat later than the rest of the scheme and is involved in the explanation of the very abrupt decrease in water ion concentration above some altitude, typically about 82 km, where atomic oxygen abruptly increases. The above scheme predicts such a decrease at the top of the water cluster layer, where $O_2^+$ production is relatively more important, because the rate of reaction (4.42) decreases as the square of the rapidly (exponentially) falling pressure and reaction (4.43) which breaks the $O_2^+ \rightarrow H_3O^+$ conversion, increases rapidly because the atomic oxygen concentration increases rapidly above ~80 km owing to the onset of photodissociation in this altitude range. Reactions (4.42) and (4.43) taken together represent an interesting ion-catalysed recombination of O with $O_2$ to produce $O_3$.

Reaction (4.44) is the kind of 'switching' reaction generally found to be fast[62] when exothermic, i.e. in which a weakly bound ligand is replaced by a

**Table 4.2** $NO^+$ **reaction sequence in moist air at 296 K**

| Reaction | Rate constant | | | | |
|---|---|---|---|---|---|
| | *Ref. 59* | *Ref. 58* | *Ref. 57* | * | |
| $NO^+ + H_2O + N_2 \rightarrow NO^+(H_2O) + N_2$ | 1.4 | 1.6 | 1.6† | 1.8 | $10^{-28}$ $cm^6$ $s^{-1}$ |
| $NO^-(H_2O) + H_2O + N_2 \rightarrow NO^+(H_2O)_2 + N_2$ | 1.2 | 1.0 | 1.1† | 1.0 | $10^{-27}$ $cm^6$ $s^{-1}$ |
| $NO^-(H_2O)_2 + H_2O + N_2 \rightarrow NO^+(H_2O)_3 + N_2$ | 1.4 | 2.0 | 1.9† | 1.0 | $10^{-27}$ $cm^6$ $s^{-1}$ |
| $NO^+(H_2O)_3 + H_2O \rightarrow H_3O^+(H_2O)_2 + HNO_2$ | 7 | 8 | 7 | 30 | $10^{-11}$ $cm^3$ $s^{-1}$ |

*French, M. A., Hills, L. P. and Kebarle, P. (1973). *Can. J. Chem.*, **51**, 456; hydration energies $\Delta H_{1,0} = 18.5$ and $\Delta H_{2,1} = 16.1$ kcal $mol^{-1}$
†Third body is NO rather than $N_2$

more tightly bound one, with near unit efficiency per collision. Reaction (4.46) is another example of this, although it is not readily predictable that (4.46) is exothermic, since the dipole moment of OH is comparable with that of $H_2O$.

Interestingly enough, as mysterious as this $NO^+ \rightarrow H_7O_3^+$ conversion process was to aeronomers (and gas phase kineticists), the basic chemistry was well known in solution inorganic chemistry. Unfortunately aeronomers knew little inorganic chemistry and inorganic chemists were not aware of this atmospheric problem so that the process required 'rediscovery'. Of course the gas-phase studies led to more specific knowledge than the solution results in which only the general conversion of $NO^+$ salts to nitrous acid in aqueous solution was known[63].

The several laboratory studies of the $NO^+$ sequence are in excellent agreement, as shown in Table 4.2. The difficulty in rationalising this $NO^+$ sequence with D-region observation is two-fold. (i) The ion $H^+(H_2O)_3$ is not the ion that appears to be the dominant one, which is $H^+(H_2O)_2$. This could conceivably be accounted for by the break-up of ambient $H^+(H_2O)_3$ into $H^+(H_2O)_2$ in the course of the rocket sampling, a process that requires less than 1 eV energy and that is actually known to occur to some extent. (ii) The timescale for the sequence is too long. Each of the steps (4.47), (4.48) and (4.49) are slow because they are three-body reactions involving a minor constituent. Ionospheric calculations then lead to the conclusion that large concentrations of $NO^+$ hydrates would be present but these have not been observed, except for a trace of $NO^+(H_2O)$.

This problem has been alleviated to some extent by showing that the reactions

$$NO^+ + CO_2 + M \rightarrow NO^+ \cdot CO_2 + M \tag{4.51}$$

and

$$NO^+ \cdot CO_2 + H_2O \rightarrow NO^+ \cdot H_2O + CO_2 \tag{4.52}$$

occur in about an order of magnitude less time than reaction (4.47). This is due to the large $[CO_2]/[H_2O]$ ratio of *ca.* 100, despite $k_{51}$ being an order of magnitude less than $k_{47}$. The switching reaction (4.52) is essentially instantaneous in the appropriate timescale. Presumably similar short cuts occur for the addition of the second and third waters by hydration to $NO^+$ but they have not yet been measured, and it will be extremely difficult to do so.

Recently it has been suggested[64] that association of $NO^+$ with $N_2$, followed by switching with $H_2O$

$$NO^+ + N_2 + M \rightarrow NO^+ \cdot N_2 + M \tag{4.53}$$

$$NO^+ \cdot N_2 + H_2O \rightarrow NO^+ \cdot H_2O + N_2 \tag{4.54}$$

might be effective. There has been some dispute about this and a large uncertainty remains. Earlier measurements[65] showed that (4.53) is either extremely slow or has a small equilibrium constant but, because of the large $N_2$ abundance, this reaction can still be important. Experimentally it is extremely difficult to measure such slow reactions and all experiments so far are at best borderline in their capability of doing this. The experimental difficulty is a result of the very weak bond, $D(NO^+ \cdot N_2) \approx 0.25$ eV $\approx T\Delta S$ at 300 K. The free energy change for the reaction $\Delta G = \Delta H - T\Delta S$ is close to zero and association is not favoured at 300 K. In other words, collisional break-up [the reverse of (4.53)] is very serious in laboratory situations and this makes it difficult to obtain a forward rate constant. It is difficult to deviate from equilibrium and it is thus easier to obtain equilibrium constants than rate constants for (4.53).

However, measuring equilibrium constants is still difficult because, owing to the weak $NO^+ \cdot N_2$ bond, almost any impurity will destroy the $NO^+ \cdot N_2$ by switching, e.g. by reaction (4.54), or by

$$NO^+ \cdot N_2 + CO_2 \rightarrow NO^+ \cdot CO_2 + N_2 \qquad (4.55)$$

or indeed by

$$NO^+ \cdot N_2 + NO \rightarrow NO^+ \cdot NO + N_2 \qquad (4.56)$$

It is extremely difficult to avoid impurities at the p.p.m. levels that are disastrous for these experiments. The atmospheric break-up problem is less severe than in room temperature laboratory studies since the D region has temperatures as low as 200 K or even lower and the equilibrium constant for (4.53) increases rapidly with decreased temperature.

At present it appears that even with $CO_2$ and $N_2$ clustering the timescale for conversion of $NO^+$ to water cluster ions is too long to agree with atmospheric observations of the relative abundances of ions. Suggestions have been made of additional reactions including

$$NO^+ \cdot H_2O + H \rightarrow H_3O^+ + NO \qquad (4.60)$$

which is exothermic but does not efficiently occur[66], and the reaction[64]

$$NO^+ \cdot H_2O + OH \rightarrow H_3O^+ + NO_2, \qquad (4.61)$$

which probably is efficient and

$$NO^+ \cdot H_2O + HO_2 \rightarrow H_3O^+ + NO + O_2 \qquad (4.62)$$

The latter two reactions have not been measured and the last one at least is beyond present measurement capability.

### 4.4.2 Terrestrial negative ion chemistry

Measurements of the negative ion composition of the ionosphere are extremely difficult to make. Negative ions are confined largely to the D region, so that all of the problems plaguing positive ion measurements in this relatively high pressure region remain. In addition, another serious problem pecular to negative ions arises. When an electric field configuration is applied to a rocket to sample the negative ions into a mass spectrometer, electrons are of course also attracted to the sampling orifice. Because of their much greater mobility, the electron current greatly exceeds the negative ion current, tending to drive the sampler negative so that negative ions are not easily sampled.

Two groups have made successful negative ion composition measurements, the group of Narcisi at AFCRL[67] and a group at the Max-Planck-Institute at Heidelberg[68]. (Earlier work reporting the observation of negative ions in the E region was almost certainly in error.) To date, these results are somewhat qualitative and are in serious disagreement. The German group has made some model calculations, using laboratory derived reaction schemes, with results which seem encouraging[69]. At this time, however, our understanding of atmospheric negative ion chemistry is in an embryonic state. In this discussion we will concentrate on the laboratory-derived reaction schemes, noting that these schemes have not yet been critically tested.

Some of the processes involved in atmospheric negative ion chemistry have been of great interest in atomic physics and in gaseous electronics for a long time. The major formation process for atmospheric negative ions

$$e^- + 2O_2 \rightarrow O_2^- + O_2 \quad (4.63)$$

has been recognised for a long time and the rate constant for this three-body attachment was measured by Chanin, Phelps and Biondi[70] in 1962. Phelps and Pack[71] also measured the reverse process, collisional detachment. The ratio of forward to reverse rate constants is the equilibrium constant. Since the equilibrium constant is determined by the energetics, this provided a value for the electron affinity of $O_2$ (0.45 eV) that has withstood the test of time, in spite of numerous claims to the contrary over the years. Recent very elegant photodetachment electron spectroscopy studies at the Joint Institute for Laboratory Astrophysics (Boulder, Colorado) have clearly resolved this problem[72].

As far back as 1958, Burch, Smith and Branscomb[73] measured $O_2^-$ photodetachment in the laboratory because of its interest as an atmospheric negative ion loss process

$$O_2^- + h\nu \rightarrow O_2 + e^- \quad (4.64)$$

However, the most important $O_2^-$ loss in the atmosphere turns out to be associative-detachment with atomic oxygen

$$O_2^- + O \rightarrow O_3 + e^- \quad (4.65)$$

which was predicted by Dalgarno[74] in 1961 and which was measured at NOAA in 1966[75], using the flowing afterglow technique. This measurement again utilises the flowing afterglow capability for reacting ions with unstable neutrals and is the only existing measurement of this reaction.

The following reaction scheme has resulted from a series of NOAA flowing afterglow studies[76–79]. The $O_2^-$ ions that survive associative-detachment can initiate a fairly complex negative ion chemistry, either by charge-transfer with ozone

$$O_2^- + O_3 \rightarrow O_3^- + O_2 \quad (4.66)$$

or by association

$$O_2^- + O_2 + M \rightarrow O_4^- + M \quad (4.67)$$

The overall reaction sequence is shown schematically in Figure 4.6.

The rapid decrease of $O_2^-$ formation rate with altitude, falling as the square of the $O_2$ density, and the rapid increase in electron detachment, due to a rapid increase in the $O/O_3$ ratio with altitude, should confine any appreciable negative ion concentration to relatively low altitudes, typically less than *ca.* 80 km, and this agrees with observation. By appreciable we mean negative ion concentrations comparable with positive ion concentrations. There will be a small concentration of negative ions even at the highest altitude, produced by radiative attachment

$$O + e^- \rightarrow O^- + h\nu \quad (4.68)$$

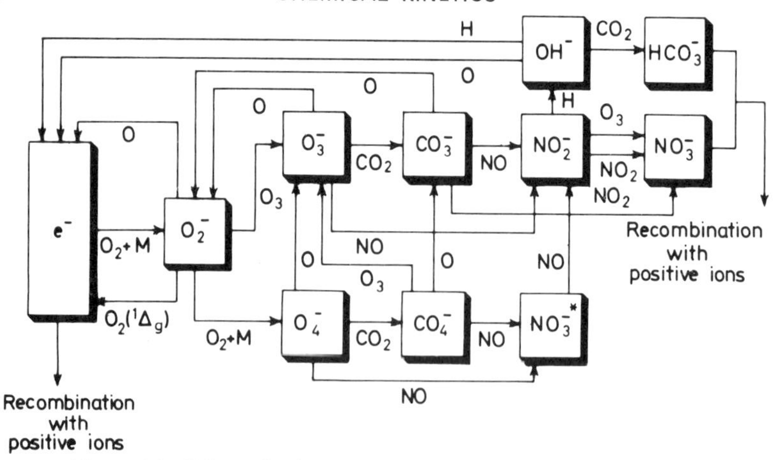


**Figure 4.6** Schematic diagram of D region negative ion chemistry

The $O^-$ production by (4.68) and loss are both proportional to [O], the loss again being associative-detachment

$$O^- + O \rightarrow O_2 + e^- \tag{4.69}$$

and a steady state treatment of these equations yields $[O^-]/n_e \approx 10^{-5}$. At high enough altitudes photodetachment of $O^-$ will become the dominant loss and then $[O^-]/n_e < 10^{-5}$. In the D region there is another weak source of $O^-$

$$e^- + O_3 \rightarrow O^- + O_2 \tag{4.70}$$

This $O^-$ will charge-transfer with $O_3$ to produce $O_3^-$ and then enter the scheme of Figure 4.6.

Both $O_3^-$ and $O_4^-$ react rapidly with the very abundant $CO_2$ in the atmosphere. The lifetimes of $O_3^-$ and $O_4^-$ are less than 0.01 s in the atmosphere and they are not expected to be observable and so far have not been. Both $O_3^-$ and $O_4^-$ are also susceptible to attack by atomic oxygen and NO, but since $[CO_2] \gg [O]$ and [NO], this is of little practical importance.

The reaction

$$O_3^- + CO_2 \rightarrow CO_3^- + O_2 \tag{4.71}$$

has an equilibrium constant greater than $1.3 \times 10^6$, which in turn leads to the present lower limit on the $CO_3$ electron affinity, $EA(CO_3) > 2.7$ eV [80].

The ion $CO_3^-$ can be destroyed by O

$$CO_3^- + O \rightarrow O_2^- + CO_2 \tag{4.72}$$

or react with NO

$$CO_3^- + NO \rightarrow NO_2^- + CO_2 \tag{4.73}$$

The $NO_2^-$ can react with H

$$NO_2^- + H \rightarrow OH^- + NO \tag{4.74}$$

but there is little atomic hydrogen in the D region so that this does not compete effectively with

$$NO_2^- + O_3 \rightarrow NO_3^- + O_2 \tag{4.75}$$

and

$$NO_2^- + NO_2 \rightarrow NO_3^- + NO \tag{4.76}$$

both of which produce the very stable negative $NO_3^-$. The electron affinity of $NO_3$ has recently been determined[81] to be $3.9 \pm 0.2$ eV, the largest established electron affinity so far.

The reaction of $CO_4^-$ with NO

$$CO_4^- + NO \rightarrow NO_3^{-*} + CO_2 \tag{4.77}$$

is a peculiar one in that the $NO_3^{-*}$ produced is not the most stable form of $NO_3^-$. This has been determined by finding that $NO_3^{-*}$ reacts with NO to produce $NO_2^-$

$$NO_3^{-*} + NO \rightarrow NO_2^- + NO_2 \tag{4.78}$$

which is just the reverse of reaction (4.76) and would therefore be endothermic for the most stable $NO_3^-$ form.

We have suggested that $NO_3^{-*}$ is the peroxy form of the ion $[O—O—N—O]^-$, while $NO_3^-$ is the nitrate ion in which all oxygens are bonded to the N atom and are equivalent. The $NO_3^{-*}$ ion can also be produced in the reaction

$$O_4^- + NO \rightarrow NO_3^{-*} + O_2 \tag{4.79}$$

It seems likely that reactions (4.77) and (4.79) are simply switching reactions, i.e.

$$O_2^- \cdot CO_2 + NO \rightarrow O_2^- \cdot NO + CO_2 \tag{4.80}$$

and

$$O_2^- \cdot O_2 + NO \rightarrow O_2^- \cdot NO + O_2 \tag{4.81}$$

where $O_2^- \cdot NO$ (or $NO_3^{-*}$) formation does not involve breaking the $O_2^-$ bond but simply represents a ligand exchange. The $NO_3^-$ produced in reactions (4.75) and (4.76) on the other hand, involves adding one O atom to the $NO_2^-$ ion and it is reasonable to suppose that this atom adds in such a way as to produce the most stable $NO_3^-$ ion.

Reactions of $NO_3^-$ that would further extend the scheme have not yet been found and probably do not occur in the atmosphere, so that $NO_3^-$ is expected to be a 'terminal' or stable atmospheric negative ion. Any stable positive or negative ion will of course hydrate in the D region, so one expects ions $NO_3^-(H_2O)_n$ to be dominant species. This prediction from the laboratory studies, which preceded observation, seemed to be substantiated by the first D-region negative ion observations[67], which did indeed find $NO_3^-$ and the hydrates to be the most conspicuous feature of the negative ion spectrum. The German results[68], while reporting $NO_3^-$, did not find its role to be so dominant, however.

The ions that have so far been observed include $NO_3^-$ and its hydrates, as expected, and also $CO_3^-$ and $CO_4^-$, which are not too surprising since the NO and O concentrations are so low (NO and O being the destructive reactants). The ions $O_3^-$ and $O_4^-$ are not observed, consistent with their short lifetimes due to the fast $CO_2$ loss, and $O_2^-$ and $NO_2^-$ are not observed, consistent with their fast loss with $O_3$.

The German group[68] reported ions at the following masses, with the tentative identification indicated in parentheses: 61 ($HCO_3^-$), 68 ($O_2^-(H_2O)_2$), and $175 \pm 2$ ($NO_3^-(HNO_3)$), as well as $CO_3^-$, $NO_3^-$, $CO_4^-$, $CO_3^-(H_2O)$ and possibly $CO_4^-(H_2O)$ and $CO_4^-(H_2O)_2$. Both groups observe $Cl^-$, which is presumably a contaminant, and Narcisi reports very small $O^-$ and $NO_2^-$

signals in the E region, while the Germans report a small D-region signal of $O_2^-$

An important loss process for the 'terminal' negative ions, and to some extent for any atmosphere negative ion, is ion–ion recombination

$$X^+ + Y^- \rightarrow \text{neutrals} \tag{4.82}$$

The rate constants for mutual neutralisation are large in all measured cases, as might be expected from the strong coulomb attractive force.

A few ion–ion recombinations have been measured in static afterglow systems and very remarkable progress has occurred in this field in recent years from the application of merged beam technique at the Stanford Research Institute[82–84]. They have so far studied the systems $H^+ + H^-$, $O^+ + O^-$, $N^+ + O^-$, $N_2^+ + O_2^-$, $O_2^+ + O_2^-$, $O_2^+ + NO_2^-$, $NO^+ + NO_2^-$, $O_2^+ + O^-$, $NO^+ + O^-$, and $Na^+ + O^-$, over the energy range as low as 0.15 eV up to several hundred volts relative kinetic energy. Thermal energy recombination coefficients are obtained by an extrapolation procedure. The range of extrapolated values for the above reactions is only from $10 \pm 4$ to $51 \pm 15 \times 10^{-8}$ cm$^3$ s$^{-1}$.

The neutral products and product states have not been determined except for the one case of $Na^+ + O^-$, where Berry and his students observed sodium D-line emission[85]. The recombination is often so exothermic that considerable neutral fragmentation could occur and for atmospheric purposes it would be interesting to know if this happens. In the D region the ultimate neutralisation probably involves reactions like

$$H^+(H_2O)_n + NO_3^-(H_2O)_m \rightarrow \text{products} \tag{4.83}$$

The products in this case may be nitric acid ($HNO_3$) and water.

Another atmospheric loss source for negative ions is photodetachment. Photodetachment has been carried out for $O^-$, $O_2^-$, $O_3^-$, $O_4^-$ and $CO_3^-$, so far as atmospherically interesting ions are concerned[73,86–90]. In all these cases photodetachment cannot compete with chemical reaction and so is only of minor concern. The $CO_3^-$ photodetachment data are suspect since the reported photodetachment threshold is clearly in error[80]. The most important D-region ions are $NO_3^-$ and its hydrates, and these ions have not as yet been studied in photodetachment.

### 4.4.3 Some special minor ion chemistry

In the above discussion we have attempted to follow the dominant reaction sequences in the D region. In addition, however, certain positive ions appear to have been observed that fall outside the scope of the presently conceived schemes. The same is true for the negative ions, as we have mentioned above, but since the negative ion observations are so sparse and since different experimenters do not agree, the problem there is less well defined. These D-region anomalies are in addition to the presence of metal ions that persist in both the D and E regions and that have not been adequately investigated.

There is often some uncertainty as to the identity of ions in the ionosphere, since only the mass is measured and this may be ambiguous.

One quite mysterious observation is that of atomic sulphur ions, which were reported by Narcisi[91] in the daytime ionosphere over Fort Churchill between 74 and 86 km in December 1967 and more recently by Zbinden *et al.*[51] around 80 km in the daytime ionosphere over Sardinia in December 1971. Since the principal sulphur isotope has a mass 32, the same as $O_2$, it is necessary to measure the mass 34 isotope and distinguish $S^+$ from $O_2^+$ by the different isotope ratios. The natural $^{34}O_2/^{32}O_2$ ratio is 0.0041 while the natural $^{34}S/^{32}S$ ratio is 0.044.

The origin of $S^+$ in the D region poses an intriguing puzzle that remains unanswered. It is known that atmospheric aerosols contain sulphur (in the form of sulphuric acid), but it is not expected that aerosols should provide a significant source of material at altitudes as high as 80 km. It is also known that sulphur is present, along with metals, in the meteors that are persistently bombarding the Earth's atmosphere. It has recently been found[92] that the reaction

$$S^+ + O_2 \rightarrow SO^+ + O \tag{4.84}$$

has a rate constant $k = 1.6 \times 10^{-11}$ cm$^3$ s$^{-1}$, which seems to place an impossible burden on any conceivable $S^+$ source in the D region. The observed ion concentrations $\sim 10^2$ cm$^{-3}$ would require a production rate $\sim 4 \times 10^4$ cm$^{-3}$ s$^{-1}$ to balance the loss rate by (4.84) and this is an ionisation rate that exceeds the total known D-region ionisation rate! It will be interesting to find the solution to this dilemma.

An ion of mass $46^+$, presumably $NO_2^+$, has been reported in the $\sim$75 km altitude region[93]. No reasonable source for this ion has yet been suggested. A reaction that might have satisfied this need is

$$NO^+(^1\Sigma^+) + O_3(^1A) \rightarrow NO_2^+(^1\Sigma_g^+) + O_2(^3\Sigma_g^-) + 1.5\text{ eV} \tag{4.85a}$$

$$\rightarrow NO_2^+(^1\Sigma^+{}_g) + O_2(^1\Delta_g) + 0.5\text{ eV} \tag{4.85b}$$

This reaction, if fast, could have been helpful in several ways: (a) as a source of $NO_2^+$; (b) as a needed additional loss process for $NO^+$; and (c) as a needed source of $O_2(^1\Delta_g)$ in auroras if the reaction occurred by channel *b*. The concentration of $O_2(^1\Delta_g)$, as determined by its i.r. emission, is known to be anomalously large in auroral conditions. Unfortunately reaction (4.85) was found[94] to be extremely slow, $k < 10^{-14}$ cm$^{-3}$ s$^{-1}$, which eliminates it from being significant.

The slowness of this exothermic reaction has been rationalised as follows. Reaction channel *a* violates spin conservation. While channel *b* conserves spin overall, it would involve $O_3$ dissociation into $O(^1D) + O_2(^1\Delta)$, which requires *ca.* 4 eV, much more than the overall reaction exothermicity, hence leading to a substantial activation energy. The energetically favoured dissociation of $O_3$ into $O(^3P) + O_2(^1\Delta)$ is spin forbidden and the $O(^3P)$ association with $NO^+(^1\Sigma)$ to produce $NO_2^+(^1\Sigma)$ would also be spin forbidden. The isoelectronic neutral reaction that is substantially more exothermic

$$CO(^1\Sigma) + O_3(^1A) \rightarrow CO_2(^1\Sigma^+{}_g) + O_2(^3\Sigma, {}^1\Delta, {}^1\Sigma) \tag{4.86}$$

is also[95] undetectably slow, $k < 4 \times 10^{-25}$ cm$^{-3}$ s$^{-1}$.

## 4.5 SUMMARY

Some general remarks concerning the ion chemistry of planetary atmospheres have been presented. The major features of the ion chemistry of the Earth's ionosphere have been briefly described. The constantly increasing sophistication of laboratory measurement technology provides a steadily increasing base of kinetic data for the understanding of planetary ionospheric chemistry. The very significant refinements in *in situ* ionospheric measurements which can be anticipated, when combined with this improved laboratory knowledge will lead to a continued increase in our knowledge in this area for the foreseeable future. It is also true that technology developed primarily for laboratory aeronomy investigations will continue to have a useful impact on other areas of chemistry.

### Note added in proof

Since this chapter was submitted a number of further developments in the field have occurred. These include the following.

An important collection of papers on ion–molecule reactions has appeared[96], including a chapter on ionospheric reactions. Recent reviews of ionospheric ion–molecule chemistry have appeared[97,98], including a substantial updating of laboratory data[98]. The rate constant for the reaction

$$O^+ + NO \rightarrow NO^+ + O \qquad (4.26)$$

has been established[99]. The thermodynamics of $NO^+$ association with $N_2$ (4.53) has now been determined[100]. Rate constants for the $O_2^+$ (a $^4\Pi_u$) ion reaction with $N_2$ (4.19) and with $O_2$, NO and other neutrals have been determined from thermal energy to *ca.* 2 eV relative kinetic energy[101,102]. The reaction of $NO^+ \cdot H_2O$ with OH (4.61) has been shown[103] to have a rate constant less than $10^{-10}$ cm$^3$ s$^{-1}$. The temperature dependences of the major F-region reactions have been measured[104] from 300 to 900 K.

In response to a suggestion that $H_2O_2^+$ ions might be present in the D-region, the appropriate production and loss reactions for $H_2O_2^+$ have been measured[105], leading to the conclusion that $H_2O_2^+$ ions are unlikely to be detectable in the ionosphere.

In response to several suggestions that ion reactions might destroy chlorofluoromethanes in the atmosphere, a number of measurements[106] have been made which show this to be unlikely.

In response to a suggestion that negative ions might provide a sink for $N_2O$ in the atmosphere, it has been shown[107] that neither $O_2^-$, $O_3^-$, $CO_3^-$ or $NO_2$ have fast reactions with $N_2O$ in spite of the fact that exothermic possibilities exist.

Increasing interest in lower atmospheric ion chemistry has led to a detailed investigation[108] of the gas phase ion chemistry of $HNO_3$.

Atmospheric ion chemistry continues to be an active field of research. It has been recognised that ambient ion measurements in the lower atmosphere (troposphere and stratosphere) would provide a powerful analytical tool for the measurement of certain important trace neutral constituents, much in the manner of chemical ionisation mass spectrometry. So far the ability to measure ambient ion composition in a high pressure atmosphere has eluded experimental capability but several groups are working on this problem.

## References


1. Kliore, A., Cain, D. L., Fjeldbo, G., Seidel, B. L. and Rasool, S. I. (1974). *Science*, **183,** 323
2. Bates, D. R. (1973). *J. Atmos. Terr. Phys.*, **35,** 1935
3. Holmes, J. C., Johnson, C. Y. and Young, J. M. (1965). *Space Research V*, 756 (Amsterdam: North Holland)
4. Narcisi, R. S. and Bailey, A. D. (1965). *J. Geophys. Res.*, **70,** 3687
5. Banks, P. M. and Kockarts, G. (1973). *Aeronomy* (New York: Academic Press) 2 volumes
6. Dubrin, J. and Henchman, M. J. (1972). *MTP International Review of Science, Physical Chemistry Series One*, Vol. 9, *Chemical Kinetics*, 213 (J. C. Polanyi, editor) (London: Butterworths)
7. McDaniel, E. W., Cermák, V., Dalgarno, A., Ferguson, E. E. and Friedman, L. (1970). *Ion Molecule Reactions* (New York: Wiley Interscience)
8. Franklin, J. L. (ed.) (1972). *Ion–Molecule Reactions* (New York: Plenum) 2 vols.
9. Friedman, L. (1968). *Ann. Rev. Phys. Chem.*, **19,** 273
10. Bowers, M. T. and Su, T. (1973). *Adv. Electronics Electron Phys.*, **34,** 223
11. Friedman, L. and Rueben, B. G. (1971). *Adv. Chem. Phys.*, **19,** 33
12. Beauchamp, J. L. (1971). *Ann. Rev. Phys. Chem.*, **22,** 527
13. Drewery, C. J., Goode, G. C. and Jennings, K. R. (1972). *MTP International Review of Science, Physical Chemistry Series One*, Vol. 5, *Mass Spectrometry* (A. Maccoll, editor) (London: Butterworths)
14. Ferguson, E. E., Fehsenfeld, F. C. and Schmeltekopf, A. L. (1969). *Adv. At. Mol. Phys.*, **5,** 1
15. Ferguson, E. E. (1968). *Adv. Electronics Electron Phys.*, **24,** 1
16. Ferguson, E. E. (1973). *At. Data Nucl. Data Tables*, **12,** 159
17. Fehsenfeld, F. C., Schmeltekopf, A. L., Dunkin, D. B., Albritton, D. L., Howard, C. J. and Ferguson, E. E. (1973). *Twenty-sixth Annual Gaseous Electronics Conference* (University of Wisconsin, October 16–19, 1973)
18. McFarland, M., Albritton, D. L., Fehsenfeld, F. C., Ferguson, E. E. and Schmeltekopf, A. L. (1973). *J. Chem. Phys.*, **59,** 6610
19. Johnsen, R. and Biondi, M. A. (1973). *J. Chem. Phys.*, **59,** 3504
20. Schaeffer, R. C. (1974). *J. Geophys. Res.*, in press
21. Kumar, S. and Hunten, D. M. (1974). *J. Geophys. Res.*, in press
20. Schaeffer, R. C. (1975). *J. Geophys. Res.*, **80,** 154
21. Kumar, S. and Hunten, D. M. (1974). *J. Geophys. Res.*, **79,** 2529
22. Hanson, W. B. and Ortenburger, I. B. (1961). *J. Geophys. Res.*, **66,** 1425
23. Donahue, T. M. (1968). *Science*, **159,** 489
24. Bates, D. R. and Lynn, N. (1959). *Proc. Roy. Soc. (London)*, **A253,** 141
25. Rapp, D. (1963). *J. Geophys. Res.*, **68,** 1773
26. Stebbings, R. B., Smith, A. C. H. and Erhardt, H. (1964). *J. Geophys. Res.*, **69,** 2349
27. Stebbings, R. F. and Rutherford, J. A. (1968). *J. Geophys. Res.*, **73,** 1035
28. Fehsenfeld, F. C. and Ferguson, E. E. (1972). *J. Chem. Phys.*, **56,** 3067
29. Bates, D. R. and Patterson, T. N. L. (1962). *Planet. Space Sci.*, **9,** 599
30. Nicolet, M. (1961). *J. Geophys. Res.*, **66,** 2263
31. Kockarts, G. (1973). *Space Science Rev.*, **14,** 723
32. Ferguson, E. E., Fehsenfeld, F. C., Dunkin, D. B., Schiff, H. I. and Schmeltekopf, A. L. (1964). *Planet. Space Sci.*, **12,** 1169
33. Sayers, J. and Smith, D. S. (1964). *Discuss. Faraday Soc.*, **37,** 167
34. Donahue, T. M. (1966). *Planet. Space Sci.*, **14,** 33
35. Giraud, A. (1971). *Mesospheric Models and Related Experiments*, 267 (G. Fiocco, editor) (Dordrecht: Reidel)

36. Danilov, A. D. and Semenov, V. K. (1970). *Space Research X*, 736 (Amsterdam: North Holland)
37. Norton, R. B., Van Zandt, T. E. and Denison, J. S. (1963). *Proceedings of the International Conference on the Ionosphere*, 26 (London: The Institute of Physics and the Physical Society)
38. Norton, R. B. and Barth, C. A. (1970). *J. Geophys. Res.*, **75,** 3903
39. McFarland, M., Albritton, D. L., Fehsenfeld, F. C., Ferguson, E. E. and Schmeltekopf, A. L. (1974). *J. Geophys. Res.*, **79,** 2005
40. Fehsenfeld, F. C., Albritton, D. L., Bush, Y. A., Fournier, P. G., Govers, T. R. and Fournier, J. (1974). *J. Chem. Phys.*, **61,** 2150
41. Rutherford, J. A., Mathis, R. F., Turner, B. R. and Vroom, D. A. (1971). *J. Chem. Phys.*, **55,** 3783
42. Rutherford, J. A., Mathis, R. F., Turner, B. R. and Vroom, D. A. (1971). *J. Chem. Phys.*, **55,** 5622
43. Rutherford, J. A., Mathis, R. F., Turner, B. R. and Vroom, D. A. (1972). *J. Chem. Phys.*, **56,** 4654
44. Rutherford, J. A., Mathis, R. F., Turner, B. R. and Vroom, D. A. (1972). *J. Chem Phys.*, **57,** 3087
45. Rutherford, J. A., Mathis, R. F., Turner, B. R. and Vroom, D. A. (1972). *J. Chem. Phys.*, **57,** 3091
46. Ferguson, E. E. and Fehsenfeld, F. C. (1968). *J. Geophys. Res.*, **73,** 6215
47. Narcisi, R. S. (1973). *Physics and Chemistry of Upper Atmosphere*, 171 (B. M. McCormac, editor) (Dordrecht, Holland: D. Reidel)
48. Fehsenfeld, F. C. (1969). *Can. J. Chem.*, **47,** 1808
49. Ferguson, E. E., Fehsenfeld, F. C. and Whitehead, J. D. (1970). *J. Geophys. Res.*, **75,** 4366, 7333
50. Hildenbrand, D. L. and Murad, E. (1970). *J. Chem. Phys.*, **53,** 3403
51. Zbinden, P. A., Hidalgo, M. A., Eberhardt, P. and Geiss, J. (1973). Paper presented at *16th Plenary Meeting of COSPAR* (Konstanz, Germany)
52. Nicolet, M. (1945). *Inst. R. Meteorol. Belg. Mem.*, **19,** 162
53. Hunten, D. M. and McElroy, B. M. (1968). *J. Geophys. Res.*, **73,** 2421
54. Fehsenfeld, F. C. and Ferguson, E. E. (1969). *J. Geophys. Res.*, **74,** 2217
55. Ferguson, E. E. and Fehsenfeld, F. C. (1969). *J. Geophys. Res.*, **74,** 5743
56. Good, A., Durden, D. A. and Kebarle, P. (1970). *J. Chem. Phys.*, **52,** 222
57. Puckett, L. J. and Teague, M. W. (1971). *J. Chem. Phys.*, **54,** 2564
58. Fehsenfeld, F. C., Mosesman, M. and Ferguson, E. E. (1971). *J. Chem. Phys.*, **55,** 2115, 2120
59. Howard, C. J., Rundle, H. W. and Kaufman, F. (1971). *J. Chem. Phys.*, **55,** 5772
60. Howard, C. J., Bierbaum, V. M., Rundle, H. W. and Kaufman, F. (1972). *J. Chem. Phys.*, **57,** 3491
61. Fehsenfeld, F. C., Schmeltekopf, A. L. and Ferguson, E. E. (1967). *J. Chem. Phys.*, **46,** 2802
62. Adams, N. G., Bohm, D. K., Dunkin, D. B., Fehsenfeld, F. C. and Ferguson, E. E. (1970). *J. Chem. Phys.*, **52,** 3133
63. Cotton, F. A. and Wilkinson, G. (1966). *Advanced Inorganic Chemistry*, 2nd ed., 344 (New York: Interscience)
64. Heimerl, J. M. and Vanderhoff, J. A. (1971). *Trans. Amer. Geophys. Union*, **52,** 870
65. Dunkin, D. B., Fehsenfeld, F. C. and Ferguson, E. E. (1971). *J. Chem. Phys.*, **54,** 3817
66. Fehsenfeld, F.C, and Ferguson. E. E. (1972). *Radio Sc.*, **7,** 113
67. Narcisi, R. S., Bailey, A. D., Della Lucca, L., Sherman, C. and Thomas, D. M. (1971). *J. Atmos. Terr. Phys.*, **33,** 1147
68. Arnold, F., Kissel, J., Krankowsky, D., Wieder, H. and Zahringer, J. (1971). *J. Atmos. Terr. Phys.*, **33,** 1169
69. Arnold, F. and Krankowsky, D. (1971). *J. Atmos. Terr. Phys.*, **33,** 1693
70. Chanin, L. M., Phelps, A. V. and Biondi, M. A. (1962). *Phys. Rev.*, **128,** 219
71. Phelps, A. V. and Pack, J. L. (1961). *Phys. Rev. Lett.*, **6,** 111
72. Celotta, R. J., Bennett, R. A., Hall, J. L., Siegel, M. W. and Levine, J. (1972). *Phys. Rev. A*, **6,** 631
73. Burch, D. S., Smith, S. J. and Branscomb, L. M. (1958). *Phys. Rev.*, **112,** 171; (1959). *ibid.*, **114,** 1952

74. Dalgarno, A. (1961). *Ann. Geophys.*, **17,** 16
75. Fehsenfeld, F. C., Ferguson, E. E. and Schmeltekopf, A. L. (1966). *J. Chem. Phys.*, **45,** 1844
76. Fehsenfeld, F. C., Schmeltekopf, A. L., Schiff, H. I. and Ferguson, E. E. (1967). *Planet. Space Sci.*, **15,** 373
77. Fehsenfeld, F. C., and Ferguson, E. E. (1968). *Planet. Space Sci.*, **16,** 701
78. Fehsenfeld, F. C., Ferguson, E. E. and Bohme, D. K. (1969). *Planet. Space Sci.*, **17,** 1759
79. Fehsenfeld, F. C. and Ferguson, E. E. (1972). *Planet. Space Sci.*, **20,** 295
80. Ferguson, E. E., Fehsenfeld, F. C. and Phelps, A. V. (1973). *J. Chem. Phys.*, **59,** 1565
81. Ferguson, E. E., Dunkin, D. B. and Fehsenfeld, F. C. (1972). *J. Chem. Phys.*, **57,** 1459
82. Aberth, W. H. and Peterson, J. R. (1970). *Phys. Rev.*, **A1,** 158
83. Peterson, J. R., Aberth, W. H., Moseley, J. T. and Sheridan, J. R. (1971). *Phys. Rev.*, **A3,** 1651
84. Moseley, J. T., Aberth, W. and Peterson, J. R. (1972). *J. Geophys. Res.*, **77,** 255
85. Weiner, J., Peatman, W. B. and Berry, R. S. (1971). *Phys. Rev.*, **A4,** 1824
86. Smith, S. J. (1960). *Proc. Fourth Int. Conf. Ionization Phenomena in Gases*, 219 (Amsterdam: North Holland)
87. Woo, S. B., Branscomb, L. M. and Beaty, E. C. (1969). *J. Geophys. Res.*, **74,** 2933
88. Byerly, R. and Beaty, E. C. (1971). *J. Geophys. Res.*, **76,** 4596
89. Burt, J. A. (1972). *J. Geophys. Res.*, **77,** 6820
90. Burt, J. A. (1972). *J. Chem. Phys.*, **57,** 4649
91. Narcisi, R. S. (1969). *Planetary Electrodynamics*, 447 (S. C. Coroniti and J. Hughes, editors) (New York: Gordon and Breach)
92. Fehsenfeld, F. C. and Ferguson, E. E. (1973). *J. Geophys. Res.*, **78,** 1699
93. Narcisi, R. S. (1973). *Physics and Chemistry of the Upper Atmosphere*, 71 (B. M. Cormac, editor) (Dordrecht, Holland: D. Reidel)
94. Fehsenfeld, F. C., Ferguson, E. E. and Howard, C. J. (1973). *J. Geophys. Res.*, **78,** 327
95. Arin, L. M. and Warneck, P. (1972). *J. Phys. Chem.*, **76,** 1514
96. (1975). *Interactions Between Ions and Molecules* (P. Ausloos, editor) (New York: Plenum)
97. Ferguson, E. E. (1975). *Atmospheres of Earth and Planets*, 197 (B. M. McCormac, editor) (Dordrecht: Reidel)
98. Ferguson, E. E. (1974). *Rev. Geophys. Space Phys.*, **12,** 703
99. Graham, E., Johnsen, R. and Biondi, M. A. (1975). *J. Geophys. Res.*, **80,** 2338
100. Johnsen, R., Huang, C. M. and Biondi, M. A. (1975). *J. Geophys. Res.*, **80,** 3374
101. Lindinger, W., Albritton, D. L., McFarland, M., Fehsenfeld, F. C., Schmeltekopf, A. L. and Ferguson, E. E. (1975). *J. Chem. Phys.*, **62,** 4101
102. Lindinger, W., Albritton, D. L., Fehsenfeld, F. C. and Ferguson, E. E. (1975). *J. Geophys. Res.*, **80,** 3725
103. Fehsenfeld, F. C., Howard, C. J., Harrop, W. J. and Ferguson, E. E. (1975). *J. Geophys. Res.*, **80,** 2229
104. Lindinger, W., Fehsenfeld, F. C., Schmeltekopf, A. L. and Ferguson, E. E. (1974). *J. Geophys. Res.*, **79,** 4753
105. Lindinger, W., Albritton, D. L., Howard, C. J., Fehsenfeld, F. C. and Ferguson, E. E. (1975). *J. Geophys. Res.*, **80,** 3277
106. Fehsenfeld, F. C., Crutzen, P. J., Schmeltekopf, A. L., Howard, C. J., Albritton, D. L., Ferguson, E. E., Davidson, J. A. and Schiff, H. I. (1976). *J. Geophys. Res.*, submitted
107. Fehsenfeld, F. C. and Ferguson, E. E. (1976). *J. Chem. Phys.*, in press
108. Fehsenfeld, F. C., Howard, C. J. and Schmeltekopf, A. L. (1975). *J. Chem. Phys.*, **63,** 2835

# 5
# Chemical Processes in the Solar System: A Kinetic Perspective

M. B. McELROY
Harvard University



## 5.1 INTRODUCTION

Our knowledge of the Earth and planets has advanced by leaps and bounds in recent years. Progress has proceeded on a broad front, with contributions from a wide variety of techniques and disciplines.

It is convenient, for present purposes, to divide the solar system into two parts: an inner region including Mercury, Venus, Earth and Mars, and an outer region containing Jupiter, Saturn, Uranus, Neptune and Pluto. Planets in the inner solar system are rocky, with densities in the range 3.95–5.52 g $cm^{-3}$. Planets in the outer solar system, with the possible exception of Pluto, are gassy or icy, with densities in the range 0.7–2.2 g $cm^{-3}$.

Excluding the Sun, eight objects in the solar system are now known to have significant gaseous envelopes. These are: in the inner solar system, Venus, Earth and Mars; in the outer solar system, Jupiter, Saturn, Uranus, Neptune and Titan, a satellite of Saturn. The criterion adopted here to assess significance is arbitrary, based simply on a consideration of surface pressures. We listed only objects with surface pressures in excess of 5 millibars*, approximately $5 \times 10^{-3}$ that of Earth. Our classification is not intended to detract from the potential scientific importance of Mercury, the Moon, Io and Ganymede, all of which are known to have thin, though no less interesting, atmospheres.

The atmospheres of Earth, Venus and Mars are dominated by species which evolved by outgassing from the interiors of these planets. It appears that the atmospheres of Mercury and the Moon were supplied in part by accretion from the solar wind, in part by decay of radiogenic elements in the crustal rocks. The atmospheres of the outer planets are almost certainly primordial, reflecting the composition of the original solar nebula. Carbon dioxied is a major constituent in the atmospheres of Venus and Mars. Hydrogen dominates the atmospheres of Jupiter, Saturn, Uranus and Neptune. Titan has vast quantities of methane, and the satellites of Jupiter may retain trace quantities of nitrogen and oxygen supplied by photochemical decomposition of ammonia and water.

Earth appears to be unique in the solar system. Significant concentrations of material are present in all three phases, solid, liquid and gas, at its surface. The composition of its atmosphere is influenced strongly by the presence of life, as are the rates for transfer of major volatile elements between the different parts of the surface system, atmosphere, ocean and sediments. Earth merits special emphasis in any review of solar system chemistry. The choice and treatment of topics in subsequent sections of this article is designed to reflect this view.

We discuss the complex cycles which influence the fate of terrestrial volatiles in Sections 5.3 and 5.4, after setting the stage with a more general discourse in Section 5.2. The photochemistry of Earth's atmospheric oxygen, with special emphasis on ozone, is treated in Section 5.5, while Sections 5.6–5.8 explore possible ways in which modern technology may influence or perturb Nature's remarkable design. Chemical processes in the atmospheres of Mars, Venus and Jupiter are discussed in Sections 5.9–5.11 and summary remarks are given in Section 5.12.

## 5.2 GENERAL REMARKS

How does one begin a review of chemical processes in the solar system? One is tempted to begin at the beginning. *In the beginning* . . . But that would be too simple. Besides, the story has already been told!

*1 mbar = $10^2$ Pa $\approx 10^{-3}$ atm.

Modern versions of the tale are more complicated. We are asked to believe that planetary bodies formed as a consequence of condensation in a primordial nebula. The nebula was possessed of angular momentum and, as time progressed, it rapidly assumed a disc-like shape. The laws of physics prevailed, and the nebula settled into a condition such that the temperature was high towards the centre, *ca.* 2000 K, and cool towards the edge, *ca.* 50 K. The composition of the nebula was similar to that of the sun, which formed later. *And then there was light . . .* Hydrogen was most abundant, followed by helium, with smaller quantities of oxygen, nitrogen, carbon, neon, iron, magnesium, and sulphur, as summarised in Table 5.1.

**Table 5.1 Abundance of elements in the solar system***

| *Element* | *Abundance*† | *Abundance*‡ |
|---|---|---|
| 1. H | $3.2 \times 10^{10}$ | 1.0 |
| 2. He | $2.2 \times 10^{9}$ | $9.9 \times 10^{-2}$ |
| 3. O | $2.1 \times 10^{7}$ | $6.8 \times 10^{-4}$ |
| 4. C | $1.2 \times 10^{7}$ | $3.7 \times 10^{-4}$ |
| 5. N | $3.7 \times 10^{6}$ | $1.2 \times 10^{-4}$ |
| 6. Ne | $3.4 \times 10^{6}$ | $1.1 \times 10^{-4}$ |
| 7. Mg | $1.1 \times 10^{6}$ | $3.3 \times 10^{-5}$ |
| 8. Si | $1.0 \times 10^{6}$ | $3.1 \times 10^{-5}$ |
| 9. Fe | $8.3 \times 10^{5}$ | $2.6 \times 10^{-5}$ |
| 10. S | $5.0 \times 10^{5}$ | $1.6 \times 10^{-5}$ |
| 11. Ar | $1.2 \times 10^{5}$ | $3.7 \times 10^{-6}$ |
| 12. Al | $8.5 \times 10^{4}$ | $2.7 \times 10^{-6}$ |
| 13. Ca | $7.2 \times 10^{4}$ | $2.3 \times 10^{-6}$ |
| 14. Na | $6.0 \times 10^{4}$ | $1.9 \times 10^{-6}$ |
| 15. Ni | $4.8 \times 10^{4}$ | $1.5 \times 10^{-6}$ |
| 16. Cr | $1.3 \times 10^{4}$ | $4.0 \times 10^{-7}$ |
| 17. P | $9.6 \times 10^{3}$ | $3.0 \times 10^{-7}$ |
| 18. Mn | $9.3 \times 10^{3}$ | $2.9 \times 10^{-7}$ |
| 19. Cl | $5.7 \times 10^{3}$ | $1.8 \times 10^{-7}$ |
| 20. K | $4.2 \times 10^{3}$ | $1.3 \times 10^{-7}$ |

*Cameron, A. G. W. (1970). *Space Sci. Rev.*, **15**, 121
†Number of atoms normalised to Si = $10^6$
‡Abundance relative to H

Condensation in the inner solar system favoured formation of bodies rich in iron, magnesium and silicon, with significant trace quantities of uranium and thorium. Radioactive decay provided a potent heat source, and, as a result, it is thought that all of the planets which formed in the inner solar system must have differentiated early in their history. Heavier elements should have settled to the core, while lighter elements would have floated to the top, forming crust and mantle. Approximately 99% of the Earth's mass is contained in its core and mantle. The mantle is composed mainly of magnesium oxide, silicon dioxide and ferrous oxide. Ferrous oxide accounts for *ca.* 10% of the total mantle mass. The core consists of two distinct layers, an inner zone which is solid, formed from an iron–nickel alloy, and an outer zone which is liquid, containing a mixture of iron and either sulphur or silicon.

Lewis[148] argued that th. Earth formed in a region of thc primordial nebula where the temperature was *ca.* 600 K. Water, which accounts for approximately 0.05% of the total Earth mass, was retained in moderate amount, bound to minerals such as tremolite. Sulphur was retained as troilite (FeS) and eventually concentrated in the liquid core. Venus, according to Lewis[148], differed from the Earth in that it formed closer to the sun, at higher temperature (~800 K), and as a consequence did not retain any significant concentrations of either $H_2O$ or S. Mars retained abundant supplies of $H_2O$. Lewis[148] estimates that the Red Planet may contain as much as 0.3% $H_2O$, with perhaps 50% FeO, concentrated in an olivine mantle. He thought it likely that Mars should have a troilite core, but argued against the presence of metallic iron.

The early history of the Earth is cloaked in mystery, We have no record of conditions which prevailed during the first billion years of the planet's existence. The oldest terrestrial rocks known to man were formed 3.8 billion years ago, and seem to suggest that the bulk chemistry of the atmosphere–ocean system has changed little since then. The planet may have retained, in its primordial atmosphere, a gaseous residue of the parent nebula. The comparative absence from today's atmosphere of cosmologically abundant elements such as $^{36}$Ar makes it clear, however, that the primordial gas, were it abundant to begin with, must have dissipated rapidly during the early phases of planetary evolution. Today's atmosphere is a secondary product, formed by outgassing from the planetary interior. The predominance of $^{40}$Ar over $^{36}$Ar is readily understood if we take into account production of the heavier isotope, due to decay of $^{40}$K. If we assume that the potassium content of the Earth is similar to that of chondritic meteorites, i.e. about 0.1% by mass, the Earth should have evolved over geologic time some $7 \times 10^{20}$ g of $^{40}$Ar, approximately 10 times the abundance of the gas now present in the atmosphere. If outgassing of $^{40}$Ar had proceeded at a similar rate on Mars, we might expect to find argon as a major component of that planet's atmosphere, contributing as much as 5 millibars to the total surface pressure. It is interesting to note that recent Soviet data appear to suggest that $CO_2$ may account for only 50% of the total Martian atmosphere. The partial pressure of $CO_2$ is *ca.* 5 millibars; argon, formed by radiogenic decay of $^{40}$K, may offer a plausible explanation for the missing atmospheric mass.

How does one account for the present composition of Earth's atmosphere–ocean system? The geologic record provides a number of important clues. It tells us in particular that Earth must have acquired near its surface, over geologic time, a rich suite of volatile compounds. These compounds, notably water, carbon dioxide, chlorine, nitrogen and sulphur, are present today with concentrations far in excess of abundances which might reasonably be attributed to the weathering of crystalline rocks. Rubey[213], in a classic paper, termed these compounds 'the distillable spirits of the Earth's solid matter'. His inventory of excess volatiles, updated somewhat to allow for more recent results, is given in Table 5.2. We can account with remarkable precision for the present composition of sedimentary rocks, the ocean and atmosphere, if we allow for weathering of major rock-forming elements such as silicon, aluminium, iron and calcium, and if we postulate in addition a steady infusion of volatiles from the mantle. Fresh volatiles may be carried to the surface by hot springs, fumaroles and volcanoes. If we assume that the major ingredients are $H_2O$, $CO_2$, HCl and $N_2$, Rubey's analysis, updated to allow for more

**Table 5.2 Concentration of volatiles now present in the atmosphere, ocean and sediments***

| | $H_2O$ | C | Cl | N | S |
|---|---|---|---|---|---|
| Present atmosphere, hydrosphere, biosphere | 14 600 | 0.4 | 276 | 40 | 13 |
| Buried in ancient sediments | 2100 | 1000 | 30 | 6 | 65 |
| Total | 16 700 | 1000 | 306 | 46 | 78 |
| Supplied by weathering of crystalline rocks | 130 | 3 | 5 | 0.6 | 30 |
| Excess volatiles | 16 600 | 997 | 300 | 45 | 48 |
| Excess volatiles, number of atoms or molecules in units of $10^{43}$ | 5 556 | 490 | 52 | 19 | 9 |
| Flux from the mantle, averaged over the total surface area of the earth and geologic time, atoms or molecules $cm^{-2}$ $s^{-1}$ | $8.1 \times 10^{10}$ | $7.1 \times 10^{9}$ | $7.6 \times 10^{8}$ | $2.8 \times 10^{8}$ | $1.3 \times 10^{8}$ |

*A comparison with estimates for the time integrated source due to weathering of crystalline rocks, and inferences as to the total concentration of volatiles which must have evolved from the crust over geologic time. Concentrations are given in units of $10^{20}$ g except where noted. Inventories are taken from Rubey[213] (1951) and Holland[105]

recent estimates of the carbon content of sediments[105,197], would imply time and spatially averaged emission rates for these species of $8.1 \times 10^{10}$, $7.1 \times 10^{9}$, $7.6 \times 10^{9}$ and $1.4 \times 10^{8}$ molecules $cm^{-2}$ $s^{-1}$, respectively. Water

**Table 5.3a Composition of sea water: elements in solution other than dissolved gases[160]**

| *Element* | *Abundance*/mg $l^{-1}$ | *Principal species* |
|---|---|---|
| Cl | $1.90 \times 10^{4}$ | $Cl^-$ |
| Na | $1.05 \times 10^{4}$ | $Na^+$ |
| Mg | $1.35 \times 10^{3}$ | $Mg^{2+}$, $MgSO_4$ |
| S | $8.85 \times 10^{2}$ | $SO_4{}^{2-}$ |
| Ca | $4.00 \times 10^{2}$ | $Ca^{2+}$, $CaSO_4$ |
| K | $3.80 \times 10^{2}$ | $K^+$ |
| Br | $6.50 \times 10^{1}$ | $Br^-$ |
| C | $2.80 \times 10^{1}$ | $HCO_3{}^-$, $H_2CO_3$, $CO_3{}^{2-}$ |
| Sr | 8.0 | $Sr^{2+}$, $SrSO_4$ |
| B | 4.6 | $B(OH)_3$, $B(OH)_2O^-$ |
| Si | 3.0 | $Si(OH)_4$, $Si(OH)_3O^-$ |
| F | 1.3 | $F^-$ |

**Table 5.3b Composition of sea water: dissolved gases[160]**

| *Gas* | *Concentration* /ml $l^{-1}$ |
|---|---|
| $O_2$ | 0–9 |
| $N_2$ | 8.4–14.5 |
| Total $CO_2$ | 34–56 |
| Argon | 0.2–0.4 |
| He + Ne | $1.7 \times 10^{-4}$ |
| $H_2S$ | 0–22 or more |

collects in the ocean. Carbon dioxide is deposited as limestone and dolomite. Chlorine accumulates in the ocean, and nitrogen remains mostly in the atmosphere. The more abundant components of the ocean and atmosphere are listed in Tables 5.3 and 5.4.

**Table 5.4 Composition of the atmosphere[84,131,208]**

| *Constituent* | *Fractional abundance* | *Comments* |
|---|---|---|
| $N_2$ | $7.81 \times 10^{-1}$ | Permanent |
| $O_2$ | $2.09 \times 10^{-1}$ | Permanent |
| Ar | $9.34 \times 10^{-3}$ | Permanent |
| $H_2O$ | $10^{-3}$–$10^{-2}$ | Variable |
| $CO_2$ | $3.3 \times 10^{-4}$ | Variable, increasing |
| Ne | $1.8 \times 10^{-5}$ | Permanent |
| He | $5.2 \times 10^{-6}$ | Radiogenic source, escaping |
| $CH_4$ | $2 \times 10^{-6}$ | Biological origin |
| Kr | $1 \times 10^{-6}$ | Permanent |
| $H_2$ | $5 \times 10^{-7}$ | Photochemical |
| $N_2O$ | $3 \times 10^{-7}$ | Biological |
| CO | $10^{-7}$ | Photochemical, industrial |
| $SO_2$ | $0$–$10^{-6}$ | Industrial |
| $O_3$ | $0$–$10^{-7}$ | Photochemical, industrial |
| Xe | $9 \times 10^{-8}$ | Permanent |
| $NO_2$ | $0$–$10^{-9}$ | Photochemical, industrial |

## 5.3 THE CARBON CYCLE

The major reservoirs for carbon are summarised in Table 5.5, and information on transfer rates is given in Table 5.6. The bulk of the Earth's surface carbon is contained in sedimentary rocks. It is present mainly as $CaCO_3$ and $CaMg(CO_3)_2$, although the sediments include also significant quantities of reduced (organic) carbon.

**Table 5.5 Carbon contents of various terrestrial reservoirs**

| *Reservoir* | *Content/g* | *Ref.** |
|---|---|---|
| Atmosphere | $7 \times 10^{17}$ | (1) |
| Land biomass | $3 \times 10^{17}$ | (2) |
| Humus | $1 \times 10^{18}$ | (2) |
| Ocean organic | $3 \times 10^{18}$ | (3) |
| Ocean biomass | $1 \times 10^{16}$ | (3) |
| Ocean inorganic: | | |
| above thermocline | $1 \times 10^{18}$ | (1) |
| below thermocline | $4 \times 10^{19}$ | (1) |
| Sediment organic | $2 \times 10^{22}$ | (1) |
| Sediment inorganic | $8 \times 10^{22}$ | (1) |

*(1) Holland, H. D. (1975). Unpublished results. (2) Craig, H. (1957). *Tellus*, **9**, 1. (3) Skirrow, G. (1965). *Chemical Oceanography*, Vol. 1 (J. P. Riley and G. Skirrow, editors) (London, New York: Academic Press)

**Table 5.6 Rates for transfer of carbon between major carbon reservoirs**

| | *Transfer process* | *Rate*/g $y^{-1}$ | *Ref.** |
|---|---|---|---|
| (a) | Land photosynthesis | $2 \times 10^{16}$ | (1, 2) |
| (b) | Ocean photosynthesis | $2 \times 10^{16}$ | (1, 3) |
| (c) | Burial of organic C in sediments | $1.2 \times 10^{14}$ | (1) |
| (d) | Burial of inorganic C as carbonate in sediments | $2.4 \times 10^{14}$ | (1) |
| (e) | River transport to ocean | $5.2 \times 10^{14}$ | (1) |
| (f) | Degassing from mantle, total C/age of earth | $2 \times 10^{13}$ | (1) |
| (g) | Degassing rate, mantle plus crust | $9 \times 10^{13}$ | (1) |
| (h) | Degassing rate for crust, (g) −(f) | $7 \times 10^{13}$ | (1) |

*(1) Holland, H. D. (1975). Unpublished results. (2) Hutchinson, G. E. (1954). *The Earth as a Planet* (G. P. Kuiper, editor) (Chicago: University of Chicago Press). (3) Koblentz-Mishke, O. J., Volkovinsky, V. V. and Kabanova, J. G. (1970). *Scientific Exploration of the South Pacific* (W. S. Wooster, editor) (Washington: National Academy of Sciences)

Carbon is cycled rapidly through the atmosphere, biosphere and upper layers of the ocean. Organic carbon is formed by photosynthesis, described symbolically by the reaction

$$CO_2 + H_2O + \text{Energy} \rightarrow CH_2O + O_2 \quad (5.1)$$

It is removed by the reverse process,

$$CH_2O + O_2 \rightarrow CO_2 + H_2O + \text{Energy} \quad (5.2)$$

which takes place during respiration and decay. According to current ideas[104,250,251], the net rate for (5.1) is slightly larger than that for (5.2). The excess organic carbon produced by (5.1) is incorporated into sedimentary rocks and eventually returned to the atmosphere as the sediments are raised and eroded*.

Carbonic acid ($H_2CO_3$), formed by combination of $CO_2$ and liquid water, plays an important role in the weathering process. Erosion of calcite ($CaCO_3$), for example, proceeds through reactions of the form

$$CO_2 + H_2O + CaCO_3 \rightarrow Ca^{2+} + 2HCO_3^- \quad (5.3)$$

The calcium and bicarbonate ions produced by (5.3) are transported to the ocean by river run-off. Reaction (5.3) proceeds to the left in the ocean, with consequent precipitation of $CaCO_3$ and release of $CO_2$.

It is of interest to consider the life history of a typical carbon atom. We assume that the atom is released from the crust at time zero, either by weathering or by degassing. Using the data in Tables 5.5 and 5.6 we estimate that $10^5$ years will elapse before the atom returns to the crust. It will return most likely as $CaCO_3$, although there is a fair chance, *ca.* 30%, that it may precipitate in organic form. Most of its $10^5$ year transit will be spent in the deep layers of the ocean, below the thermocline†. Some 1900 years will be spent in the atmosphere, 800 years will be taken up as part of the land biomass, most

*The reader may be interested to note that the forces of erosion and uplift must strike an approximate balance on a geologic timescale. According to Leopold *et al.*[143], erosion at the present rate would eliminate all topography in a period as short as 44 million years. It is clear from the geologic record that erosive processes must have played a key role in shaping the Earth's surface relief for at least 3 billion years.

†A region of the ocean characterised by a high degree of static stability. The thermocline effectively separates the warm surface waters from the cold waters which prevail at depth. The thermocline is encountered typically at depths of a few hundred metres.

probably incorporated in the wood tissue of trees. The atom will spend on average 2800 years as a component of humus, 2800 years as inorganic carbon above the thermocline, 7700 years as marine organic carbon, and a cumulative total of 27 years as part of the ocean's thinly populated biosphere. It will cycle many times between ocean and atmosphere, biosphere and soil, before returning to the sediment. The residence time in the sediment is approximately $3 \times 10^8$ years, and the representative atom may pass through the weathering–erosion–precipitation cycle some 15 times over the 4.5 billion year history of the Earth.

Many of the numbers in Tables 5.4 and 5.5 are uncertain, and the present state of knowledge scarcely justifies a detailed quantitative analysis of the carbon cycle. Several conclusions can be drawn, however, which serve to severely limit our thinking on the possible paths for evolution of carbon.

It is clear that the rates for photosynthesis, reaction (5.1), and respiration–decay, reaction (5.2), must balance to a fairly high degree of accuracy over both long and short timescales. Suppose that the respiration–decay process were temporarily suspended, for one reason or another. Photosynthesis could exhaust all of the inorganic carbon of the surface ocean–atmosphere reservoir in a time as short as 40 years. Transport of $CO_2$ from the deep ocean, however, would replenish the inorganic carbon content of surface waters and allow photosynthesis to continue with undiminished vigour for much longer periods*. The total reservoir of marine inorganic carbon could take up the slack for $10^3$ years. Its store of inorganic carbon would then be exhausted and life as we know it would terminate. Photosynthesis would cease and could not resume until weathering of sedimentary rocks, or various outgassing processes, had replenished the reservoir of dissolved $CO_2$. At present rates of supply a period of *ca.* $1.7 \times 10^5$ years would be required to restore the budget of inorganic carbon to its original level.

It is clear that the life cycle has maintained a relatively steady pace, at least over the past $10^5$ years. It follows that the rates for reactions (5.1) and (5.2), $R_1$ and $R_2$, must have balanced to better than 1 part in $10^2$ over this period. We may conclude that

$$R_1 = (1 \pm 6 \times 10^{-3})R_2, \quad \Delta t \approx 10^5 \text{ years} \tag{5.4}$$

An imbalance of $6 \times 10^{-3}$ would suffice to produce or destroy the total oxygen content of today's atmosphere in $2 \times 10^6$ years, and could account for the total organic carbon content of the sediments in $8 \times 10^7$ years. If we assume that photosynthesis has continued at more or less today's rate for the past $5 \times 10^8$ years, a reasonable assumption based on studies of the geochemical characteristics of ancient sediments[88,210], we must conclude that photosynthesis has balanced respiration and decay to better than 1 part in $10^3$ on this timescale, i.e. that

$$R_1 = (1 \pm 10^{-3})R_2, \quad \Delta t \approx 5 \times 10^8 \text{ years} \tag{5.5}$$

*Bolin and Eriksson[21] estimate that the time constant for complete mixing of ocean water should be *ca.* 600 years. The inorganic carbon contained in the atmosphere–surface ocean reservoirs represents approximately 1 part in 24 of the total ocean–atmosphere budget of inorganic carbon. It follows that exchange with deep ocean water can replenish surface carbon on a timescale of 25 years, somewhat less than the timescale for depletion of the element by uncompensated photosynthesis.

Atmospheric oxygen may be formed by burial of organic carbon represented symbolically by

$$CO_2 \rightarrow C + O_2 \tag{5.6}$$

or by decomposition of water

$$2H_2O \rightarrow 4H + O_2 \tag{5.7}$$

in which the excess H is lost by evaporation to space. Much of the discussion which remains in this section will be devoted to an attempt to define the relative roles of (5.6) and (5.7). Reaction (5.6) is balanced to some extent by the weathering of reduced carbon in exposed sediments, described by

$$C + O_2 \rightarrow CO_2 \tag{5.8}$$

Reaction (5.7), on the other hand, proceeds irreversibly. Capture of H atoms from the interplanetary medium is entirely negligible in the present context.

If the oxygen budget were determined primarily by escape we should expect an even tighter balance between $R_1$ and $R_2$ than that given by (5.4) and (5.5). Equations (5.4) and (5.5) define maximum permissible bounds on the difference between $R_1$ and $R_2$. It is conceivable, and indeed consistent with all current constraints, to suppose that burial of organic carbon over geologic time has led to *no* net change in the oxidation state of the ocean–atmosphere system. The reduced carbon content of the sediment in this case would simply reflect the oxidation state of the primitive gases which evolved from the crust and mantle and we would have

$$R_1 = (1 \pm 2 \times 10^{-5})R_2, \quad \Delta t \approx 5 \times 10^8 \text{ years} \tag{5.9}$$

a condition which seems only slightly more demanding than the essential constraint summarised by (5.5).

Reaction (5.7) takes place primarily in the stratosphere, mesosphere and thermosphere, above 30 km, in a complex sequence of chemical steps which we can summarise most readily by considering the net flux of various species through a suitably chosen reference surface. It is convenient to focus attention on a level near the tropopause, at *ca.* 18 km. Global mean values for the flux of various key gases may be readily estimated using data presented by Wofsy *et al.*[269]. We find upward fluxes for $CH_4$, CO and $O_2$, balanced by downward fluxes for $CO_2$, $H_2O$ and $H_2$. The flow rates $\phi$ (molecules cm$^{-2}$ s$^{-1}$) are given by

$$\phi_{CH_4}(18 \text{ km}) = +4.5 \times 10^9 \tag{5.10}$$

$$\phi_{CO}(18 \text{ km}) = +4.5 \times 10^9 \tag{5.11}$$

$$\phi_{O_2}(18 \text{ km}) = +1.1 \times 10^9 \tag{5.12}$$

$$\phi_{CO_2}(18 \text{ km}) = -9 \times 10^9 \tag{5.13}$$

and

$$\phi_{H_2O}(18 \text{ km}) = -9 \times 10^9 \tag{5.14}$$

with $\phi_{H_2}(18 \text{ km})$ satisfying the relation

$$4\phi_{CH_4} + 2\phi_{H_2O} + 2\phi_{H_2} = +2 \times 10^8 \tag{5.15}$$

The number on the right-hand side of (5.15) is numerically equal to the escape rate for H [151].

For most planets the H escape rate is limited by the rate at which the gas may be supplied to the escape region[112]. The escape rate approaches a limiting magnitude defined by the maximum rate at which H can diffuse through the stratosphere and mesosphere. This limiting flux for the Earth is given[112] approximately by

$$\phi_H = \frac{1}{\mathscr{H}}\{1.8 f_{CH_4} b_{CH_4} + 0.8 f_{H_2O} b_{H_2O} + 1.2 f_{H_2} b_{H_2}\} \tag{5.16}$$

where the quantities $f_i$ denote mixing ratios for $CH_4$, $H_2O$ and $H_2$, $\mathscr{H}$ is the scale height of the atmosphere, and the coefficients $b_i$ may be expressed in terms of the coefficients $D_{ix}$ describing diffusion of a gas $i$ through a background gas $x$ of number density $n_x$:

$$b_i = D_{ix}\, n_x \tag{5.17}$$

In general, $D_{ix}$ and $n_x$ are strong functions of altitude. The coefficients $b_i$, however, depend only on temperature and are essentially constant over the height range of interest for present purposes.

As may be inferred immediately from the data in equations (5.10)–(5.15), the hydrogen atoms which escape from the earth are supplied by upward diffusion of $CH_4$. Methane is produced by bacteria, which flourish under

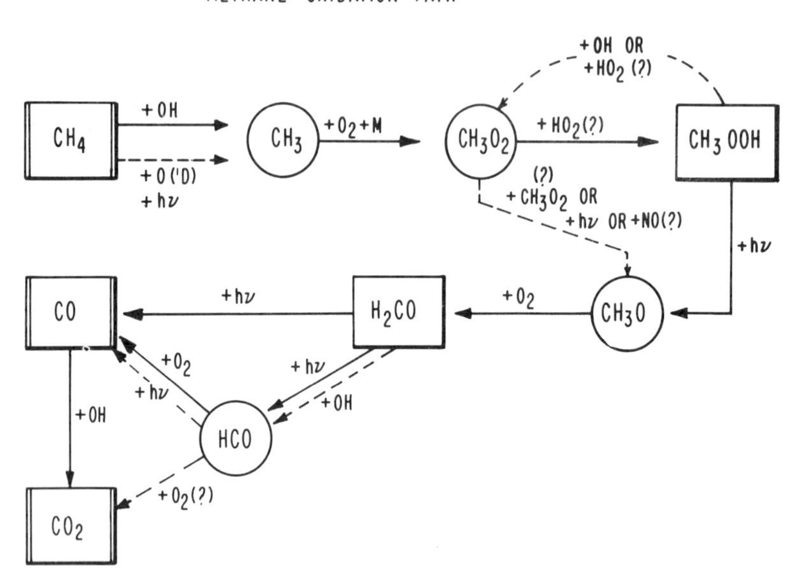


**Figure 5.1** Classification of C species in $CH_4$ oxidation path. The square bordered with a double line represents a lifetime of greater than 1 month (observable gas). The square represents a lifetime of 1 day (observable in rain). The circle represents a lifetime of less than 10 min (reactive intermediate) (After Wofsy *et al.*[269], by courtesy of the American Geophysical Union)

anaerobic conditions at the bottom of lakes, ponds, marshes and rice fields. Methanobacteria are to be found also in the rumina of various herbivorous mammals. Oxidation of $CH_4$ in the contemporary atmosphere is initiated by the reaction

$$OH + CH_4 \rightarrow H_2O + CH_3 \quad (5.18)$$

Further chemistry, illustrated by Figure 5.1, leads to production of $H_2CO$ and $H_2$, through the reactions

$$CH_3O + O_2 \rightarrow H_2CO + HO_2 \quad (5.19)$$

and

$$h\nu + H_2CO \rightarrow H_2 + CO \quad (5.20)$$

respectively. Molecular hydrogen produced above *ca.* 50 km flows up, fuelling the escape of H, which is released ultimately by

$$O + H_2 \rightarrow OH + H \quad (5.21)$$

and by

$$OH + H_2 \rightarrow H_2O + H \quad (5.22)$$

followed by

$$h\nu + H_2O \rightarrow H + OH \quad (5.23)$$

Molecular hydrogen formed below 50 km flows down, to be removed mainly by (5.22), with most of the loss taking place in the troposphere below 18 km. The equilibrium concentration of OH results from a balance of production by

$$h\nu + O_3 \rightarrow O(^1D) + O_2 \quad (5.24)$$

followed by

$$O(^1D) + H_2O \rightarrow 2OH \quad (5.25)$$

and loss by (5.18), with additional contributions due to

$$OH + CO \rightarrow H + CO_2 \quad (5.26)$$

The concentration of $H_2$ is set mainly by a balance between (5.20) and (5.22).

The escape rate of H is a fairly critical function of the altitude at which the $H_2$ flux changes sign, a level established primarily by competition between escape and oxidation as the effective sinks for $H_2$. Flux reversal in the contemporary atmosphere takes place at *ca.* 50 km. As a consequence the escape process operates today with relatively low efficiency. For every $3 \times 10^5$ molecules of $CO_2$ reduced by (5.1), $6 \times 10^3$ carbon atoms are released to the atmosphere as $CH_4$, supplying four hydrogen atoms which can escape to space, liberating one molecule of $O_2$. Escape of hydrogen over geologic time, $4.5 \times 10^9$ years, at the rate indicated by (5.15), could supply an amount of oxygen equal to $1.8 \times 10^{21}$ g, approximately 50% more than the amount now present in the atmosphere. On the other hand, the rate at which oxygen is produced by the escape process (5.7) is significantly less than the rate at which it is removed by weathering of reduced sedimentary carbon. It would appear that the oxygen balance, at least for the present atmosphere, is maintained mainly by reactions (5.6) and (5.8). Reaction (5.8) may have played a more important role in the past when the level of oxygen was lower.

It could be important also in the present where it could act to stabilise oxygen over relatively short timescales, as we discuss below.

We consider the effects of a small decrease in oxygen. This exercise will serve to model either a situation which might have prevailed in the past, or a present-day circumstance which could be induced, for example, by a temporary increase in the rate for weathering of reduced sedimentary material. We would expect an increase in the rate of anaerobic decay, with consequent rise in the production of $CH_4$ and CO. The concentration of OH would be reduced, and reaction (5.22) would be less effective as a sink for $H_2$. As a result, a relatively larger fraction of the $H_2$ produced by (5.20) would flow up, supplying escape of H. The concentration of $CH_4$ could change in response to the perturbation in a time period which might be as short as five years, and the concentration of $H_2$ would respond essentially instantaneously, in about two years, maintaining equilibrium with $CH_4$. The H escape rate would increase rapidly. The time constant for growth of oxygen would be reduced accordingly and could become comparable in magnitude with the value associated with (5.7). It appears therefore that reaction (5.8) could be important for oxygen on timescales less than $10^5$ years. In timescales in excess of $10^5$ years we must look for other mechanisms to account for the stability of oxygen.

On the longer timescales one might expect temporary imbalances between the rates for supply and removal of carbon. There would be no reason to expect precise equilibrium between the rates for weathering of sedimentary rocks and the rates for burial of fresh sedimentary material. In particular it would be difficult to envisage any inherent geophysical or geochemical factors which might force the oxidation state of the ocean–atmosphere system to remain static in face of fluctuating rates for the supply and loss of reduced material. It seems clear that one must look to the biological domain for simple interpretations of the balances described by equations (5.5) and (5.6).

Consider the response of the biospheric system to a small fluctuation in the level of atmospheric oxygen. There should be a corresponding change in the concentration of oxygen dissolved in the ocean. An increase in atmospheric $O_2$, induced for example by a transitory increase in the rate at which reduced carbon was incorporated in sediments, would lead to a concomitant increase in marine $O_2$, with a consequent increase in the biospheric consumption of reduced carbon, thus limiting the effect of the original perturbation. A change in atmospheric and oceanic $O_2$ would lead to a change in the marine nutrient cycle which, as we shall see, could serve to further stabilise the oxidation state of the atmosphere–ocean system.

Decomposition of marine organic material in the presence of adequate amounts of $O_2$ may be described by the stoichiometric relation[206]

$$(CH_2O)_{106}(NH_3)_{16}H_3PO_4 + 138O_2 \rightarrow 106CO_2 + 122H_2O + 16HNO_3 + H_3PO_4 \quad (5.27)$$

When all or most of the available oxygen is consumed, decomposition of organic material may continue using nitrate, either by

$$(CH_2O)_{106}(NH_3)_{16}H_3PO_4 + 84.8HNO_3 \rightarrow 106CO_2 + 42.4N_2 + 148.4H_2O + 16NH_3 + H_3PO_4 \quad (5.28)$$

or by

$$(CH_2O)_{106}(NH_3)_{16}H_3PO_4 + 94.4HNO_3 \rightarrow 106CO_2 + 55.2N_2 + 177.2H_2O + H_3PO_4 \quad (5.29)$$

Denitrification takes place mainly in regions of the ocean where the concentration[195] of dissolved $O_2$ is less than 0.1 ml $l^{-1}$ and seems to proceed to completion[206] before reduction of sulphate commences. Reactions (5.28) and (5.29) appears to represent major sinks for fixed marine nitrogen[41,47], which is supplied in part by drainage from continents[39], in part by rainfall[115]. Reactions (5.28) and (5.29) could deplete the oceans' supply of useful nitrogen, present primarily as nitrate, in a period as short as $10^3$ years if these reactions were to constitute a major path for oxidation of organic material. Removal of fixed nitrogen would be limited only by the rate at which $NO_3^-$ would be supplied from depth and the reserve of nitrate in surface waters could be used up in as little as 25 years if respiration were to switch from (5.27) to either (5.28) or (5.29).

A decrease in the concentration of atmospheric $O_2$ would cause a decrease in the concentration of marine $O_2$, and oxidation of organic matter would tend to shift from the oxygen cycle (5.27) to the nitrogen cycles (5.28) and (5.29). The concentration of dissolved N, a limiting nutrient for much of the marine biosphere, would be lowered accordingly, with consequent effects on the biological productivity of the ocean. Fewer fish* would consume less food, allowing more reduced carbon to fall to the ocean bottom, where it could be incorporated in sediments, liberating oxygen which would be released to the atmosphere, where it could remove the perturbation which was imposed at the beginning of this exercise. A decrease in the productivity of the ocean would lead also to a decrease in the biological demand for phosphorus,

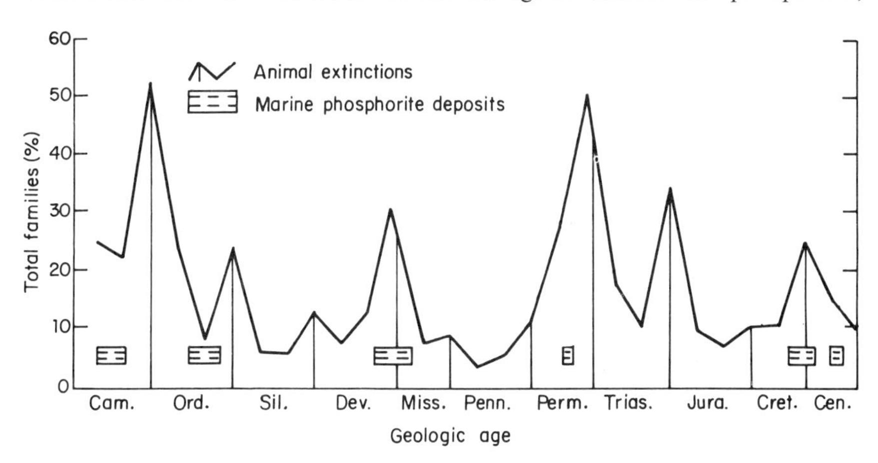


**Figure 5.2** Episodes of extinction in the animal kingdom, and approximate ages of extensive marine phosphorite deposits. The last appearances of animal families through geologic time are given in percentages. Main episodes of extinction occurred near the close of the Cambrian, Devonian, Permian, Triassic and Cretaceous periods. The size of the symbol designating the age of phosphorite deposits is not intended as an estimate of their relative economic importance nor of their amount of total phosphorus (After Piper and Codispoti[195], by courtesy of the American Association for the Advancement of Science)

*We have in mind of course the totality of organisms which might be expected to play a role in decay. A change in the population density of these organisms would influence the rate at which organic material would be incorporated in sediments at the ocean bottom. Marine bacteria known to play a pivotal role in the decay cycle have a relatively high nitrogen requirement[278]. It seems reasonable to suppose that consumption of organic material would be affected relatively more rapidly than its source, consistent with the argument outlined here.

which might then accumulate in surface waters, and eventually precipitate as an inorganic salt. The interconnections between marine oxygen, nitrogen and phosphorus were discussed recently by Piper and Codispoti[195]. Figure 5.2, taken from their paper, shows a possible correlation between the age of major phosphoritic deposits and periods of rapid evolutionary change in the characteristics of the ocean biomass.

It would appear that the marine nutrient cycle, directly affected by the availability of oxygen, could regulate the biological productivity of the ocean, thus controlling the rate at which reduced material was incorporated in sediments, ensuring a close balance between production and loss of reduced gases as required by (5.5). The ocean could respond to a small imbalance in the rates for reactions (5.1) and (5.2) by modest fluctuations in the size of the ocean biomass — more or less fish, feast or famine. With larger imbalances there could be mass extinction of major families of marine life. On a geologic timescale, an increase in oxygen due to (5.7) should be accompanied by an increase in the concentration of dissolved N. There should be an associated growth in the size of the marine biosphere which could adjust, in an evolutionary sense, to the changing characteristics of the environment.

The discussion to date has emphasised processes which might play a major role in the contemporary biosphere. We have sought to identify reactions which might serve to regulate carbon and to maintain the stability of the atmosphere over periods as long as $5 \times 10^8$ years. How did the system arrive at its present configuration? By what remarkable combination of circumstances did the Earth evolve from the anonymity of the inanimate nebula to the present pulsating exchange of biological, physical and chemical diversity. The view which follows is personal, speculative and to some extent anthropomorphic. I bring no special qualifications to the task. To paraphrase Rubey, I am neither oceanographer, geochemist, paleontologist nor biologist. The subject, however, is one on which all men are both equally knowledgeable and equally ignorant. The choice of epithet depends on the prejudice and patience of the reader.

We will suppose that the state of the Earth some hundreds of millions of years after its formation was basically the same as today, except for the absence of living things. Approximately 70% of the total surface area was covered by ocean. The ocean had accumulated gradually, fed by the gases which evolved from the crust and mantle. We assume that the chemical composition of emerging volatiles was then similar to today. In order of importance we might expect $H_2O$, $CO_2$, CO, HCl, $N_2$ and $H_2$ with fluxes perhaps comparable with the values given in Table 5.2. The ratios, by number, for $H_2$ relative to $H_2O$ and for CO relative to $CO_2$, could be of order

$$\frac{H_2}{H_2O} = 0.02 \tag{5.30}$$

and

$$\frac{CO}{CO_2} = 0.06 \tag{5.31}$$

if we adopt Fudali's[79] estimate for gases in equilibrium with melts of basaltic composition at 1200 °C. Escape of hydrogen, reaction (5.7), would be regulated by the rate at which water vapour could be transferred across the cold tropical tropopause. The escape route of H, and as a consequence the production of oxygen, in the prebiotic Earth would take place at rates similar to today*. These rates would be too slow to balance the emission of reduced gases from the crust and mantle, and it is difficult to escape the conclusion that the early Earth must have evolved a reducing atmosphere.

Carbon monoxide would have been removed by reactions such as

$$h\nu + H_2O \rightarrow H + OH \tag{5.32}$$

followed by

$$CO + OH \rightarrow CO_2 + H \tag{5.33}$$

with the H atoms recombining either through

$$H + O_2 + M \rightarrow HO_2 + M \tag{5.34}$$

followed by

$$H + HO_2 \rightarrow H_2 + O_2 \tag{5.35}$$

or through

$$H + H + M \rightarrow H_2 + M \tag{5.36}$$

As a consequence we might expect that the prebiotic atmosphere and ocean should have been enriched in $H_2$. The concentration of $O_2$ would have been much smaller than today, although $O_2$ would have played a very important role, limiting the effective rate for production of $O_2$ through reactions such as (5.34) followed by

$$OH + HO_2 \rightarrow H_2O + O_2 \tag{5.37}$$

The reducing environment of the early Earth would have provided ideal conditions for the chemical synthesis of large organic molecules. These molecules could have formed either under the influence of ultraviolet solar radiation, or at the expense of energy released by the lightning bolts of primitive storms. Ultraviolet radiation and lightning could have supplied a veritable shower of organic manna which would have rained down on the primeval ocean, providing essential nurture for the original citizens of the sea. It seems probable that these hardy fellows were simple unicellular heterotrophs, procaryotic antecedents of the bacteria and blue-green algae, dependent on manna and the vagary of ocean currents for growth and survival. It seems reasonable that the procaryotic heterotrophs should have evolved in such a manner as to reduce their dependence on manna. The development of the eucaryotic cell, the antecedent of plants, animals, protozoa, fungi and most algae, with its ability to respire and to extract energy from sunlight, would seem like an obvious step along the inexorable path to nutritional autonomy. The eucaryotic cell would have been preceded most likely by simpler photosynthetic procaryotes and it seems probable that the purple and green bacteria of today's environment are sturdy survivors of these early anaerobic organisms. The original photosynthetic bacteria would have been obliged to use reduced materials for growth, i.e. $H_2S$, $H_2$ or various organic compounds. In this sense, like their non-photosynthetic procaryotic relatives, they would have relied on abiotic synthesis and marine dynamics for food. They would have lacked the ability to influence their environment, to exert any measure of control over their destiny, although, by taking advantage of energy available in the diffuse sunlight which reached their

*Larger escape rates for H could apply during the earliest period of atmospheric evolution. The atoms would be formed by thermospheric reactions involving $O^+$, $N_2^+$ and $H_2$, This path would be short-circuited when the oxygen level of the atmosphere, supplied mainly by photolysis of $H_2O$, grew to the point where $O_2^+$ and $NO^+$ should become major components of the lower ionosphere.

habitat, they would have clearly advanced at least one step up the evolutionary ladder. They would also have lacked the ability to use water as a photosynthetic electron donor. As a consequence they could not have evolved oxygen, and the oxidation state of the ocean and atmosphere would have been relatively unchanged by their development and evolution.

The green and purple bacteria exhibit a number of remarkable characteristics which merit attention in the present context. The major absorption bands for their photosynthetic functions are located in the far red and infrared regions of the spectrum. They can flourish, for example, beneath a dense cover of algae and can live in either fresh water or marine environments. They have the capability to fix nitrogen and can be induced in some circumstances to form $H_2$ at the expense of either organic or inorganic electron donors. It is hard to escape the conclusion that photosynthetic bacteria must have played a key role in the early biosphere. Their survival to the present day is a notable tribute to their durability, and is all the more remarkable, given the simplicity of their unicellular structure.

It would appear that the growth of the biosphere during the period described above would have been limited ultimately by the supply of reduced material. Meteorology, with its ability to synthesise organic material using lightning and u.v. radiation as catalysts, would have ruled the world. It would seem logical that the world should have sought to free itself from this bondage, and rational that it should do so by developing an organism with the ability to grow using nutrients readily available in its environment, e.g. $CO_2$ and $H_2O$. Reaction (5.1) provides a natural mechanism. Its implementation, however, required an enormous evolutionary step, the development of a eucaryotic cell which could photosensitise the abstraction of H from $H_2O$. The development of this cell represented probably the largest single step on the evolutionary path leading to the growth of higher life forms.

The first oxygen-releasing autotrophs would have been faced with a puzzling dilemma. Oxygen is poisonous to all forms of life in the absence of oxygen-mediating enzymes. Did the original photo-autotrophs evolve with full protective armour or, as seems more likely, did they dispose of their oxygen waste by attaching it to suitable oxygen acceptors? Vulnerable photo-autotrophs in today's environment dispose of their oxygen by attaching it to sulphide ions which are converted in the process to sulphate. Cloud[42] made the interesting suggestion that ferrous iron might have played the role of oxygen acceptor for the early photo-autotrophs. His suggestion provides a ready explanation for the remarkable banded iron formations* which characterise sedimentary rocks with ages in excess of about $1.8 \times 10^9$ years.

The oxygen level of the atmosphere and ocean would have grown slowly after the onset of water based photosynthesis. The photosynthesis rate would, presumably, have been much less then than today and the supply of iron would have placed important restrictions on the growth of the marine biosphere. These restrictions would have disappeared with the evolution of

*The banded iron formations are layers of sedimentary rocks which were deposited, apparently in open bodies of water, on a near global scale. The layers consist mainly of silicious material and are alternately rich and poor in iron. Iron is present mainly as haematite, indicating that the medium in which the sediments were formed must have contained a source of oxygen.

advanced oxygen-mediating enzymes, an obvious next step in the evolutionary cycle. These enzymes would have released the photosynthetic organisms from obligatory dependence on iron, an element which would have become increasingly rare as the oxygen level grew. Life would have been free to proliferate in the ocean, with rapid growth in oxygen, as reduced carbon was incorporated in sediments. If we assume that the rate at which oxygen was released by burial of reduced carbon should have exceeded the rate at which oxygen was used to oxidise reduced gases emitted from the mantle by a factor $X$ and if we assume that the oxidation of reduced gases took place at mean rates implied by the data in Tables 5.5 and 5.6, then we may estimate that $O_2$ would have grown to 1% of its present level in approximately $10^6 \times (X-1)^{-1}$ years. If production had exceeded loss by a factor of 1.01, oxygen would have grown to 1% of its present level in $10^8$ years. If we assume that the burial rate for reduced carbon in the past was related to the total photosynthesis rate in approximately the same manner as today, i.e. smaller by a factor of *ca.* $3 \times 10^{-3}$, then oceanic photosynthesis, at approximately 0.1% of today's rate, would have sufficed to raise atmospheric oxygen to 1% of its present value in $10^8$ years. The size of the ocean biosphere, and the rate of photosynthesis, might have been limited in this period by the supply of nutrients such as nitrogen and phosphorus. The supply of nitrogen would have increased as photosynthetic organisms spread to land, allowing a profusion of nitrogen-fixing blue-green algae and, ultimately, symbiotic bacteria to develop and enhance the source of nitrogen. Life would have switched from a fermentative to an oxidative metabolism, as the level of oxygen grew past the Pasteur point[17,215]. Escape of H supplied by $CH_4$ could have been important in this era as a source of oxygen, as the rate for photosynthesis approached its present value, and as the concentration of oxygen converged on its current amount. The system could have settled down to a steady state in as little as $5 \times 10^8$ years after the level of photosynthesis was established at its present rate. At that time burial of reduced carbon would have been balanced by the weathering of reduced sedimentary rocks, and the stabilising forces discussed earlier would have been in full bloom.

## 5.4 THE NITROGEN CYCLE

The bulk of Earth's volatile nitrogen, approximately $4 \times 10^{21}$ g, resides in the atmosphere as $N_2$. The remainder, approximately $6 \times 10^{20}$ g, is distributed between sediments, ocean, soil and biosphere, with concentrations as given in Table 5.7.

This section is concerned with the manner in which nitrogen is exchanged between these various reservoirs. It is clear that biological processes must play a critical role. Indeed, in the absence of life one might expect a major shift in nitrogen from the air to the sea. The oceans, at the current epoch, contain much less nitrate than one might anticipate on the basis of chemical equilibrium. The deficit reflects, in a simple way, the importance of the biospheric demand for nitrogen. Nitrogen is an essential nutrient and yet, in the form in which it may bc utilised by the biosphere, it is in remarkably limited supply. Before the element can be incorporated into the tissue of living

**Table 5.7 Nitrogen content of various terrestrial reservoirs**

| *Reservoir* | *Content/g* | *Ref.* |
|---|---|---|
| Atmosphere | $4 \times 10^{21}$ | (1) |
| Land biomass | $1 \times 10^{16}$ | (2) |
| Humus | $6 \times 10^{16}$ | (3) |
| Soil inorganic | $1 \times 10^{16}$ | (4) |
| Ocean biomass | $8 \times 10^{14}$ | (5) |
| Ocean organic | $2 \times 10^{17}$ | (5) |
| Ocean nitrate | $6 \times 10^{17}$ | (6) |
| Ocean ammonia + nitrite | $1 \times 10^{16}$ | (6) |
| Ocean $N_2$ | $2 \times 10^{19}$ | (6) |
| Sediments | $6 \times 10^{20}$ | (1) |

*(1) Holland, H. D. (1975). Personal communication. (2) Table 5.5, assuming a C-to-N ratio of 30. The C-to-N ratio of trees is typically in the range 20–40. (3) Table 5.5, assuming a C-to-N ratio of 15; Chet, I. (1975). Personal communication. (4) Delwiche, C. C. (1970). *The Biosphere* (San Francisco; W. H. Freeman) as corrected to take account of a numerical error in earlier source material. (5) Table 5.5, assuming a C-to-N ratio of 12; Mitchell, R. (1975). Personal communication. (6) Vaccaro, R. F. (1965). *Chemical Oceanography*. Vol. 1 (J. P. Riley and G. Skirrow, editors) (London, New York: Academic Press)

things it must be fixed, i.e. it must be transformed, from the relatively abundant though chemically inert form $N_2$, to more useful compounds such as $NH_3$ and $NO_3^-$. The transformation requires a significant investment of energy. We must supply 226 kcal mol$^{-1}$ in order to break the N—N bond. Some of this energy is returned during the formation of $NH_3$ and $NO_3^-$. The expenditure is impressive nonetheless. The global production of nitrogen fertiliser in 1974, $4 \times 10^7$ metric tons, had a market value of \$8 billion. A chemical plant with the capacity to produce $10^3$ metric tons per day requires a capital investment of \$100 million[98]. By way of comparison, as discussed below, the annual production of fixed nitrogen by natural processes exceeds 100 million metric tons and is achieved by relatively simple micro-organisms. These organisms draw on sunlight, both visual and infrared, for their energy requirements. They operate at temperatures near 30 °C, in contrast to chemical fertiliser plants which require temperatures in excess of 500 °C.

The bulk of the natural fixation results from a symbiotic relation between certain bacteria and seed plants, neither of which acting alone has the capability to transform $N_2$. The classic example of symbiosis is the association between bacteria of the genus *Rhizobium* and plants of the family *Leguminosae*. The bacteria are apparently stimulated by various compounds released into the root zone of the legumes by excretion. The bacteria then invade the plant, entering it through the root hair, which tends to curl or otherwise deform in response to the release of a microbial product usually identified as indoleacetic acid. A small fraction of the infected root hairs, typically less than 5%, develop nodules. Narrow hypha-like infection tubes are formed, which converge on the central portions of the incipient nodule, where the bacteria are eventually released into the cytoplasm of the host plant. Injection of bacteria into the host plant triggers a period of rapid cell division in the plant's surrounding cellular wall. The final structure consists of a central core containing the rhizobia and a surrounding cortex which includes the vascular system of the plant.

The role of the plant in symbiotic nitrogen fixation is as yet not well understood. The relationship between plant and bacterial agent is often quite specific. A particular bacterial strain may be highly effective in certain hosts but may tend towards parasitic behaviour in others. Nodules of legumes actively metabolising $N_2$ are distinctly red in colour, and as a consequence easily recognisable. The red colouration results from the presence of haemoglobin. Synthesis of this compound apparently requires the symbiotic association of plant and bacteria. Its role in the fixation process remains obscure, however, despite early work by Hamilton *et al.*[96] which appears to indicate that the haemoglobin of soybean nodules has the ability to oxidise $N_2$. Biological fixation appears to require trace quantities of iron, calcium and molybdenum. Large quantities of carbohydrates are consumed during the synthesis of relatively small quantities of fixed nitrogen[216]. Synthesis apparently involves two distinct enzymes, nitrogenase and hydrogenase, which promote reactions of nitrogen and hydrogen, respectively, and there are reasons to believe that $NH_3$, or a similar species involving N—H bonds, is the primary fixation product.

Once fixed, nitrogen is transported rapidly from the nodule into the remainder of the plant. A healthy leguminous crop can be remarkably productive. Fixation rates for nitrogen as large as 50 kg per acre per year are common in temperate areas, and yields as much as 2–3 times larger can be realised in well managed pastures. These rates may be compared with estimates for the globally averaged fixation of nitrogen which range between 1 and 2 kg per acre per year, as discussed below. The alfalfas, clovers and lupines are particularly important. Peas, beans and peanuts are relatively ineffective[2].

Nitrogen fixation may be effected also by certain free-living bacteria, and by blue-green algae. The important bacteria include members of the heterotrophic classes *Azotobacter* and *Clostridium*, in addition to the photosynthetic purple sulphur, green sulphur and non-sulphur purple bacteria, *Thiorhodaceae*, *Chlorobacteriaceae* and *Athiorhodaceae*, respectively. *Azotobacters* are strict aerobes. They are highly sensitive to pH and are essentially absent from soil with pH lower than 6.0. *Clostridia* are more common and probably play a larger role. They flourish under anaerobic conditions and may be present in typical arable land with a concentration as large as $10^5$ cells per gram, with highest population densities observed in the vicinity of plant root systems. Grasslands are particularly effective. The extensive root network of this medium provides an abundant supply of carbohydrate, an essential nutrient to feed the voracious appetites of the bacteria. Their appetites are so large in fact that heterotrophic bacteria can make at most a minor contribution to the global production of fixed nitrogen. This conclusion is consistent with results from a variety of experiments in which agricultural soils were inoculated with large concentrations of active *Azotobacter*, without significant enrichment in the fixed nitrogen content of the medium[2].

The role of the purple and green bacteria was described earlier in connection with the discussion on the origin and evolution of life. These organisms are remarkably effective in their ability to fix nitrogen. They require, however, rather special conditions for growth and proliferation. Their contribution to the global budget of fixed nitrogen is consequently small, although they may play an important role in localised media. Their activity in deep, horizontally stratified, meromictic lakes is particularly noteworthy. These lakes are permanently cold, relatively stagnant and anaerobic below depths of *ca.* 20 m. The photosynthetic bacteria develop in a narrow band within the anaerobic zone. They multiply, feeding on nutrients released from the sediments below, drawing energy from sunlight at wavelengths transmitted by the oxygen evolving photosynthetic organisms above. They are able to function beneath a relatively dense layer of surface algae, in cold waters devoid of oxygen, conditions which would be exceedingly inhospitable for most other lifeforms. They have the ability to form molecular hydrogen using a variety of organic and inorganic materials as electron donors. Production of $H_2$ is inhibited by the presence of $N_2$, suggesting that the enzymatic system involved in $N_2$ fixation may serve a dual function. In the absence of $N_2$ the reducing power derived from the electron donor and the ATP produced by photophosphorylation are used apparently to make $H_2$. During the fixation of $N_2$ this capability is employed to reduce $N_2$ to $NH_3$.

The blue-green algae are by far the most important of the non-symbiotic nitrogen fixers. They can grow in soil cultures totally devoid of fixed nitrogen, and are evidently very primitive, as discussed in Section 5.3. They have the ability to synthesise organic materials from $CO_2$ and $H_2O$, in contrast to the photosynthetic bacteria which must rely on preformed organic nutrient. They flourish proliferously in aquatic media, although they can develop also in well drained fields and are often the first plants to return to areas which may have suffered heavy denudation. The algae can play an important role in the nitrogen cycle of rice fields and other localised media. They are important also on a global scale. They are particularly effective in the tropics where the fixation rate may be as high as 30 kg per acre per year.

Nitrogen fixed by non-symbiotic processes is absorbed directly into the body tissue of living organisms. Nitrogen fixed by symbiotic activity is first incorporated into the cellular structure of the bacterial agents which regulate fixation. It is then transferred to the host plant, where it may be used to synthesise the proteins and nucleic acids essential to growth. The organic materials synthesised by plants and algae may be assimilated by animals and various micro-organisms, and in the process the organic structures may be converted to more or less complex forms. Animals are somewhat wasteful in their use of organic material. A significant fraction of the fixed nitrogen absorbed in their daily food supply may be released as a component of animal excrement. Ammonia is the primary compound evolved by invertebrates. Nitrogen may appear in a variety of forms in the excreta of vertebrates. Uric acid, $C_5H_4N_4O_3$, predominates for reptiles and birds. Urea, $CO(NH_2)_2$, is more common for mammals.

The organic materials of plants and animals may be converted back to inorganic form, as the plants and animals die, and as their body stuff is worked over by various classes of micro-organic life. Organic nitrogen is released primarily as ammonium, which in turn may be converted to nitrate, or assimilated directly by new generations of plants. Nitrate is the most abundant form of fixed inorganic nitrogen, not only in the soil but also in the sea. The formation of nitrate from ammonium by micro-organic activity is termed nitrification. The conversion of organic nitrogen to ammonium is called ammonification. Conversion of organic to inorganic material is referred to generally as mineralisation and denitrification is the word used to describe the microbial reduction of nitrate and nitrite, which results in release of $N_2$ and $N_2O$. Denitrification completes the cycle which began with the fixation of atmospheric $N_2$.

Significant concentrations of fixed nitrogen are released to the atmosphere during the various transformations described above. In soils with pH above 8.0 [261], the reaction

$$NH_4^+ + OH^- \rightleftarrows NH_3 + H_2O \tag{5.38}$$

may proceed to the right, with consequent emission of $NH_3$. The pH tends to rise during decomposition of organic material, and large concentrations of $NH_3$ may be formed in a layer of rotting plant and animal residue, even under conditions where the underlying substrate may be highly acidic. Thin biological films, frequently observed in local regions of the sea surface, may provide additional sources of $NH_3$[129]. These films are formed by decay of organic material, and it seems probable that the pH in the films should be significantly higher than that in bulk seawater. Release of $NH_3$ would be raised accordingly. Junge[129] describes observational evidence to support this view.

Just as ammonification tends to raise the pH of the medium in which it takes place, nitrification acts to lower pH. This result may be readily understood if we note that the first step in nitrification involves the bulk reaction

$$2NH_4^+ + 3O_2 \rightarrow 2NO_2^- + 4H^+ + 2H_2O \tag{5.39}$$

Nitrite is unstable at pH below about 5.5. In this case, (5.39) should be followed by reactions leading to release of NO. It seems unlikely, however, that these reactions should contribute significantly to the global budget of NO. Oxidation of ammonium proceeds slowly under acidic conditions. The global production of nitrite in media with pH below 5.5 is almost certainly small, and unlikely to affect the overall function of the global nitrogen cycle.

Denitrification leads primarily to production of $N_2$ and $N_2O$, although it can provide also a small source of NO. The growth of micro-organisms involved in denitrification is seldom limited by the supply of nitrate. Most active denitrifying species can grow under both aerobic and anaerobic conditions. They can take part in a variety of functions affecting both soil and marine budgets of fixed nitrogen. These functions include, for example, proteolysis, the bacteriological decomposition of proteins, and ammonification. The function of the bacteria during denitrification is simply summarised by reactions (5.28) and (5.29). When sufficient supplies of oxygen are available, the bacteria may switch for respiration to the aerobic cycle (5.27).

The quantity of $N_2$ and $N_2O$ evolved in any given situation is influenced by a variety of environmental factors. Moisture content and pH are particularly important. Volatilisation of nitrogen is essentially absent at moisture levels lower than *ca.* 60% of the water holding capacity of the host soil. The dependence of gas release on soil water content reflects in a simple way the important role which water plays in regulating the flow of $O_2$ to sites of biological activity in the soil. The bacteria active in denitrification have a low tolerance to hydrogen ions. As a consequence, denitrification can take place to a significant extent only in media with pH larger than about 5.5. The pH of the decay medium influences not only the rate of denitrification, but also the composition of evolved gases. Production of $N_2O$ is pronounced in soils with pH lower than about 6.5. Under such circumstances $N_2O$ may account for as much as 50% of the total nitrogenous gas release. Production of $N_2$ dominates in soils with pH higher than about 6.5, while there may be some production of NO if denitrification takes place in soils with pH near 5.0 [2].

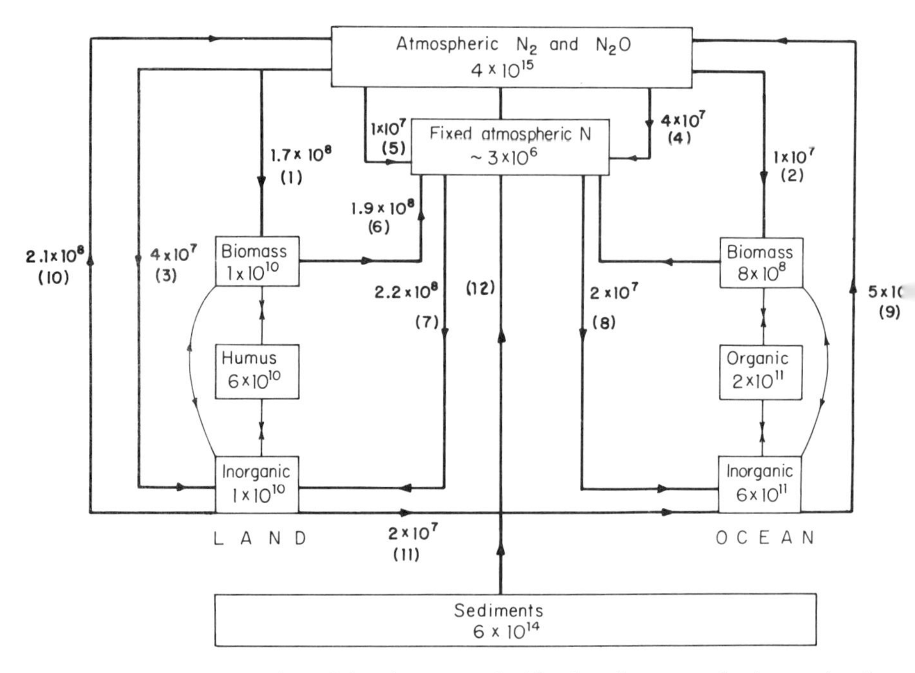


**Figure 5.3** Representation of the nitrogen cycle. Numbers in rectangular boxes give the abundance of N measured in metric tons. Transfer rates are given in tons per year, and the numbers in parentheses define the transfer reactions as follows: (1) land biological fixation; (2) marine biological fixation; (3) fixation in chemical fertiliser; (4) fixation due to combustion; (5) fixation due to lightning; (6) transfer due to volatilisation of $NH_3$; (7) transfer due to rain over land; (8) transfer due to rain over sea; (9) marine denitrification; (10) land denitrification; (11) transfer from land to sea in river run-off; (12) transfer due to raising of sediments, with a comparable amount due to deposition

Major components of the global nitrogen cycle are shown schematically in Figure 5.3. We expect that there should be significant transfer of fixed nitrogen from the land to ocean biomass. The bulk of nitrogen fixation takes place on land as discussed above. Denitrification may take place either in the soil or sea and, as seen earlier, the productivity of land and marine biospheres are similar. The transfer of fixed nitrogen from the land to sea can take place in two ways. The first and most obvious involves the transport of nitrate and dissolved organic material in rivers. Emery *et al.*[72] estimate a nitrogen source due to rivers of magnitude $1.9 \times 10^7$ metric tons per year. The second contribution to the marine budget of fixed nitrogen involves the volatilisation of $NH_3$ by land biota, its oxidation in the atmosphere to NO, $NO_2$ and $HNO_3$, and the eventual rain out of some fraction of these species on the sea, mainly as $NO_3^-$ and $NH_4^+$. The oxidation of atmospheric $NH_3$ was discussed by McConnell[163]. He suggested that the oxidation was initiated by reaction with OH

$$OH + NH_3 \rightarrow H_2O + NH_2 \tag{5.40}$$

and argued that the subsequent chemistry should lead mainly to production of $NO_x$ ($HNO_2 + HNO_3 + NO + NO_2 + NO_3 + N_2O_5$). Using measurements of $NH_4^+$ and $NO_3^-$ in rain[7,73,125,131] we estimate that marine rainfall should deliver a net source of fresh, fixed nitrogen to the marine biosphere of magnitude $2 \times 10^7$ metric tons per year, essentially the same as the source associated with rivers. This computation is consistent with McConnell's[163] estimate of $1.9 \times 10^8$ metric tons per year for the total global source of fixed atmospheric nitrogen. Abiological processes contribute approximately $4 \times 10^7$ metric tons per year to this budget[98]. The remainder, approximately $1.5 \times 10^8$ metric tons per year, may be attributed to the biogenic production of $NH_3$.

Our estimate for the release to the atmosphere of biogenically produced $NH_3$ may be compared with Hardy and Havelka's[98] value for the rate at which nitrogen was fixed by all processes, natural and anthropogenic, in 1974: $2.6 \times 10^8$ metric tons. This figure may be broken down as follows. Biological processes accounted for $1.8 \times 10^8$ metric tons, of which approximately $9 \times 10^7$ metric tons were contributed by fixation in agricultural soils. Lightning supplied an additional $10^7$ metric tons. Industrial processes were responsible for $8 \times 10^7$ metric tons, divided more or less equally between the manufacture of nitrogenous fertiliser and the consumption of fossil fuel. Man's contribution to the global production of fixed nitrogen was therefore at least as large as $8 \times 10^7$ metric tons, 30% of the total, and could have been as much as $10^8$ metric tons (38%) if we were to make the assumption that man's intervention had increased the globally averaged yield of fixed nitrogen by a relatively modest factor of 30%.

In assigning values to the various rates in Figure 5.3, we assumed that rain and river run-off were the major contributions to the marine budget of fixed nitrogen. *In situ* production by marine algae was set somewhat arbitrarily equal to $10^7$ metric tons per year. The contribution of $N_2O$ to the denitrification cycle was computed using data presented by Bates and Hays[10]. We assumed that the relative rates for production of $N_2O$ and $N_2$ were similar, irrespective of the site for denitrification, soil or sea. On a globally averaged basis, with the numbers in Figure 5.3, $N_2O$ accounts for approximately 8% of the total sink for fixed nitrogen, a result which appears plausible in light of the earlier discussion, which seemed to imply that $N_2O$ should be a minor product of denitrification in media with pH in excess of 6.5.

Our estimate for the marine source of nitrogen released from the ocean as $N_2O$, $4 \times 10^6$ metric tons per year, is markedly less than the value reported by Hahn[94], $9 \times 10^7$ metric tons per year. On the other hand, Hahn's result is based on relatively uncertain extrapolation of a regrettably limited data base. He used measurements of the concentration of $N_2O$ dissolved in waters of the North Atlantic, and sought to estimate the flux, $\phi$, to the atmosphere using the approximate relation

$$\phi = D/Z(C - P\alpha) \tag{5.41}$$

where $D$ was a diffusion coefficient, $Z$ a diffusion length, $C$ the concentration of gas in solution, $P$ its pressure in the overlying atmosphere, and $\alpha$ its solubility. The quantity $D/Z$, an effective diffusion velocity, was estimated empirically using data presented by Broecker and Peng[24]. The concentrations $C$ were measured by Hahn[94] during three cruises, the first in April 1969 covering the latitude band 0.6° N to 60° N along the 30° W meridian, the second in June 1970 covering a relatively limited geographic area near 63° N 15° W, and the third in June 1971 near 40° N extending from 11° W to 43° W. On this basis Hahn computed a nitrogen, $N_2O$, flux from the North Atlantic of $7 \times 10^6$ metric tons per year. Extrapolation to global scale, assuming a somewhat larger value for $D/Z$, but without data for $C$, gave the result for the global flux noted above, $9 \times 10^7$ metric tons per year. Hahn concluded that his result for the North Atlantic should be correct to within an order of magnitude. The extrapolation to global scale could introduce even larger errors, and would be particularly serious if the concentration of $N_2O$ in surface waters should be found to show significant seasonal variation. We may note in this context that nitrification and denitrification in soil are distinctly seasonal phenomena, with peak production of nitrate, at temperate latitudes, taking place in spring and fall[2]. These remarks serve to emphasise the uncertainties in the present preliminary attempts to define the nitrogen cycle of the sea. Hahn's pioneering efforts should be extended to other seasons and locations. As we shall see in Section 5.6, the study of the biogeochemistry of nitrous oxide may develop in the next few years into a subject of great social, as well as scientific, significance*.

We may estimate, in approximate fashion, the global rate for nitrification, by combining the value for the photosynthetic production of carbon given in Table 5.6, $4 \times 10^{16}$ g per year, with a reasonable guess for the abundance of nitrogen in representative organic tissue. The data in Table 5.7 suggest a value of order 15 for the ratio C to N, and lead therefore to an estimate of $3 \times 10^{15}$ g per year for the net rate at which nitrogen is transformed globally by nitrification. This result may be compared with the value derived earlier for denitrification, *ca.* $2 \times 10^{14}$ g per year. It would appear that anaerobic decay, leading to formation of $N_2$ and $N_2O$, takes place on average about 6% of the time during the typical cycle of fixed nitrogen from organic to inorganic form. This result is intriguingly close to the value derived earlier for the fractional concentration of $CH_4$ evolved during volatilisation of organic carbon, 2%. The similarity between the two numbers is reassuring. Common sense, however, suggests that we should continue to retain a healthy degree of scepticism, to recognise explicitly the skeletal nature of our understanding and to resist the temptation to play down the importance of the vast areas where hard experimental data are either absent or inadequate.

We have attempted in these sections to develop a global view of the carbon, nitrogen and oxygen cycles. Our aim was to try to define the wood but at the same time to maintain the trees in perspective. It is clear that more remains to be done, and that a satisfactory understanding of any aspect of the environment may require contributions from many, if not all, traditional spheres of intellectual activity. The following sections are devoted to a discussion of ozone. Ozone is a trace constituent of the atmosphere, accounting for less than 1 part in $10^6$ of the total atmospheric mass. It plays an exceed-

*A more detailed critique of present uncertainties in our understanding of the marine nitrogen cycle is given by McElroy *et al.*[290].

ingly important role in the biospheric environment, however. In the absence of ozone, the surface of the Earth would be bathed in solar radiation with wavelengths as short as 2400 Å. Even a small change in ozone could have a potentially serious environmental impact*. It is now clear that man has the ability to modify ozone in a variety of subtle ways. He can do so by flying aircraft in the stratosphere, with his use of Freon as a refrigerant and as a propellant for aerosol sprays, by fumigating his fields with methyl bromide, and, perhaps, by his application of fertilisers to stimulate the productivity of the world's agricultural economy. The chemistry of ozone in the normal atmosphere is discussed in Section 5.5. The variety of potential perturbations are discussed in Sections 5.6–5.8. As the reader may note, much of the discussion on perturbations is based on work of relatively recent vintage. Some, indeed, has yet to appear in the professional literature. It seemed important, however, that the brew should be presented in as complete a fashion as possible. Several of the bouquets in Sections 5.6–5.8 are exposed here for the first time. We recognise that new science, like freshly bottled wine, is often best consumed before it has the opportunity to age. The reader should beware: today's vintage could become tomorrow's vinegar.

## 5.5 ATMOSPHERIC OZONE

The chemistry of atmospheric ozone was discussed first by Sydney Chapman, in a classic paper published in 1930. He proposed that ozone would be formed in a series of reactions initiated by photolysis of $O_2$ in the Herzberg continuum below 2400 Å;

$$h\nu + O_2 \rightarrow O + O \tag{5.42}$$

The atoms formed in (5.42) may react with $O_2$ to form $O_3$,

$$O + O_2 + M \rightarrow O_3 + M \tag{5.43}$$

where M denotes an appropriate third body. Ozone may be recycled by

$$h\nu + O_3 \rightarrow O + O_2 \tag{5.44}$$

followed by (5.43). Odd oxygen is removed by

$$O + O_3 \rightarrow O_2 + O_2 \tag{5.45}$$

*The environmental impact of a change in ozone has been discussed by a number of official bodies during the past several years, most recently by the Climatic Impact Assessment Program (CIAP) of the U.S. Department of Transportation. In a report to the Congress of the United States, made public in December 1974, CIAP concluded that a variety of organisms, including agricultural and wild plants, phytoplankton, insects, toad embryos and larvae, were sensitive to u.v. B-radiation now reaching the surface of the Earth. They noted that many of these organisms had at best a limited ability to affect the repair of biological tissue which might be damaged owing to conceivable changes in the level of u.v. B-radiation. They concluded that a reduction in ozone by $x\%$ might cause an increase of $2x\%$ in the incidence rate for non-melanomic skin cancer for a population with the ethnic mix characteristic of the United States. There would be a corresponding increase of $10^4x$ in the annual number of new cases of the disease in the U.S. Non-melanomic skin cancer is rarely fatal, although it may be unpleasant and costly. Fair skinned Caucasians are especially vulnerable, an aspect of the problem which could have political as well as sociological implications beyond the scope of this article.

Chapman's scheme gives a good first order description for atmospheric $O_3$. It tends to predict too much $O_3$ at low latitudes, too little at high latitudes. The discrepancies between theory and observation may be removed, at least in part, if one allows for transport. Most of the world's ozone is manufactured and destroyed by chemical processes, in a latitude band between about 30° N and 30° S, at high altitudes, above *ca.* 30 km. At higher latitudes and lower altitudes, chemical processes play a less important role, and the distribution of ozone is set mainly by dynamics. The relative importance of chemistry and dynamics at any given location may be assessed from a consideration of relevant time constants.

The time constant for chemistry is given by

$$t_c(r) = \frac{[O_3]_r}{2 \times J_1[O_2]_r} \tag{5.46}$$

where $[O_3]_r$ and $[O_2]_r$ define the number densities of $O_3$ and $O_2$ at $\boldsymbol{r}$, and $J_1$ gives the first-order rate constant for reaction (5.42). The rate constant $J_1$ is a function of position and time, given by

$$J_1 = \int (\pi F)_\lambda [O_2]_r Q^1{}_\lambda \exp(-\tau_\lambda) d\lambda \tag{5.47}$$

where $(\pi F)_\lambda$ defines the solar flux (photons cm$^{-2}$ s$^{-1}$ Å$^{-1}$) at wavelength $\lambda$ at the top of the atmosphere, $Q^1{}_\lambda$ is the cross section (cm$^2$) for absorption of photons at $\lambda$ by $O_2$, and $\tau_\lambda$ is an optical depth, defined by

$$\tau_\lambda = \int \{[O_2]_s Q^1{}_\lambda + [O_3]_s Q^2{}_\lambda\} ds \tag{5.48}$$

Here $Q^2{}_\lambda$ is the cross section for absorption of solar radiation at wavelength $\lambda$ by $O_3$, and the integration is carried out along the optical path $\boldsymbol{s}$ between $\boldsymbol{r}$ and the sun. For solar zenith angles $\theta$ less than *ca.* 87°, the integration over optical path $\boldsymbol{s}$ may be transformed to an integral over height $z$ and evaluated directly, to yield

$$\tau_\lambda = \frac{1}{\cos\theta}[\{O_2\}_r Q^1{}_\lambda + \{O_3\}_r Q^2{}_\lambda] \tag{5.49}$$

where $\{O_2\}_r$ and $\{O_3\}_r$ are vertical column densities (molecules cm$^{-2}$) for $O_2$ and $O_3$ above $\boldsymbol{r}$. If $t_c$ is long compared with the time for a significant change in solar zenith angle $\theta$, then the expression for $t_c$, equation (5.46), should be adjusted to allow for an average over time-varying solar conditions.

The relatively large values for $t_c$ which apply at high latitudes and low altitudes result from a combination of small values for $\cos\theta$ and large values for $\{O_2\}_r$ and $\{O_3\}_r$. The column density of $O_3$, as illustrated by Figure 5.4, varies from a value of order $5 \times 10^{18}$ cm$^{-2}$ near the equator to a value approaching $1.6 \times 10^{19}$ cm$^{-2}$ near the poles. The variation of $O_3$ with latitude is determined primarily by dynamics. A small amount of $O_3$, approximately 1% of the total global source, is allowed to leak out of the region of the atmosphere in which chemistry plays a dominant role. The gas is then transported, both north and south. It behaves as an essentially inert constituent of the atmosphere and the concentration above any selected high latitude station is set by a balance between supply by local winds and slow

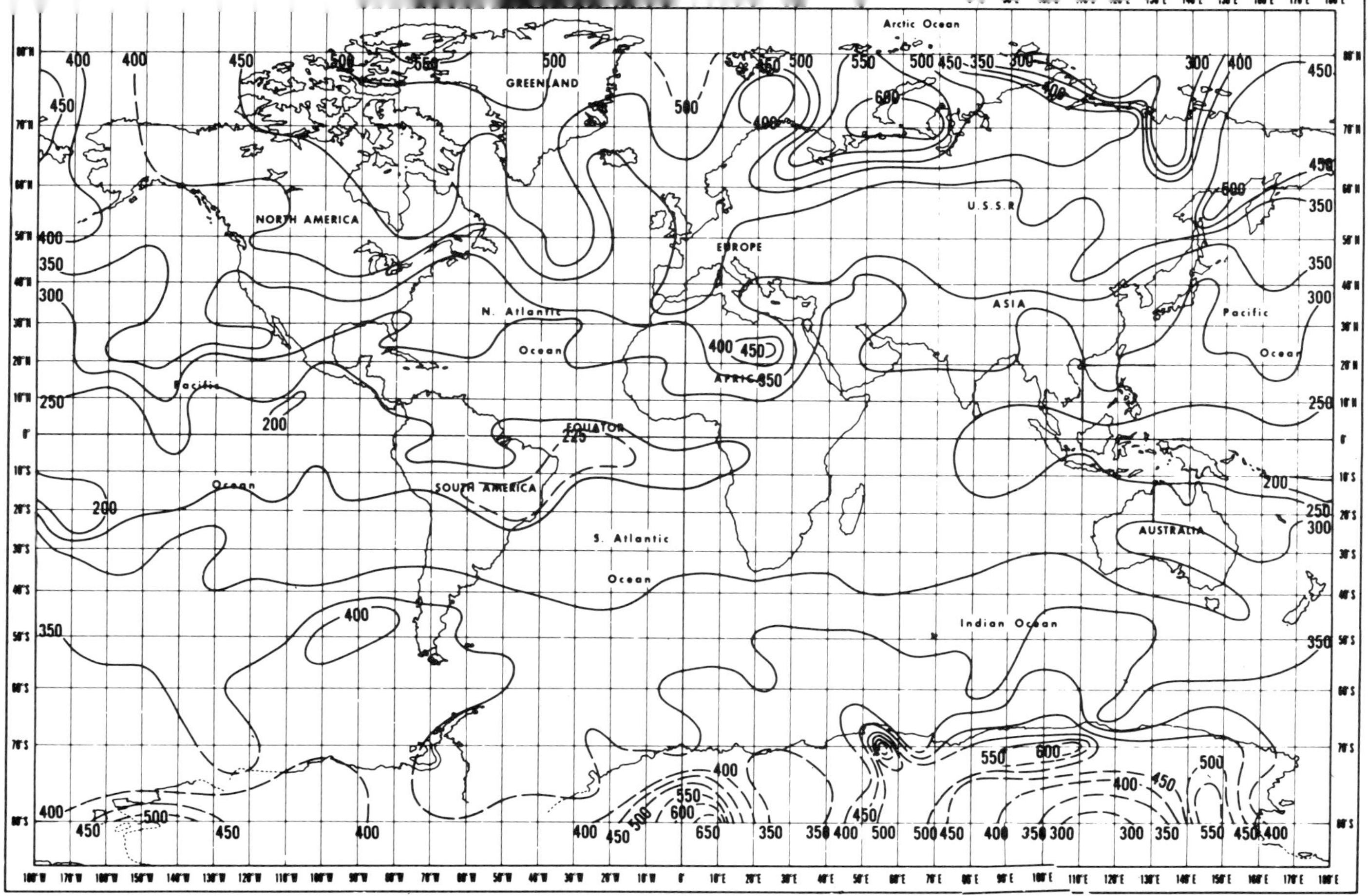


Figure 5.4 Total ozone content ($10^{-3}$ cm STP), 28 April 1969 (After Prabhakara *et al.*[200], by courtesy of NASA)

loss by chemistry. The general features of the ozone distribution in Figure 5.4 may be reproduced with a relatively simple dynamical model. The model tends to overestimate, however, the concentration of $O_3$ at all latitudes and altitudes, if we restrict attention to the simple chemical scheme proposed by Chapman[32]. The discrepancies at high altitude may be removed using catalytic hydrogen chemistry introduced by Bates and Nicolet[11]. Catalysis by nitric oxide plays a major role in the removal of $O_3$ at lower altitudes[50,122], and may account for as much as 80% of the total sink for odd oxygen, as discussed below.

Catalysis of oxygen recombination by hydrogen may take place either through

$$H + O_3 \rightarrow OH + O_2 \tag{5.50}$$

followed by

$$OH + O \rightarrow O_2 + H \tag{5.51}$$

or through

$$OH + O_3 \rightarrow HO_2 + O_2 \tag{5.52}$$

followed by

$$HO_2 + O \rightarrow OH + O_2 \tag{5.53}$$

These reactions are important above 50 km, and account for a reduction by about a factor of 2 in the density of ozone calculated at 60 km. Reactions involving $HO_2$, for example

$$HO_2 + O_3 \rightarrow OH + 2O_2 \tag{5.54}$$

followed by

$$OH + O_3 \rightarrow HO_2 + O_2 \tag{5.55}$$

may play a role at lower altitudes. Catalysis by $HO_2$, however, is in general small compared with catalysis by NO, which takes place through the reaction

$$NO + O_3 \rightarrow NO_2 + O_2 \tag{5.56}$$

followed by

$$NO_2 + O \rightarrow NO + O_2 \tag{5.57}$$

Reactions (5.50) + (5.51), (5.52) + (5.53) and (5.56) + (5.57) are equivalent to the net reaction

$$O + O_3 \rightarrow 2O_2 \tag{5.58}$$

Reactions (5.54) + (5.55) correspond to the bulk reaction

$$2O_3 \rightarrow 3O_2 \tag{5.59}$$

The odd hydrogen compounds H and OH are formed by photolysis of $H_2O$:

$$h\nu + H_2O \rightarrow OH + H \tag{5.60}$$

and by

$$h\nu + O_3 \rightarrow O(^1D) + O_2 \tag{5.61}$$

or

$$h\nu + O_2 \rightarrow O(^1D) + O \tag{5.62}$$

followed by

$$O(^1D) + H_2O \rightarrow 2OH \tag{5.63}$$

and

$$O(^1D) + H_2 \rightarrow OH + H \tag{5.64}$$

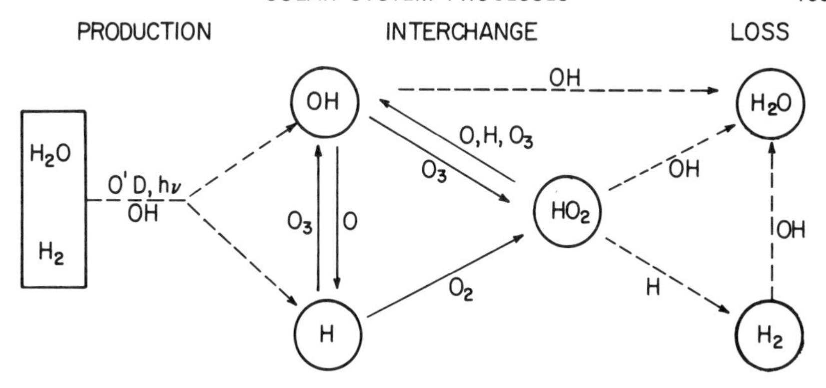


**Figure 5.5** Production and loss process for major forms of odd hydrogen (After McConnell and McElroy[165], by courtesy of the American Meteorological Society)

The major sinks for odd hydrogen, as shown schematically in Figure 5.5, are

$$OH + HO_2 \rightarrow H_2O + O_2 \tag{5.65}$$

and

$$H + HO_2 \rightarrow H_2 + O_2 \tag{5.66}$$

with (5.63) predominant, except at the highest altitudes.

Oxides of nitrogen may be formed by photochemical decomposition of $N_2O$, by oxidation of $NH_3$, or by a variety of reactions involving charged particles. Decomposition of $N_2O$ takes place primarily by the reaction

$$h\nu + N_2O \rightarrow N_2 + O(^1D) \tag{5.67}$$

with additional contributions due to

$$O(^1D) + N_2O \rightarrow 2NO \tag{5.68}$$

and

$$O(^1D) + N_2O \rightarrow N_2 + O(^3P) \tag{5.69}$$

Reaction (5.68) accounts for a NO source[173] of approximately $1.8 \times 10^5$ metric tons per year, roughly 1% of the total $N_2O$ sink*. Most of the $NO_x$ formed by (5.68) is released directly into the stratosphere, where it may be involved immediately in the chemistry of odd oxygen. The height integrated source of $NO_x$ due to oxidation of $NH_3$ is almost certainly larger than that due to (5.68). Oxidation of $NH_3$, however, proceeds mainly in the troposphere, by reactions such as (5.40), followed by

$$NH_2 + O_3 \rightarrow NH_2O + O_2 \tag{5.70}$$

$$NH_2O + O_2 \rightarrow HNO + HO_2 \tag{5.71}$$

*In order to facilitate comparison with the data in Section 5.4, results for the production of $NO_x$, and for the removal of $N_2O$ and $NH_3$, are quoted here in terms of equivalent mass units of nitrogen. Reaction (5.68) makes a small contribution, of order 0.07%, to the global budget of fixed nitrogen. As may be seen, a relatively small fraction of the total nitrogen released by microbiological activity at the Earth's surface is eventually involved in the catalytic sink for $O_3$.

and

$$HNO + O_2 \rightarrow HO_2 + NO \quad (5.72)$$

McConnell[163] points out that the oxidation of $NH_3$ could lead to a net sink for stratospheric $NO_x$. Laboratory studies by Gordon *et al.*[87] suggest that the reactions

$$NH_2 + NO \rightarrow N_2 + H_2O \quad (5.73)$$

and

$$NH + NO \rightarrow N_2 + OH \quad (5.74)$$

may be fast. In this case, removal of $NH_2$ could proceed by (5.73) rather than (5.70), if the rate constant for (5.70) were less than *ca.* $10^{-14}$ $cm^3$ $s^{-1}$. Further work, to define the fate of stratospheric $NH_3$, would be useful. The uncertainties in present models, however, are unlikely to seriously affect our discussion of $O_3$ chemistry. The concentration of $O_3$ is determined mainly by dynamics at altitudes below 25 km, where $NH_3$ might be expected to play a serious role in the chemistry of $NO_x$.

Galactic cosmic rays[22,214,253] and solar proton events[52] may contribute to the stratospheric budget of $NO_x$, by reactions such as

$$\begin{aligned} e^- + N_2 &\rightarrow N^+ + N + 2e^- \quad (5.75)\\ &\rightarrow 2N + e^- \end{aligned}$$

and

$$\begin{aligned} N^+ + O_2 &\rightarrow NO^+ + O\\ &\rightarrow N + O_2{}^+ \quad (5.76) \end{aligned}$$

followed by

$$NO^+ + e^- \rightarrow N + O \quad (5.77)$$

and

$$N + O_2 \rightarrow NO + O \quad (5.78)$$

According to current models, the contribution of galactic cosmic rays to the global budget of stratospheric $NO_x$ could be as large as 10% that of $N_2O$. The time-averaged contribution due to solar protons could be comparable with that from galactic cosmic rays, according to Crutzen *et al.*[52]. Cosmic rays and solar protons could affect both the spatial and temporal distribution of $O_3$. The associated production of $NO_x$ would be confined to relatively high latitudes, above about 60°, and the $NO_x$ catalysis would have to be transported to lower latitudes before it could have a direct influence on $O_3$. The production of $NO_x$ due to cosmic rays would take place at a relatively steady rate, although it should tend to maximise during periods of low solar activity, a consequence of the well-known Forbush decrease in galactic radiation, as discussed, for example, by Simpson[227]. Production of $NO_x$ due to solar protons would be somewhat more erratic. Most of the production during the 1960s might be attributed to two major storms, the first lasting from the 12th to the 16th November 1960, the second taking place during the period 2–5 September, 1966. Production of $NO_x$ by solar protons during the early part of the present decade should be dominated by the storm which occurred during the period 2–10 August, 1972. A careful examination of the detailed observational data for $O_3$ might throw some light on the role of energetic particle events in stratospheric chemistry. As emphasised by Ruderman and Chamberlain[214], modulation of $O_3$ by cosmic rays might be expected to exhibit cyclical behaviour, with a period similar to that for magnetic activity on the sun, about 11 years. Maximum production of $NO_x$ would take place during solar minimum. Maximum destruction of $O_3$, and consequently minimum concentrations of the gas, would be expected to occur somewhat later, with the delay defined by the details of the stratospheric wind field[214]. A critical examination of $O_3$ records taken during 1974 and 1975 might reveal some effects of the proton storm in 1972, an event of unusual strength and persistence.

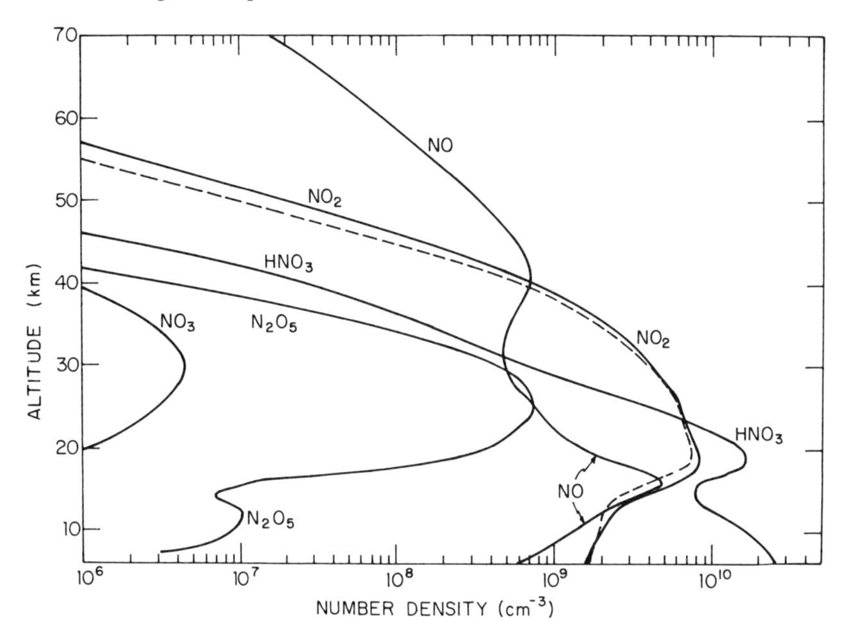


**Figure 5.6** Calculated vertical distributions of individual $NO_x$ constituents in the normal atmosphere, 30° N latitude, solar declination +12°. The dashed curve shows $NO_2$ calculated using a small rate for the reaction of NO with $O_3$ (After McElroy *et al.*[177], by courtesy of the American Meteorological Society)

Concentrations of major forms of stratospheric $NO_x$, computed for a typical mid latitude station, are shown in Figure 5.6. We may note that $HNO_3$ is the dominant component at low altitudes. Nitrogen dioxide is important between 25 and 40 km, and NO is the major constituent above 40 km. The height distribution of individual $NO_x$ species, $i$, may be found by solving the relevant continuity relations

$$\frac{\partial n_i}{\partial t} + \nabla \cdot {}_i\phi = P_i - L_i \tag{5.79}$$

where $n_i$ denotes density, $\phi_i$ flux, and $P_i$, $L_i$ chemical production and loss rates, respectively. Chemical exchange between various forms of $NO_x$ can take place with relative ease at all altitudes, as illustrated by Figure 5.7, and it is customary to sum (5.79) over species $i$, introducing an analogous relation for total $NO_x$. The flux $\phi$ may be evaluated in an approximate fashion, using the concept of eddy diffusion, as discussed, for example, by

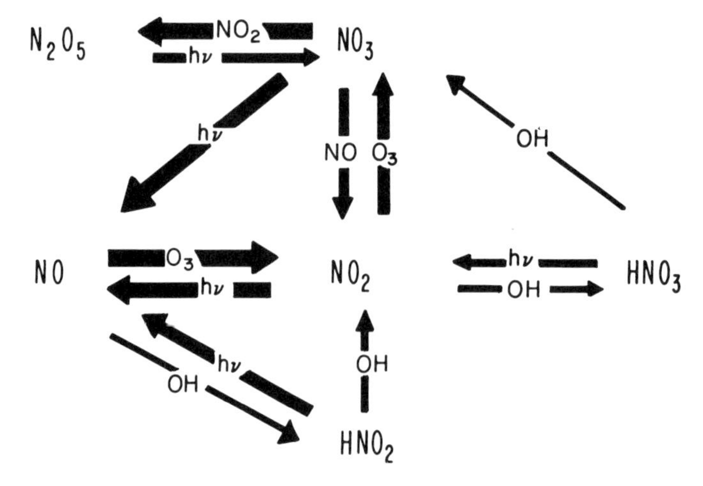


**Figure 5.7** The odd nitrogen cycle in the stratosphere. Width of arrows provide an indication of the rate ($s^{-1}$) of various reactions (After McConnell and McElroy[165], by courtesy of the American Meteorological Society)

Lettau[144], Newell[186], Colegrove *et al.*[48] and Godiksen *et al.*[85]. If we assume that the major variations in $\phi$ are confined to the vertical dimension $z$, and if we neglect the explicit dependence of $n$ on $t$, then (5.79) may be simplified to give

$$-\frac{\mathrm{d}}{\mathrm{d}z}\{K(z)\ N(z)\frac{\mathrm{d}f}{\mathrm{d}z}\} = P - L \tag{5.80}$$

where $K(z)$ is an effective diffusion coefficient ($cm^2\ s^{-1}$), $N(z)$ denotes total density at $z$, $f$ is the mixing ratio of $NO_x$ and $P$, $L$ are the production and loss rates, respectively. A comprehensive discussion of various approaches to the selection of $K$ is given by Johnston *et al.*[124] (see also Hunten[113]). The results in Figure 5.6, and in subsequent figures, were obtained with a profile for $K$ chosen to reproduce observed concentrations of $CH_4$, CO and $N_2O$[264]. This profile gives reasonable agreement with data on carbon-14 analysed by Johnston *et al.*[124], and the results for $NO_x$ in Figure 5.6 are in satisfactory accord with observational data summarised by McConnell and McElroy[165], Schiff[217], McConnell[164] and Wofsy[263].

Results for the height distribution of various forms of odd hydrogen are shown in Figure 5.8, which also includes data for O and $O(^1D)$. The metastable $O(^1D)$ is particularly important. It provides the major source for odd hydrogen throughout the Earth's lower atmosphere. It serves to regulate the concentration of OH, which in turn plays a dominant role in the chemistry of $NO_x$, as shown in Figure 5.7. It provides the first step in the chemical chain which leads to the oxidation of hydrocarbons in the troposphere, by reactions similar to (5.18)–(5.20). It may control the eventual sink for gases such as CO, and thus influence directly the quality of the air we breathe. It is a remarkable fact that the bulk of the chemistry of the Earth's lower atmosphere may be governed by a gas, $O(^1D)$, present with a concentration

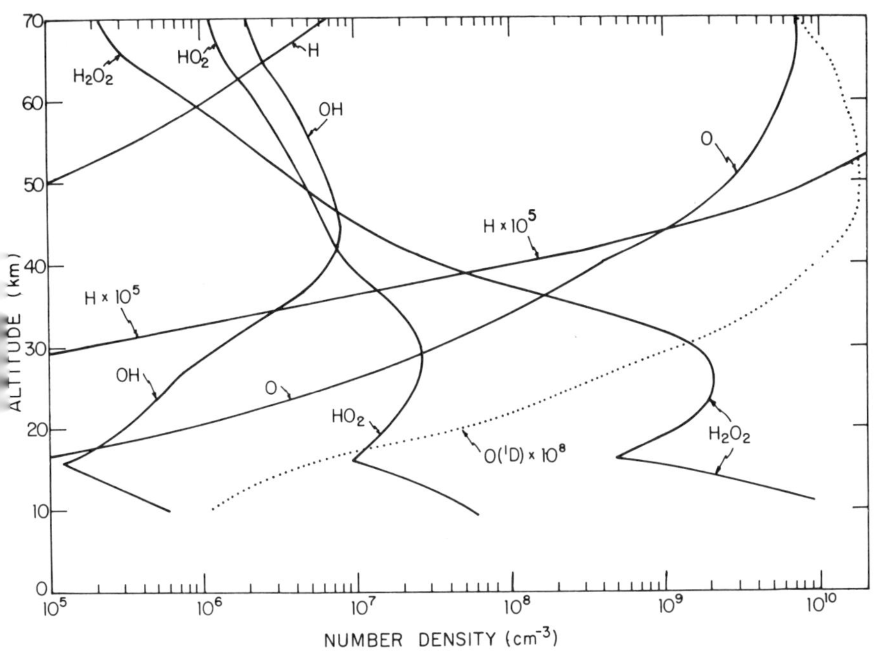


**Figure 5.8** Calculated number densities (24 h average) for hydrogenous radicals, $H_2O_2$, O($^3$P) and O($^1$D) for the normal atmosphere, 30° N latitude, solar declination +12°. The tropopause is at 16 km (After McElroy *et al.*[177], by courtesy of the American Meteorological Society)

of less than 1 part in $10^{20}$. Its abundance is set by a balance between photolysis, reactions (5.59) and (5.60), and quenching,

$$O(^1D) + N_2 \rightarrow O(^3P) + N_2 \tag{5.81}$$

and

$$O(^1D) + O_2 \rightarrow O(^3P) + O_2 \tag{5.82}$$

Most models assume that the rate constants for (5.81) and (5.82) are independent of temperature. The results in Figures 5.6 and 5.8, for example, were obtained with $k_{81} = 4.2 \times 10^{-11}$ cm$^3$ s$^{-1}$ and $k_{82} = 6 \times 10^{-11}$ cm$^3$ s$^{-1}$, values implied by room temperature measurements reported by Noxon[190], Lowenstein[157], and Gauthier and Snelling[81]. In view of the obvious importance of odd hydrogen, it would seem useful to extend the laboratory studies to lower temperatures, to obtain data over the range of temperatures of interest for the stratosphere, from *ca.* 180 K to 300 K. Additional work on the sink for odd hydrogen, reaction (5.65), would also be valuable. Most current models take $k_{65} = 2 \times 10^{-10}$ cm$^3$ s$^{-1}$, following measurements reported by Hochanadel *et al.*[103], although Kaufman[135] has raised questions regarding the interpretation of their data. The ambiguities may be resolved in the next few months by direct measurements of atmospheric OH. As this article goes to press, several groups, led by J. G. Anderson at the University of Michigan and by D. D. Davis at the University of Maryland, have instrumentation at advanced states of development, ready for a space-age attack on the problem of stratospheric chemistry. Anderson[4] has already reported preliminary measurements for the concentration of stratospheric O. His earlier data[3] for upper stratospheric and mesospheric OH are in satisfactory accord with theory, as shown in Figure 5.9.

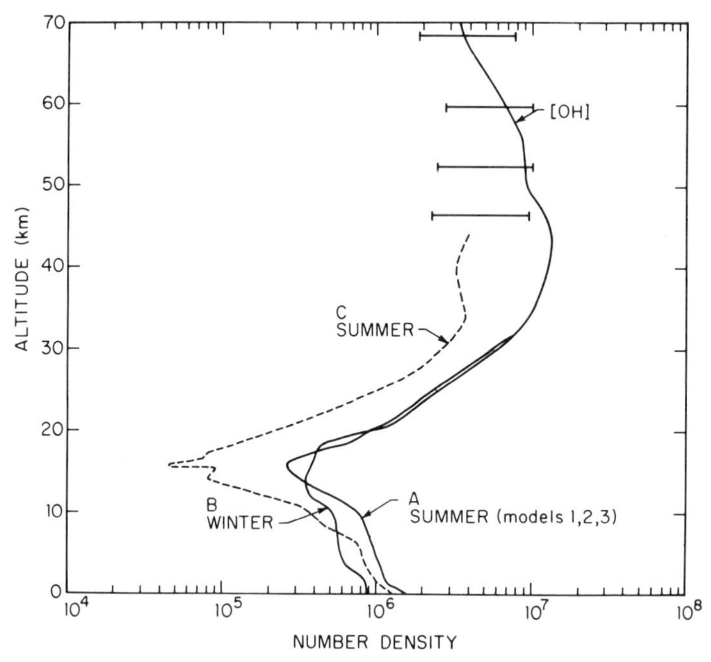


**Figure 5.9** Calculated OH number density with various models (From Wofsy *et al.*[269], by courtesy of the American Geophysical Union)

Observed and calculated profiles for $O_3$ are shown in Figure 5.10. The computations shown here were carried out for late summer conditions, a solar declination of 12°, at a latitude of 30° N. Curves A and D allow for the full range of $HO_x$–$NO_x$ chemistry as defined by Table 5.8. They differ in the choice of rate constant for the reaction of NO with $O_3$, a distinction of relatively minor importance in the present context. Reactions involving $NO_x$ were excluded in the computations summarised by curve B. Reactions involving $NO_x$ and $HO_x$ were omitted in curve C. The results in Figure 5.10 clearly attest to the importance of $HO_x$ catalytic chemistry at high altitudes, as suggested by Bates and Nicolet[11], and to the importance of $NO_x$ catalytic chemistry at low altitude, as proposed by Crutzen[50] and Johnston[122].

The various source and sink terms for odd oxygen are summarised in Figure 5.11. It seems clear that $NO_x$ chemistry must dominate below 40 km. The Chapman reaction, (5.45), has a major influence between 47 and 54 km. Reactions (5.51) and (5.53) are important above 54 km. As is evident from Figure 5.10, the theoretical model, with relatively few adjustable parameters, gives excellent agreement with experimental data, obtained using a variety of techniques, both chemical[101] and optical[100,120,138]. The agreement between theory and observation shown in Figure 5.10 provides an important measure

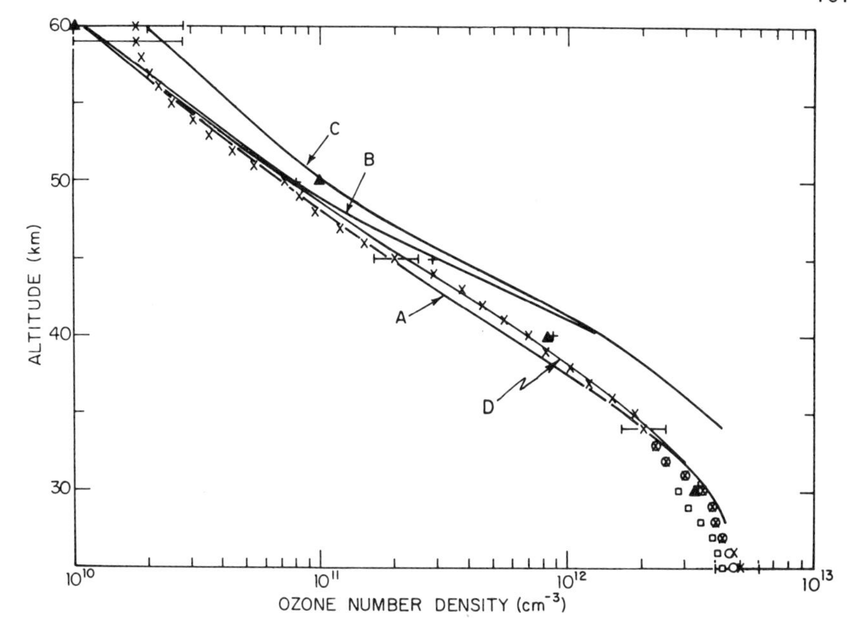


**Figure 5.10** Calculated ozone profiles for the normal atmosphere, 30° N latitude, +12° solar declination. Curve A was calculated with the complete chemical model of Table 5.4 and computed $NO_x$ profiles; reactions involving $NO_x$ were omitted in curve B; reactions involving $NO_x$ and $HO_x$ were omitted in curve C. Curve D is the complete calculation, as in curve A, but with the lower rate for reaction $NO + O_3 \rightarrow NO_2 + O_2$. Observations shown are as follows: ×, (with error bars), chemiluminescent rocketsonde, and ○, Mast–Brewer electrochemical ballonsonde, at Wallops Island (38° N), 16 September 1968[101]; ▲, rocketsonde solar u.v. absorption from White Sands (32° N), 14 June 1949[75,120]; +, rocketsonde solar u.v. absorption from Pt. Mugu, California (34° N), 18 June 1970[138]; □, Mast–Brewer balloonsondes from Tallahassee, Florida (30.4° N), 1966 summer average[100]

**Table 5.8 Chemical model**

| *Reaction* | *Rate* | *Ref.* |
|---|---|---|
| 1. $O_2 + h\nu \rightarrow O + O$ | $4 \times 10^{-9}$ | 173 |
| 2. $O_3 + h\nu \rightarrow O(^1D) + O_2(^1\Delta)$ | $4.6 \times 10^{-3}$ | 92 |
| $\rightarrow O(^3P) + O_2$ | $1.7 \times 10^{-4}$ | 166 |
| 3. $O + O_2 + M \rightarrow O_3 + M$ | $1.1 \times 10^{-34} \exp(500/T)$ | 121, 128, 218 |
| 4. $O + O + M \rightarrow O_2 + M$ | $2.8 \times 10^{-34} \exp(710/T)$ | 29 |
| 5. $O + O_3 \rightarrow 2O_2$ | $1.3 \times 10^{-11} \exp(-2140/T)$ | 289 |
| 6. $H_2O + h\nu \rightarrow H + OH$ | $1.0 \times 10^{-6}$ | 165 |
| 7. $HO_2 + h\nu \rightarrow O + OH$ | $3.9 \times 10^{-4}$ | 165 |
| $\rightarrow H + O_2$ | 0 | 165 |
| 8. $H_2O_2 + h\nu \rightarrow 2OH$ | $5.7 \times 10^{-5}$ | 165 |
| 9. $O(^1D) + H_2O \rightarrow H_2 + O_2$ | $0.6 \times 10^{-10}$ | |
| 10. $O(^1D) + H_2O \rightarrow 2OH$ | $2.1 \times 10^{-10}$ | 18, 193, 224, 280 |
| 11. $O(^1D) + H_2 \rightarrow OH + H$ | $1.3 \times 10^{-10}$ | 192, 273, 280 |
| 12. $O + OH \rightarrow O_2 + H$ | $5.0 \times 10^{-11}$ | 133, 134 |
| 13. $O + HO_2 \rightarrow OH + O_2$ | $6.0 \times 10^{-11}$ | 103, 133, 134 |

**Table 5.8**—*continued*

| Reaction | Rate | Ref. |
|---|---|---|
| 14. $O + H_2O_2 \rightarrow O_2 + H_2O$ | $1.5 \times 10^{-12} \exp(-2222/T)$ | 1, 281 |
| 15. $O + H_2O_2 \rightarrow OH + HO_2$ | $1.5 \times 10^{-12} \exp(-2222/T)$ | 1, 281 |
| 16. $O + H_2 \rightarrow OH + H$ | $5.3 \times 10^{-11} \exp(-5100/T)$ | 133, 134 |
| 17. $OH + O_3 \rightarrow HO_2 + O_3$ | $1.3 \times 10^{-12} \exp(-950/T)$ | 165, 281 |
| 18. $HO_2 + O_3 \rightarrow OH + O_3$ | $1.3 \times 10^{-12} \exp(-1820/T)$ | 282, 283 |
| 19. $OH + OH \rightarrow H_2O + O$ | $2.6 \times 10^{-12}$ | 65, 133, 134 |
| 20. $O_2(^1\Delta) + M \rightarrow O_2 + M$ | $4.4 \times 10^{-19}$ | 279 |
| 21. $OH + H_2O_2 \rightarrow H_2O + HO_2$ | $4.1 \times 10^{-13} T^{\frac{1}{2}} \exp(-1200/T)$ | 90, 91 |
| 22. $OH + HO_2 \rightarrow H_2O + O_2$ | $2.0 \times 10^{-10}$ | 103 |
| 23. $H + OH \rightarrow H_2 + O$ | $3.0 \times 10^{-11} \exp(-4180/T)$ | 133, 134 |
| 24. $OH + H_2 \rightarrow H_2O + H$ | $6.7 \times 10^{-12} \exp(-2010/T)$ | 90, 91 |
| 25. $HO_2 + HO_2 \rightarrow H_2O_2 + O_2$ | $9.5 \times 10^{-12}$ | 165 |
| | $3.0 \times 10^{-12}$ | |
| 26. $H + HO_2 \rightarrow H_2 + O_2$ | $3.0 \times 10^{-11} \exp(-330/T)$ | 218 |
| 27. $H + HO_2 \rightarrow 2OH$ | $10^{-10} \exp(-330/T)$ | 218 |
| 28. $H + HO_2 \rightarrow H_2O + O$ | 0 | 218 |
| 29. $H + O_3 \rightarrow OH + O_2$ | $2.6 \times 10^{-11}$ | 133, 134 |
| 30. $H + O_2 + M \rightarrow HO_2 + M$ | $1.8 \times 10^{-32} \exp(340/T)$ | 140 |
| 31. $H + H_2O_2 \rightarrow H_2 + HO_2$ | $5.0 \times 10^{-12} \exp(-2100/T)$ | 1 |
| 32. $H + H_2O_2 \rightarrow OH + H_2O$ | $5.0 \times 10^{-13} \exp(-2100/T)$ | 1 |
| 33. $NO_2 + h\nu \rightarrow NO + O$ | $4.1 \times 10^{-3}$ | 95 |
| $\rightarrow N + O_2$ | 0 | |
| 34. $N_2O + h\nu \rightarrow N_2 + O(^1D)$ | $7.4 \times 10^{-7}$ | 173, 287 |
| $\rightarrow NO + N$ | $\leqslant 7.4 \times 10^{-9}$ | |
| 35. $NO + h\nu \rightarrow N(^4S) + O(^3P)$ | $4.0 \times 10^{-6}$ | 28 |
| 36. $HNO_3 + h\nu \rightarrow OH + NO_2$ | $8.6 \times 10^{-5}$ | 286, 288 |
| $\rightarrow H + NO_3$ | 0 | |
| $\rightarrow O + HNO_2$ | 0 | |
| 37. $HNO_2 + h\nu \rightarrow OH + NO$ | $1.3 \times 10^{-3}$ | 165 |
| $\rightarrow H + NO_2$ | 0 | |
| $\rightarrow O + HNO$ | 0 | |
| 38. $N_2O_5 + h\nu \rightarrow NO_2 + NO_3$ | $7.7 \times 10^{-5}$ | 126 |
| $\rightarrow 2NO_2 + O$ | 0 | |
| 39. $NO_3 + h\nu \rightarrow NO + O_2$ | $1.0 \times 10^{-2}$ | 221 |
| $\rightarrow NO_2 + O$ | | |
| 40. $O(^1D) + N_2 \rightarrow O + N_2$ | $1.2 \times 10^{-11} \exp(253/T)$ | 293 |
| 41. $O(^1D) + N_2O \rightarrow 2NO$ | $3.0 \times 10^{-10} \exp(430/T)$ | 89, 273, 293 |
| 42. $N + O_2 \rightarrow NO + O$ | $1.5 \times 10^{-11} \exp(-3580/T)$ | 218 |
| 43. $N + O_2(^1\Delta) \rightarrow NO + O$ | $\leqslant 3.0 \times 10^{-15}$ | 258 |
| 44. $N + NO \rightarrow N_2 + O$ | $2.2 \times 10^{-11}$ | 218 |
| 45. $O + NO_2 \rightarrow NO + O_2$ | $9.1 \times 10^{-12}$ | 59 |
| 46. $NO + O_3 \rightarrow NO_2 + O_2$ | $1.5 \times 10^{-12} (\exp(-1330/T)$ | 284 |
| 47. $NO_2 + O_3 \rightarrow NO_3 + O_2$ | $1.31 \times 10^{-13} \exp(-2475/T)$ | 277 |
| 48. $NO + NO_3 \rightarrow 2NO_2$ | $1.0 \times 10^{-11}$ | 165 |
| 49. $NO_2 + NO_3 \rightarrow NO + O_2 + NO_2$ | $2.2 \times 10^{-13} \exp(-1850/T)$ | 165 |
| 50. $2NO_3 \rightarrow 2NO_2 + O_2$ | $5.0 \times 10^{-15} \exp(-3000/T)$ | 284 |
| 51. $N + NO_2 \rightarrow N_2O + O$ | $7.9 \times 10^{-12}$ | 218 |
| 52. $N + NO_2 \rightarrow 2NO$ | $6.1 \times 10^{-12}$ | 218 |
| 53. $N + NO_2 \rightarrow O_2 + N_2$ | $1.8 \times 10^{-12}$ | 218 |
| 54. $N_2O_5 + H_2O \rightarrow 2HNO_3$ | 0 | |
| 55. $OH + HNO_3 \rightarrow H_2O + NO_3$ | $8.9 \times 10^{-14}$ | 291 |
| 56. $HO_2 + NO \rightarrow OH + NO_2$ | $2.0 \times 10^{-11} \exp(-606/T)$ | 285, 292 |
| 57. $N + O_3 \rightarrow NO + O_2$ | $2.0 \times 10^{-12} T^{\frac{1}{2}} \exp(-1200/T)$ | 218 |
| 58. $O + HNO_3 \rightarrow OH + NO_3$ | $\leqslant 1.0 \times 10^{-11} \exp(-1860/T)$ | 165 |

**Table 5.8**—*continued*

| *Reaction* | *Rate* | *Ref.* |
|---|---|---|
| 59. $O + HNO_3 \rightarrow HNO_2 + O_2$ | 0 | 165 |
| 60. $H + HNO_3 \rightarrow H_2O + NO_2$ | 0 | 165 |
| 61. $H + HNO_3 \rightarrow H_2O + NO_3$ | $1.0 \times 10^{-12} \exp(-1180/T)$ | 165 |
| 62. $H + HNO_3 \rightarrow OH + HNO_2$ | $1.0 \times 10^{-11} \exp(-1180/T)$ | 165 |
| 63. $N + OH \rightarrow NO + H$ | $6.8 \times 10^{-11}$ | 29 |
| 64. $N + O \rightarrow NO + h\nu$ | $1.0 \times 10^{-17}$ | 188 |
| 65. $OH + HNO_2 \rightarrow H_2O + NO_2$ | $5.0 \times 10^{-13}$ | 165 |
| 66. $O + HNO_2 \rightarrow OH + NO_2$ | $1.6 \times 10^{-11} \exp(-1860/T)$ | 165 |
| 67. $NO + O \rightarrow NO_2 + h\nu$ | $6.4 \times 10^{-17}$ | 188 |
| 68. $NO + O + M \rightarrow NO_2 + M$ | $4.0 \times 10^{-33} \exp(965/T)$ | 218 |
| 69. $NO_2 + O + M \rightarrow NO_3 + M$ | $5.0 \times 10^{-31}$ | 218 |
| 70. $OH + NO_2 + M \rightarrow HNO_3 + M$ | (function of $z$) | 294 |
| 71. $OH + NO + M \rightarrow HNO_2 + M$ | $\dfrac{1.2 \times 10^{-11} \exp(806/T)}{2.3 \times 10^{20} + M}$ | 165 |
| 72. $NO_2 + NO_3 \underset{r}{\overset{f}{\rightleftarrows}} N_2O_5^*$ | $1.0 \times 10^{-12}$ (f) | 165 |
| | $\dfrac{1.0 \times 10^{3} + M}{2.6 \times 10^{19} + M} + 1.0 \times 10^{7}$ (r) | |
| 73. $N_2O_5 + M \underset{r}{\overset{f}{\rightleftarrows}} N_2O_5^* + M$ | $0.9 \times 10^{-5} \exp(-9700/T)$ (f) | |
| | $3.7 \times 10^{-11}$ (r) | |
| 74. $H_2O + NO + NO_2 \rightarrow 2HNO_2$ | $10^{-34}$ | 256 |
| 75. $ClO + h\nu \rightarrow Cl + O$ | $2.2 \times 10^{-3}$ | 71, 265 |
| 76. $HCl + h\nu \rightarrow H + Cl$ | $2.0 \times 10^{-7}$ | 209 |
| 77. $Cl_2 + h\nu \rightarrow Cl + Cl$ | $1.6 \times 10^{-3}$ | 82 |
| 78. $ClOO + h\nu \rightarrow Cl + O_2$ | $1.9 \times 10^{-3}$ | 123, 265 |
| 79. $OH + HCl \rightarrow H_2O + Cl$ | $2 \times 10^{-12} \exp(-313/T)$ | 295 |
| 80. $Cl + H_2 \rightarrow HCl + H$ | $2 \times 10^{-11} \exp(-2164/T)$ | 16, 80, |
| 81. $H + HCl \rightarrow H_2 + Cl$ | $1 \times 10^{-11} \exp(-1605/T)$ | 257 |
| 82. $O + HCl \rightarrow OH + Cl$ | $1.88 \times 10^{-11} \exp(-3573/T)$ | 270 |
| 83. $Cl + O_3 \rightarrow ClO + O_2$ | $1.84 \times 10^{-11}$ | 45 |
| 84. $Cl + Cl + M \rightarrow Cl_2 + M$ | $1.6 \times 10^{-33} \exp(880/T)$ | 102, 259 |
| 85. $Cl + O_2 + M \rightarrow ClOO + M$ | $1.7 \times 10^{-33}$ | 123, 187 |
| 86. $ClOO + M \rightarrow Cl + O_2 + M$ | $1.3 \times 10^{-11} \exp(-1000/T)$ | 46, 123 |
| 87. $ClO + ClO \rightarrow ClOO + Cl$ | $1.3 \times 10^{-12} \exp(-1150/T)$ | 43, 46, 123 |
| 88. $Cl + ClOO \rightarrow ClO + ClO$ | $1.44 \times 10^{-12}$ | 123 |
| 89. $Cl + ClOO \rightarrow Cl_2 + O_2$ | $1.5 \times 10^{-10}$ | 123 |
| 90. $ClO + NO \rightarrow Cl + NO_2$ | $1.7 \times 10^{-11}$ | 45 |
| 91. $O + ClO \rightarrow Cl + O_2$ | $5.3 \times 10^{-11}$ | 15 |
| 92. $ClO + ClO + M \rightarrow Cl_2 + O_2 + M$ | $5 \times 10^{-32}$ | 123 |
| 93. $H + Cl_2 \rightarrow HCl + Cl$ | $3.5 \times 10^{-11}$ | 233 |
| 94. $O + Cl_2 \rightarrow ClO + Cl$ | $1.36 \times 10^{-11} \exp(-1560/T)$ | 43 |
| 95. $ClO + O_3 \rightarrow ClOO + O_2$ | 0 | 265 |
| 96. $H + ClO \rightarrow HCl + O$ | $1.0 \times 10^{-11}$ | 265 |
| 97. $Cl + HO_2 \rightarrow HCl + O_2$ | $1.0 \times 10^{-11}$ | 265 |
| 98. $ClOO + O_3 \rightarrow ClO + 2O_2$ | 0 | 265 |
| 99. $Cl + CH_4 \rightarrow HCl + CH_3$ | $5 \times 10^{-11} \exp(-1791/T)$ | 44, 254 |
| 100. $HBr + h\nu \rightarrow H + Br$ (b) | $6.7 \times 10^{-6}$ | 209 |
| 101. $CH_3Br + h\nu \rightarrow CH_3 + Br$ | $1.8 \times 10^{-5}$ | 58 |
| 102. $HBr$ + rain, aerosol $\rightarrow$ removed from atmosphere | $\begin{cases} 1.7 \times 10^{-6} & 0 \leqslant Z \leqslant 5\ \text{km} \\ 0 & 5\ \text{km} < Z \end{cases}$ | 268 |
| 103. $Br + O_3 \rightarrow BrO + O_2$ | $1.2 \times 10^{-12}$ (300 K) | 254 |
| | $1.0 \times 10^{-10} \exp(-1320/T)$ | |
| | $2.5 \times 10^{-11} \exp(-900/T)$ | |
| 104. $NO + BrO \rightarrow NO_2 + Br$ | $2.0 \times 10^{-11}$ | 254 |
| 105. $Br + HO_2 \rightarrow HBr + O_2$ | $1.0 \times 10^{-10}$ | 61 |

**Table 5.8**—*continued*

| Reaction | Rate | Ref. |
|---|---|---|
| | $1.0 \times 10^{-11}$ <br> 0 | 268 |
| 106. $BrO + O \rightarrow Br + O_2$ | $8.0 \times 10^{-11}$ | 254 |
| 107. $BrO + BrO \rightarrow 2Br + O_2$ | $6.4 \times 10^{-12}$ | 254 |
| 108. $BrO + h\nu \rightarrow Br + O$ | Rate unknown in the atmosphere, probably slower than $k_2$ according to Watson[254] | |
| 109. $OH + HBr \rightarrow H_2O + Br$ | $3.7 \times 10^{-11} \exp(-600/T)$ | 254 |
| 110. $OH + CH_3Br \rightarrow H_2O + CH_2Br$ (c) | $5.5 \times 10^{-12} \exp(-1900/T)$ <br> $1.0 \times 10^{-11} \exp(-1500/T)$ | 268 |
| 111. $Br + H_2O_2 \rightarrow HBr + HO_2$ | $1.0 \times 10^{-11} \exp(-1200/T)$ | 268 |

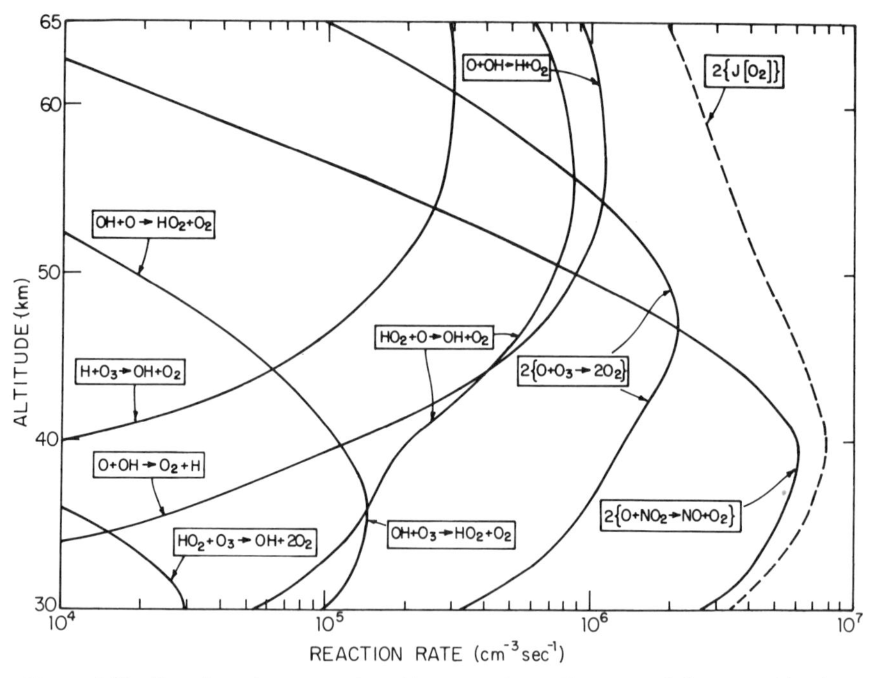


**Figure 5.11** Reactions important in odd oxygen loss. The sum of the recombination reaction rates equals the production rate at every altitude, consistent with the assumption of photochemical equilibrium (After McElroy *et al.*[177], by courtesy of the American Meteorological Society)

of support for the model calculations described in Sections 5.6–5.8. It would appear that now, finally, almost 50 years after Chapman's pioneering work in 1930, we have a model which may be expected to give a satisfactory, first-order, account of stratospheric $O_3$. More complex models are available, or in course of development, which will allow a more detailed description of the global distribution of $O_3$. For a preliminary account of the physical basis for these models the reader is referred to an excellent discussion by Cunnold *et al.*[54].

## 5.6 PLANES, BOMBS AND FERTILISER

This section is devoted to an elaboration of possible perturbations to $O_3$, which might be induced by a variety of anthropogenically derived sources of stratospheric $NO_x$. We shall emphasise models for the reduction of $O_3$ by high-flying supersonic transports (SSTs), although we shall also discuss potential effects of NO formed in the fireballs of nuclear explosions. Nitric oxide, in the normal state of the stratosphere, is produced primarily by photochemical oxidation of $N_2O$. Nitrous oxide is a product of microbiological denitrification and we shall introduce here the possibility that increased application of nitrogenous fertilisers could trigger an increase in the global production of $N_2O$, with ultimately serious consequences for stratospheric $NO_x$ and $O_3$.

The temperature field was taken as given in the computations described in Section 5.5. It is important for applications to the perturbed atmosphere, however, that we should demonstrate a capability to model temperature as well as $O_3$. The temperature of the stratosphere is set in part by dynamics, in part by radiation, with the latter dominant above about 35 km according to Blake and Lindzen[20]. Dissociation of $O_3$, reaction (5.44), is a major heat source, with most of the photon energy in (5.44) converted to heat at heights

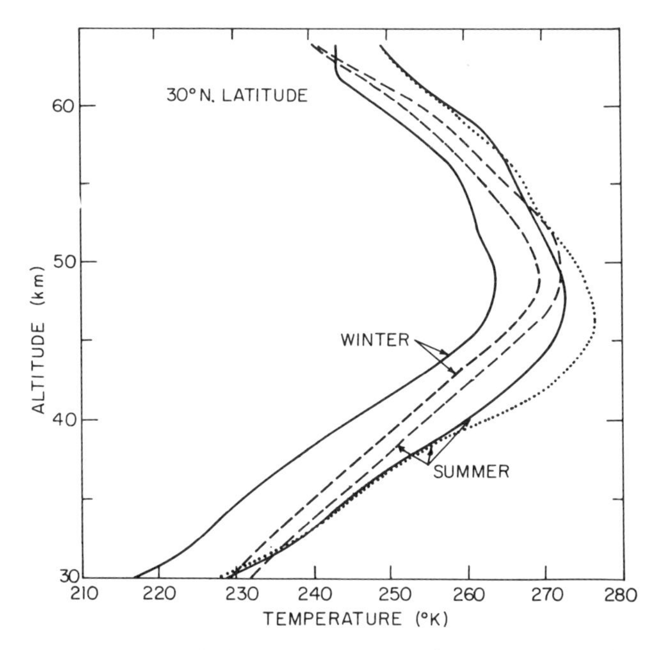


**Figure 5.12** Calculated temperature profiles (solid curves) for the normal atmosphere at solar declinations of +12° (summer) and −12° (winter). Dashed curves are from U.S. Standard Atmosphere Supplements (1966) for July and January. Dotted curve is temperature calculated with a small rate for NO + $O_3$ (From McElroy *et al.*[177], by courtesy of the American Meteorological Society)

of interest here, either immediately by (5.44) and subsequent superelastic collisions, or eventually by reactions involving $O(^1D)$, (5.81) and (5.82), $O_2(^1\Delta_g)$, and various vibrationally excited states of $O_2$, $N_2$ and $CO_2$. The energy absorbed in (5.44) is released ultimately as i.r. radiation, by vibrationally excited states of $CO_2$ and $O_3$. The temperature tends to rise above 20 km, a consequence of the heat absorbed in (5.44). It reaches a maximum near 45 km, and declines at higher altitudes, in the mesosphere, as the atmosphere becomes optically thin at i.r. wavelengths. Computed values for temperature are compared with observation over a limited height range in Figure 5.12. The agreement is satisfactory, and suggests that the thermal model used here, taken from Blake and Lindzen[20] and Dickinson[64], should be adequate for present purposes.

There are a variety of factors which act to stabilise $O_3$ in the perturbed atmosphere. We shall attempt to list these factors explicitly. They are taken into account, as fully as possible, in the numerical models described later in this review. There is a popular belief, fostered in part by several of the industries which may be affected by the issues raised here, that the atmosphere has an unlimited capacity for autorepair. This view is apparently not supported by the analysis described here. The atmosphere's capacity for autorepair is finite. Man's capacity for self delusion is evidently less constrained!

Consider a perturbation in the high altitude concentration of stratospheric $NO_x$. An increase in $NO_x$ should lead to a decrease in $O_3$. There should be a related decrease in temperature, and a corresponding increase in the rate

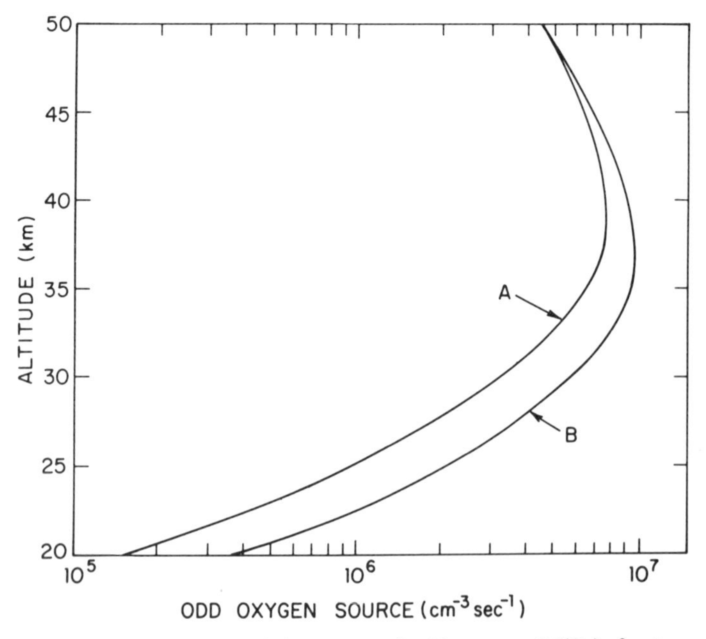


**Figure 5.13** Variation of the source of odd oxygen, $2J(O_2)$, for two levels of stratospheric $NO_x$: curve A, normal atmosphere; curve B, perturbed model f (From McElroy *et al.*[177], by courtesy of the American Meteorological Society)

constant for (5.43). This should lead to a reduction in O and tend to suppress the magnitude of the original perturbation in $O_3$. The importance of temperature–ozone coupling was emphasised first by Blake and Lindzen[20] and was incorporated fully in the models developed by McElroy *et al.*[177].

Consider a decrease in the column density of $O_3$ above some representative stratospheric level, *z*. A decrease in $O_3$ will lead to an increase in the transmission of radiation at wavelengths below 2400 Å, with a consequent increase in the production of odd oxygen due to dissociation of $O_2$ in the Herzberg continuum. The effect is illustrated in Figure 5.13. Here A refers to the normal atmosphere: B denotes a perturbed condition, in which the vertical column density of $O_3$ was reduced by about 26% owing to a globally averaged $NO_x$ input, at 20 km, of $4 \times 10^8$ molecules $cm^{-2}$ $s^{-1}$ [177]. Maximum production of odd oxygen for the normal atmosphere, under the conditions modelled in Figure 5.13, takes place at 39 km. Peak production in model B is at 37 km and is *ca.* 13% larger than the source at the maximum in model A. The production at 28 km, near the base of the chemical zone, is $2.5 \times 10^6$ $cm^{-3}$ $s^{-1}$ in the normal condition, model A, which may be compared with the value $4.1 \times 10^6$ $cm^{-3}$ $s^{-1}$ computed for the same level in model B.

A perturbation in $O_3$ may have an effect also on the natural source of stratospheric $NO_x$. As discussed in Section 5.5, $N_2O$ is removed mainly by

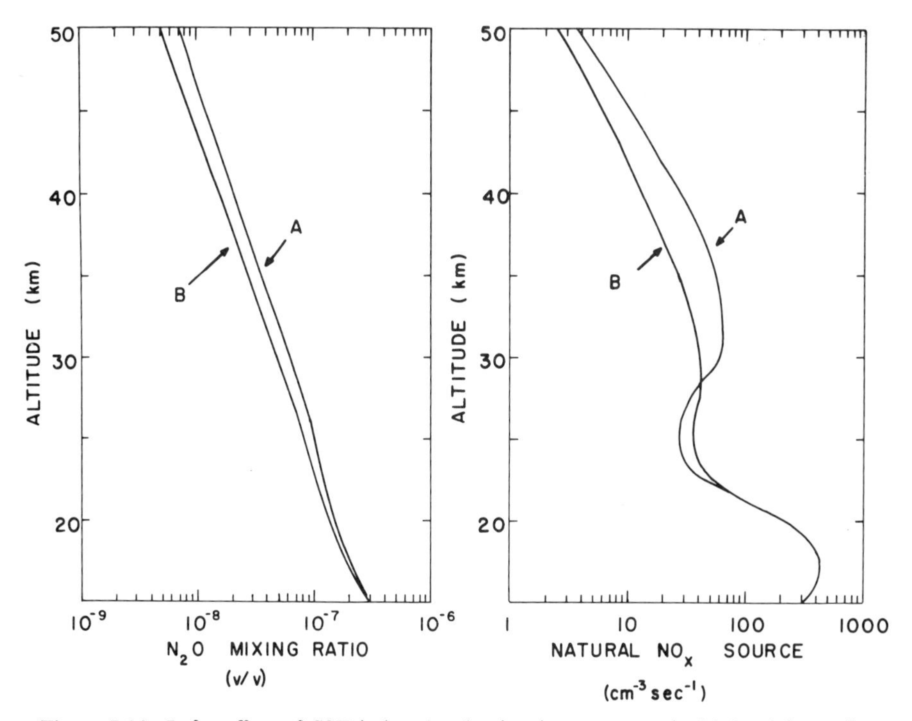


**Figure 5.14** Left—effect of SST induced reduction in ozone on the $N_2O$ mixing ratio: curve A, normal atmosphere; curve B, perturbed model f. Right—variation of natural $NO_x$ production induced by SST injection of $NO_x$: curve A, normal atmosphere; curve B, model f (From McElroy *et al.*[177], by courtesy of the American Meteorological Society)

(5.67) with a small contribution from (5.68)*. Dissociation of $N_2O$ is energetically possible at wavelengths below 3370 Å. The cross section, however, reaches a maximum at relatively short wavelengths, near 1300 Å. The rate constant for (5.67) is height dependent and affected to some extent by $O_3$, at least at the lower altitudes, i.e. below 30 km. With the profile B in Figure 5.13 the rate constant for (5.67) has a time and spatially averaged value of $6 \times 10^{-10}$ $s^{-1}$ at the Earth's surface, which may be compared with the value $5 \times 10^{-10}$ $s^{-1}$ derived for the unperturbed condition. The $N_2O$ profile, and $NO_x$ source, are altered accordingly, as shown in Figure 5.14. A reduction in $O_3$ at high altitudes leads to a reduction in the high-altitude concentration of $O(^1D)$, with a corresponding reduction in the production of $NO_x$. A reduction in high altitude $O_3$, however, allows deeper penetration of u.v. sunlight, with a corresponding increase in the rates for production of $O(^1D)$ and $NO_x$ at low altitudes, as illustrated by Figure 5.14. The height integrated rates for production of $NO_x$ in models A and B are $8.9 \times 10^7$ $cm^{-2}$ $s^{-1}$ and $8.2 \times 10^7$ $cm^{-2}$ $s^{-1}$, respectively. The corresponding values for $NO_x$ production above 28 km are $5.8 \times 10^7$ $cm^{-2}$ $s^{-1}$ and $4.3 \times 10^7$ $cm^{-2}$ $s^{-1}$. The production of $NO_x$ due to oxidation of stratospheric $NH_3$ takes place primarily at low altitudes, and is relatively unaffected by a change in the concentration of $O_3$.

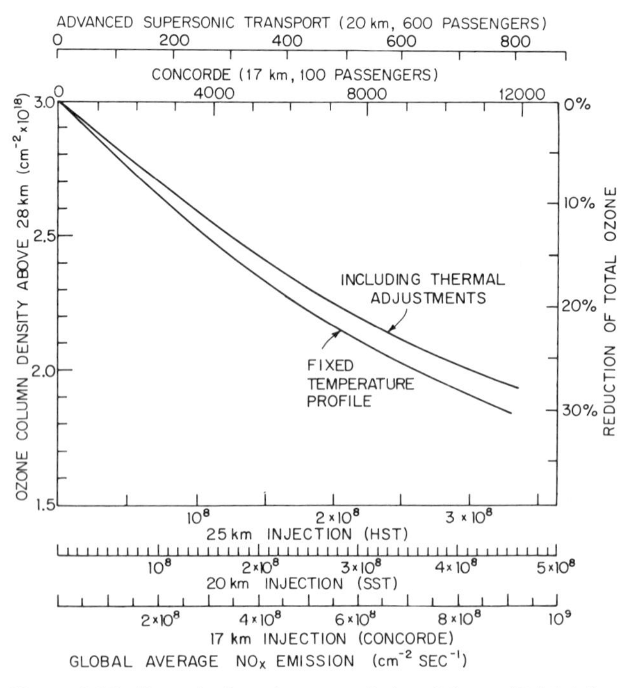


**Figure 5.15** Perturbations in ozone induced by artificial injections of $NO_x$ (After McElroy *et al.*[177], by courtesy of the American Meteorological Society)

*A growing body of evidence, summarised by McElroy *et al.*[290], suggests that there may be an important additional sink for $N_2O$ such that the effective lifetime for the gas in the atmosphere may be as short as 10 years. If true, this evidence would imply an upward revision in the estimate given below for the fractional yield of $N_2O$ in denitrification to perhaps as large as 50%. It would exacerbate possible stratospheric problems associated with agricultural practice, but would not seriously impact the more qualitative aspects of the discussion given here.

A summary of diverse models for the perturbed stratosphere is shown in Figure 5.15. The vertical axis to the left gives the column density of $O_3$ above 28 km, as a function of various anthropogenic sources of $NO_x$. The magnitude and nature of the anthropogenic source is defined by the several scales on the horizontal axis to the bottom of the figure. We show results for $NO_x$ injections centred at 25, 20 and 17 km. Models with the source at 17 km were designed to simulate possible effects of the Anglo–French Concorde, and should be appropriate also for the Soviet TU 144. Results with the source centred at 20 km would have applied to the supersonic transport (SST) considered by the United States during the 1960s. The source at 25 km is included in order to model possible effects of a conceptual hypersonic transport (HST). We assumed that aircraft emissions were distributed over a relatively narrow height range, and used a triangular shape function to model the input of exhaust gases. The triangular function was assigned a vertical extent of 2 km at the half intensity points for Concorde, TU 144 and HST. A somewhat broader dispersion function, with a half-width of 4 km, was adopted for the SST. The numbers of Concordes and SSTs associated with a given injection of $NO_x$ are indicated by the horizontal scales to the top of the figure. We assumed that the aircraft were operational at cruise altitude for an average of 7 hours per day, and used emission data presented by Broderick *et al.*[23] to effect the scale change from $NO_x$ input to numbers of operational aircraft*. The reduction in $O_3$, expressed as a percentage of the total global concentration of $O_3$, is shown by the vertical axis to the right of Figure 5.15.

We assumed, in computing total global concentrations of $O_3$ under perturbed conditions, that densities at lower altitudes should scale linearly in response to a change in the concentration of $O_3$ at the lower boundary of the photochemical zone. We located the lower boundary at the height for which the chemical time constant, $t_c$, had a value equal to 15 days. This procedure may be justified rigorously, to the extent (a) that the dynamics of the lower atmosphere can be considered independent of $O_3$; (b) the chemical loss rate for $O_3$ is first order in the concentration of $O_3$; and (c) we can ignore sources of odd oxygen at lower altitudes. These conditions are apparently satisfied to an accuracy adequate for present purposes. Results from a more comprehensive dynamical study of a hypothetical flight profile by Cunnold *et al.*[53] are in excellent agreement with the data shown here, as discussed by McElroy and Wofsy[174].

The possibility that nuclear bombs might provide a significant source of stratospheric $NO_x$, and that nuclear weapons testing in the atmosphere could have had a detectable effect on $O_3$, was raised first by Foley and Ruderman[78]. The initial explosion of the nuclear device takes place at temperatures above $10^5$ K. Approximately one-third of the energy released by the bomb is carried away by the blast wave which forms almost instantaneously. Approximately one-third of the energy is emitted, either to space or to the

*Broderick *et al.* assumed that Concorde, carrying 100 passengers at a cruise speed of $2.1 \times 10^3$ km h$^{-1}$ for 7 h d$^{-1}$, would emit a globally averaged quantity of $NO_x$ equivalent to $7.74 \times 10^4$ molecules cm$^{-2}$ s$^{-1}$. Their data for the SST assume an advanced version of the aircraft under development earlier, an SST with the capacity to transport 600 passengers at a cruise speed of $3.2 \times 10^3$ km h$^{-1}$, operational for an average of 7 h d$^{-1}$, with an equivalent emission index of $5.73 \times 10^5$ molecules $NO_x$ cm$^{-2}$ s$^{-1}$.

ground, at wavelengths for which the intervening atmosphere is relatively transparent. The remaining energy is used up in the formation of the fireball, which in its early stages of development has[123] a central temperature near 6000 K. The fireball cools by expansion, and by the entrainment of cold surrounding air. It tends to rise and, if the initial energy is high enough, it may reach the stratosphere, carrying with it a significant quantity of $NO_x$ formed while the ambient temperatures are above 2000 K. Johnston *et al.*[123] estimate that a $y$ megaton nuclear bomb would release between $0.17y \times 10^{32}$ and $y \times 10^{32}$ molecules of NO. Weapons testing in the atmosphere between 1952 and 1962 accounted for approximately 512 megatons, according to Foley and Ruderman[78], a total production of NO of between $8 \times 10^{33}$ and $5 \times 10^{34}$ molecules, using the release rates calculated by Johnston *et al.*[123]. These numbers correspond to a globally averaged production of between $5 \times 10^6$ and $3 \times 10^7$ molecules $cm^{-2}$ $s^{-1}$, in which we have assumed that release of $NO_x$ took place at a relatively steady rate over the 10-year interval of the tests. Nuclear testing in the 1950s and early 1960s could have provided a source of stratospheric $NO_x$ equivalent to that from a fleet of between 65 and 390 Concordes, or between 9 and 52 advanced design SSTs, with the use patterns and emission indices discussed above. A major nuclear war could trigger the release of as much as $10^3$ megatons, and if we assume that the associated production of $NO_x$ should be averaged over a period of about three years, a typical residence time for the stratosphere[177], we may calculate a globally averaged $NO_x$ source of between $3 \times 10^7$ and $2 \times 10^8$ molecules $cm^{-2}$ $s^{-1}$. With the data in Figure 5.15, the reduction in $O_3$ due to weapons testing would be between 0.5 and 5%. A major nuclear war could lead to a decrease in global $O_3$ by as much as 20%, with larger effects locally, perhaps as much as 30%, if the war were concentrated in a single hemisphere, as seems probable with the present disposition of the world's nuclear powers*. The effects on $O_3$ would persist for several years after the last blast and Hampson[97] was led to describe the effect of a nuclear war on $O_3$ as the ultimate doomsday weapon. This view seems somewhat extreme. It is hard to believe that a 20–30% reduction in $O_3$, for a period of less than five years, could have effects in any way comparable with those which would immediately follow the release of the world's nuclear arsenal. Yet the $O_3$ issue has become a factor of significance in the debate on nuclear strategy. F. C. Iklé, director of the U.S. Arms Control and Disarmament Agency, warned publicly that the reduction in $O_3$ following a nuclear war could 'shatter the ecological structure that permits man to

*The uncertainty in Johnston *et al.*'s estimate for the $NO_x$ production per megaton may be somewhat larger than the measure quoted in their paper. Crutzen[51] points out that a realistic treatment of chemical processes in the fireball should allow for reactions involving $H_2O$ which would act to lower the production of $NO_x$ and reduce accordingly the effects on $O_3$. On the other hand, many of the nuclear tests, and a fair fraction of the explosions expected in a nuclear war, would take place in marine environments, and one might expect the fireball to entrain a significant quantity of sea salt. Chlorine and bromine released by high-temperature reactions in the fireball could contribute to the reduction of stratospheric ozone, and, as noted in Sections 5.7 and 5.8, these elements are more effective than $NO_x$ as catalysts for recombination of $O_3$. Johnston *et al.*[123] thought that $O_3$ records obtained during the 1960s showed a direct observational effect which could be attributed to earlier atmospheric tests. Their conclusions have been challenged, however, by Angel and Korshover[6].

remain alive on this planet'. The Defense Department, while conceding that all-out nuclear war could reduce $O_3$ over temperate regions by as much as 50–75%, suggested that the effects of an environmental perturbation of such magnitude would be relatively harmless. The *New York Times*, in an editorial published November 12, 1974, under the title, 'If in doubt — gamble' termed the Defense Department's attitude 'preposterous' and concluded that 'either to continue pouring the commercial spray gas into the air in the hope that it may not prove lethal after all or to plan for a type of war that can expose the entire world to something far worse than nuclear fall-out — that is folly. It is comparable to a child skipping through a minefield, on the theory that he won't necessarily step on a mine and if he does it won't necessarily prove fatal'. As we shall see below, the minefield may be nurtured by nitrogenous fertiliser, and agriculturally enhanced production of $N_2O$, rather than nuclear war, may provide the ultimate threat to $O_3$.

As was noted in Section 5.4, man's influence on the rate for global fixation of nitrogen in 1974 was somewhere between 30 and 40% of the total production, estimated by Hardy and Havelka[98] at $260 \times 10^6$ metric tons. Averaged over suitably long timescales, the rate for denitrification must balance the rate for fixation. Any imbalance would lead to a major change in the fixed nitrogen content of the various reservoirs summarised in Table 5.7. The change in soil nitrogen could take place in as little as 200 years, and could affect the land and ocean biomasses in times as short as 38 and 16 years, respectively. The time constant for modification of fixed nitrogen in the ocean above the thermocline would be about 300 years, which may be compared with the time of the order of 800 years required to change the $CO_2$ content of the atmosphere and surface-ocean due to the consumption of fossil fuels*.

Man's influence on the global budget of fixed nitrogen has grown rapidly in the past 25 years. Nitrogenous fertilisers played a major role in this growth and it is clear that it must continue. Fertiliser is an essential element in the fight to produce a food reserve adequate to feed the world's expanding population. Nitrogenous fertiliser accounted for less than 1% of the total production of nitrogen fixed by all processes, natural and anthropogenic, in 1950. It contributed approximately 15% to the total fixation in 1974 and Hardy and Havelka[98] estimate that the fertiliser industry may grow to an annual production of *ca.* $200 \times 10^6$ metric tons by the turn of the century. They point out that growth is limited by factors which depend primarily on the availability of capital to finance the construction of new plants, rather than on demand. It seems clear that man's intervention in the nitrogen cycle must at least double the fixation rate by the turn of the century. There should be a corresponding increase in the rate for denitrification, and the concen-

*In carrying out the computations summarised here we assumed that a relatively small fraction of the world's total supply of fixed nitrogen, $5 \times 10^7$ metric tons per year, was delivered ultimately to the sea. The time constant for change in the abundance of fixed marine nitrogen should be adjusted downward if a relatively larger fraction of the global source of N were allowed to reach the sea. We adopted a value of $5 \times 10^9$ metric tons per year for the release of carbon due to consumption of fossil fuels, roughly 1% of the total global release of carbon due to respiration and decay, with the data summarised in Section 5.3.

tration of $O_3$ could be reduced on a relatively short timescale (less than 100 years) by as much as 20% due to the increased microbiological source of $N_2O$*.

It is difficult to assign a precise timescale to the perturbation in $O_3$ which might be expected due to the application of fertiliser. According to the arguments presented in Section 5.4, the land operates as an essentially closed cycle. We would expect a steady state to be reached in a few hundred years if the application of fertiliser were to stabilise at some constant rate. If we make the reasonable assumption that the active part of the nitrogen cycle may be confined to the upper layers of the soil, equilibrium could be established more rapidly, and one might expect a significant change in the release of $N_2O$, and in the concentration of $O_3$, by the middle of the next century. It seems clear that this matter needs further attention. It raises problems of enormous complexity, affecting agriculture, soil science, microbiology, public health, oceanography, atmospheric chemistry and meteorology. Its resolution will require combinations of scientific and technical talents not readily available in any existing institutional framework. We hope that the discussion presented here may serve to at least focus the issue. It can be addressed in a serious fashion, and fortunately we have time to prepare a response appropriate to the concerns raised here, should they be substantiated by further work.

## 5.7 SPRAY CANS, REFRIGERATORS AND PAPER PULP

It has been known, since early work by Norrish and Neville[189], that the recombination of $O_3$ may be catalysed by trace quantities of chlorine. The key reactions, analogous to (5.56) + (5.57) for $NO_x$, are

$$Cl + O_3 \rightarrow ClO + O_2 \tag{5.83}$$

and

$$ClO + O \rightarrow Cl + O_2 \tag{5.84}$$

Chlorine atoms may be removed from the atmosphere, by

$$Cl + H_2 \rightarrow HCl + H \tag{5.85}$$

and by

$$Cl + CH_4 \rightarrow HCl + CH_3 \tag{5.86}$$

with the latter especially important at lower altitudes. Hydrogen chloride

*In making estimates for the reduction in $O_3$ we assumed that the rate for production of $N_2O$ might be taken to scale linearly in response to a change in fixation. Accordingly the surface source of $N_2O$ was set equal to $2.4 \times 10^9$ molecules $cm^{-2}\ s^{-1}$ and the computations were carried out in the manner described by McElroy *et al.*[177]. In practice there could be a net increase in denitrification following excess application of fertiliser. The factors discussed in Section 5.3, which act to stabilise $O_2$ in the normal environment, will tend to promote a shift in the decay cycle from aerobic to anaerobic processes as man's agricultural intervention leads to a net increase in the productivity of the biosphere. The relative yields of $N_2$ and $N_2O$ evolved during denitrification are influenced primarily by the pH of the decay medium, as discussed in Section 5.4. We assumed that the mean pH of the Earth was unaltered by the application of fertiliser and retained a constant ratio for the fractional quantity of $N_2O$ evolved during denitrification. It is clear that a change in the average pH of the Earth due to anthropogenic activity, were it to take place on a short timescale, could have effects on the biospheric environment over and above those discussed here.

may be recycled by

$$OH + HCl \rightarrow H_2O + Cl \tag{5.87}$$

and by

$$h\nu + HCl \rightarrow H + Cl \tag{5.88}$$

and will be removed ultimately from the atmosphere by a variety of heterogeneous processes, including rain and washout.

The chemistry of atmospheric chlorine was discussed first by Wofsy and McElroy[265], Crutzen[51] and Stolarski and Cicerone[235]. Wofsy and McElroy[265], following Duce[67], concluded that marine aerosols were a major source for gaseous chlorine. They inferred an annual yield of $2 \times 10^8$ metric tons for HCl, using data presented by Junge[130], Duce[67], Buat-Menard and Chesselet[26], Chesselet *et al.*[34] and Buat-Menard[25]. Production of HCl due to industrial activity should account for approximately $10^7$ tons $y^{-1}$ and there could be additional release of similar magnitude owing to volcanoes and hot springs[74].

Wofsy and McElroy[265] estimated a global mean lifetime for gaseous chlorine of less than 10 days. This result may be compared with lifetimes of less than a day derived by Chesselet *et al.*[34] from measurements taken over rural France. Wofsy and McElroy[265] argued that Chesselet's data could be reconciled with their result for the average global condition if one assumed that about 15% of the Earth's land mass was effective as a sink for HCl. The mixing ratio (v/v) for all forms of gaseous chlorine appears to be variable, with observed values[130] in the range $1–6 \times 10^{-9}$.

Surface sources of HCl play a minor role in the budget for stratospheric chlorine. The heterogeneous loss processes discussed above are exceedingly efficient, and, as a consequence, only a small fraction of the gas released at ground level manages to penetrate to the stratosphere. It appears that the chlorine budget of the stratosphere may be dominated by less reactive species, such as $CFCl_3$, $CF_2Cl_2$, $CH_3Cl$, $CCl_4$ and $CHCl_3$. The potential importance of the chlorofluoromethanes $CFCl_3$ and $CF_2Cl_2$ was noted first by Molina and Rowland[180]. According to current models for the chemistry of atmospheric chlorine, these gases, known commonly by their duPont trade names Freon 11 and Freon 12*, are removed primarily by photolysis

$$h\nu + CFCl_3 \rightarrow CFCl_2 + Cl \tag{5.89}$$

*Commercially viable procedures for the production of chlorofluoromethanes were developed first by Midgely and Henne[178], who are responsible also for the numbering system used to describe the various substituted methanes and ethanes. The numbers were incorporated in the registered trademarks of Kinetic Chemicals Inc., and were acquired later by duPont. The numbering system was donated to the industry by duPont, who retained however the original trade name, Freon, which remains the registered trademark of the duPont Company. The first digit on the right denotes the number of fluorine atoms in the compound. The second digit from the right is one more than the number of hydrogen atoms. The third digit from the right is one less than the number of carbon atoms. If the compound contains only one carbon atom this digit is omitted. The more common chlorofluorocarbons, denoted by their duPont names, are Freon 11 ($CFCl_3$), Freon 12 ($CF_2Cl_2$), Freon 22 ($CHF_3Cl$), Freon 113 ($C_2F_3Cl_3$) and Freon 114 ($C_2F_4Cl_2$). In the subsequent discussion we shall refer to the compounds interchangeably as chlorofluoromethanes and Freons. The word Freon has acquired a significance over and above its duPont connection. It is used here with a capital F in deference to that company's trademark. A list of alternate trade names is given in the *Science and Technology of Aerosol Packaging* (1974).

and

$$h\nu + CF_2Cl_2 \rightarrow CF_2Cl + Cl \tag{5.90}$$

Their lifetimes in the atmosphere, as computed, for example, by Wofsy, McElroy and Sze[267], using cross sections measured by Rowland and Molina[212], are 45 and 68 years, respectively. If we assume that anthropogenic release of Freon should continue indefinitely at rates which prevailed in 1972, $2.2 \times 10^5$ metric tons for $CFCl_3$ and $3.5 \times 10^5$ metric tons for $CF_2Cl_2$, then the Freon industry could supply and maintain a budget of atmospheric chlorine as large as $2.2 \times 10^7$ metric tons, larger by about a fraction of 4 than the concentration of chlorine now present in the atmosphere, supplied mainly by decomposition of marine aerosols. It seems clear that industrial production and release of Freons could lead to significant change in the total budget of atmospheric chlorine. The change in stratospheric chlorine would be particularly remarkable. The concentration of chlorine in rain would remain relatively constant, and should continue to be controlled primarily by HCl of marine origin.

The Freon industry has grown rapidly in the past 20 years, with the growth most striking for Freon 11. According to Rowland and Molina[212], approximately 85% of the current production of $CFCl_3$ is used to provide propellants for aerosol sprays. Almost 2 billion of these products were marketed last year in the United States, nearly eight cans for every man, woman and child in the country. Approximately 8% of the total production of $CFCl_3$ was used as a foaming agent in the manufacture of products as diverse as coffee cups and living-room cushions. Freon 12, the most widely used of the chlorofluoromethanes, was employed mainly as a refrigerant and, to a lesser extent, as an industrial solvent. Almost 93% of the world's supply of chlorofluoromethanes in 1955 could be attributed to production within the continental United States. The source today is more widely dispersed. Howard and Hanchett[108] estimate that approximately 47% of the total production of fluorocarbons in 1973 came from the United States, with a similar amount from Europe, and most of the remainder produced in Japan[162].

Approximately 60% of the total Freon source is used to propel aerosol sprays, 25% is used as a coolant in refrigerators and air conditioners, 10% is employed as a foaming agent, and 5% is consumed as an industrial solvent. The Freon industry enjoyed an average growth of 8.7% per year between 1948 and 1972. The growth rate was somewhat larger in the 1960s. Production of Freon 11 advanced in this period at an annual rate of 22% per year, and R. L. McCarthy, technical manager of the Freon Products Laboratory of the duPont Company, believes that a growth of 10% per year would represent a reasonable projection for future demand. The reduction in $O_3$, computed by Wofsy, McElroy and Sze[267], with various conceptual models for the future consumption of Freon, is shown in Figure 5.16. The gross features of chlorine chemistry are illustrated by Figure 5.17.

Model A in Figure 5.16 assumes that production and release of Freon continues indefinitely at rates which prevailed in 1972. Models B, C and D assume initial growth rates of 10% per year, with growth allowed to proceed without interruption in model D. Production of Freon is assumed to end abruptly in 1978 with model B, but to continue until 1995 with model C. A

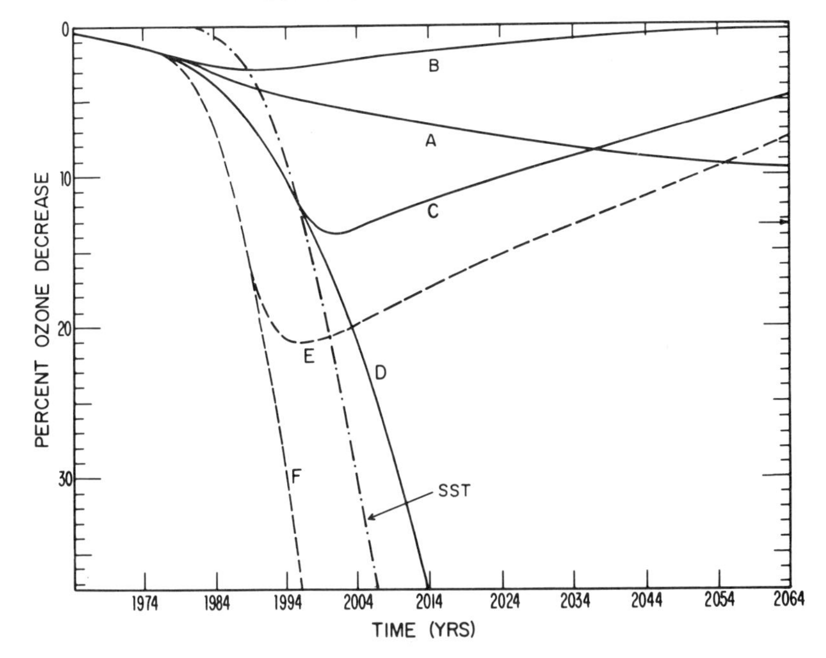


**Figure 5.16** Reductions in global ozone computed for six models of Freon use. Emissions of $CF_2Cl_2$ and $CFCl_3$ were assumed to be $3.5 \times 10^5$ and $2.2 \times 10^5$ metric tons, respectively, in 1972. The growth rates for each of the Freons are taken as 10% per year (7 year doubling) for models B, C and D; 22% per year (3.5 year doubling) for E and F, with production held constant for A. In models D and F, growth continues indefinitely; in models B, C and E, Freon emissions are assumed to cease abruptly in 1978, 1995 and 1987, respectively. For comparison we show ozone reductions due to SST operations, using Grobecker's 1974 estimate for the upper bound on NO injection by future fleets. The arrow indicates steady-state conditions approached by model A at long times (After Wofsy *et al.*[267], by courtesy of the American Association for the Advancement of Science)

growth rate of 21% per year is adopted for models E and F, and production is assumed to terminate in 1987 with model E. The height distribution and time variability of $O_3$ was found by numerical solution of a set of time dependent one-dimensional diffusion equations, similar to (5.79), with $\phi_i(Z)$ set equal to $K(Z)N(Z)\,\mathrm{d}f_i/\mathrm{d}Z$, where all symbols have the significance discussed earlier. We allowed for time-dependent vertical flow of $N_2O$, $NO_x$, $CH_4$, CO, $O_3$, $CF_2Cl_2$, $CFCl_3$ and odd chlorine, taken as the sum of HCl, ClO, ClOO, Cl and $Cl_2$. The computations illustrated in Figure 5.16 were carried out with the full chemical model given in Table 5.8, with procedures described in Section 5.5. It seems clear that an unconstrained Freon industry could introduce serious problems for atmospheric $O_3$. The problem is particularly serious in light of the long time constants which prevail for atmospheric Freon. A decision reached as early as 1978, model B, to eliminate Freon use would still allow for a reduction in $O_3$ by as much as 3% by the year 1990. If the decision to suspend Freon production were postponed until 1995, model C, the reduction in $O_3$ could exceed 10%, and would be appreciable for as much as 200 years after termination of Freon use. Rates

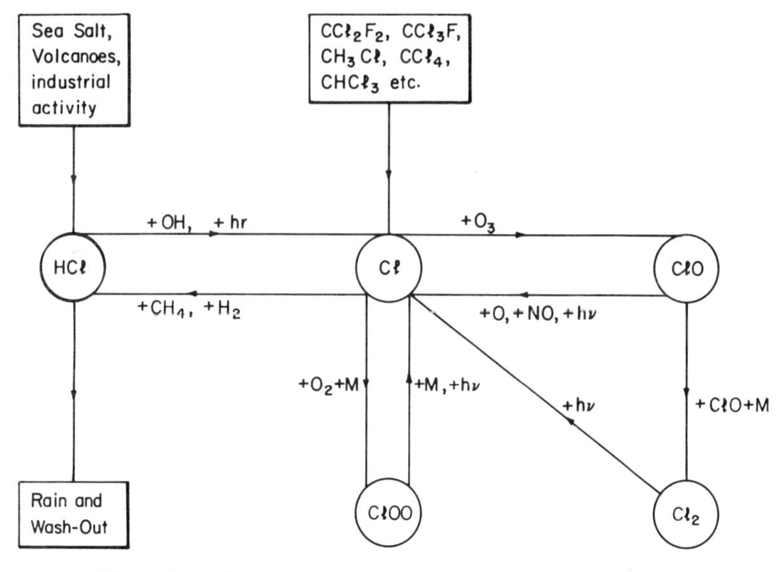


**Figure 5.17** Schematic representation of chlorine chemistry

for the production and loss of odd oxygen in 1995 are shown for model C in Figure 5.18a. As may be seen from the figure, the catalytic effects of odd chlorine could be comparable with those for $NO_x$ by 1995. Concentrations

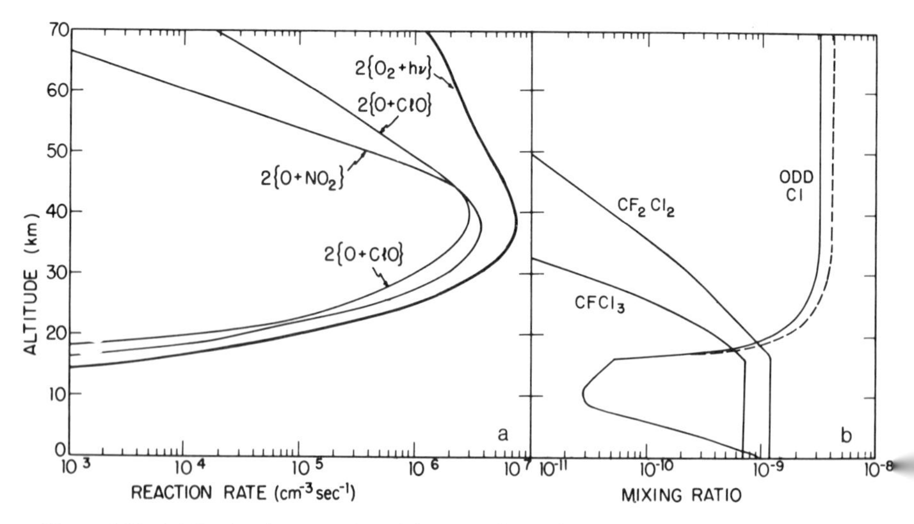


**Figure 5.18** (a) Production rates for odd oxygen from photolysis of $O_2$ (bold line) together with loss rates due to reactions of O with ClO and $NO_2$ (model C, 1995). Above 45 km, removal of odd oxygen is dominated by reactions of O with $O_3$, OH and $HO_2$. (b) Vertical profiles of $CFCl_3$ (Freon 11), $CF_2Cl_2$ (Freon 12), and odd chlorine are given for model C in 1995, In that year Freon production is assumed to end, and the dashed curve shows odd chlorine concentrations calculated for 2001. Model C predicts a stratospheric concentration of HCl consistent with Farmer's 1974 upper limit in 1973 (After Wofsy *et al.*[267], by courtesy of the American Association for the Advancement of Science)

of $CFCl_3$, $CF_2Cl_2$ and odd chlorine, derived for the same model year, are shown in Figure 5.18b*.

Most studies of atmospheric Freon have focused on possible effects which the gas might have on $O_3$[36,51,180,267]. Ramanathan[204] has raised an additional issue which may provide further cause for concern. He points out that the compounds $CFCl_3$ and $CF_2Cl_2$ have a number of strong bands which could interact with planetary radiation at i.r. wavelengths. Freons in the atmosphere could act like the glass in a greenhouse, and Ramanathan[204] estimates that the surface temperature of the Earth could be raised by as much as 0.9 K if the concentrations of $CFCl_3$ and $CF_2Cl_2$ in the atmosphere were increased from their present values, near 0.1 p.p.b., to *ca.* 2 p.p.b. He argues that the increase in surface temperature should be linearly related to the mixing ratios of $CFCl_3$ and $CF_2Cl_2$ for mixing ratios less than *ca.* 5 p.p.b. and drew attention to the possible radiative effects of $CCl_4$, $CHCl_3$ and $CH_2Cl_2$. Ramanathan's calculations were based on a simple globally-averaged radiative-convective model for thermal structure. As he points out, this model neglects a variety of possible feedback mechanisms. A more comprehensive dynamical study would appear desirable, in light of the obvious economic and agricultural implications of even a small change in surface temperature.

**Table 5.9 Measurements of halomethanes[275]**

| *Species* | *Mixing ratio* (*v*/*v*) | *Ref.* |
|---|---|---|
| | ($10^{-12}$) | |
| $CCl_2F_2$ | 102–115 | 153 |
| $CCl_3F$ | 60–80 | 260 |
| | 80–90 | 153 |
| $CH_3Cl$ | 400 | 154 |
| | $550 \pm 50$ | 205 |
| $CH_3I$ | 1.2 | 153 |
| $CCl_4$ | $71 \pm 7$ | 156 |
| | $75 \pm 8$ | 260 |
| | 111–138 | 153 |
| $CHCl_3$ | 19–27 | 153 |
| | 20 | 205 |
| $C_2HCl_3$ | $1.5 \pm 1.2$ (Southern Hemisphere) | 154 |
| | $15 \pm 12$ (Northern Hemisphere) | |
| $C_2Cl_4$ | $<1.5$ (Southern Hemisphere) | 154 |
| | $28 \pm 9$ (Northern Hemisphere) | |
| $CH_3CCl_3$ | $24 \pm 3$ (Southern Hemisphere) | 154 |
| | $65 \pm 17$ (Northern Hemisphere) | |

*The results in Figures 5.17 and 5.18 should be revised to allow for more recent data on the reaction of Cl with $CH_4$ and $O_3$. A preliminary analysis suggests that the reduction in $O_3$ computed for any given model year should be reduced by about a factor of 2. This change is equivalent to a small shift in the horizontal (time) axis in Figure 5.17 and does not, in any sense, *remove* the concern that Freon could pose a serious threat to atmospheric $O_3$.

The atmosphere, in addition to $CFCl_3$ and $CF_2Cl_2$, contains small, though detectable, concentrations of $CH_3Cl$, $CCl_4$ and $CHCl_3$, as measured by Lovelock[153,154], Wilkniss *et al.*[260] and Rasmussen[205]. A summary of various measurements is given in Table 5.9. Methyl chloride appears to be produced mainly by marine biological activity[154,205], and there is evidence for analogous sources of methyl bromide[205,268] and methyl iodide[156,275]. Carbon tetrachloride is used extensively as an intermediate in the manufacture of $CFCl_3$ and $CF_2Cl_2$. An uncertain amount of the gas is inadvertently released to the atmosphere, and could account[181] for a significant fraction of the observed concentration of $CCl_4$, although there appears to be strong empirical support for a natural source in addition to the anthropogenic contribution[154]. Chloroform, used widely as an anaesthetic for almost 50 years after its discovery in 1831, and more recently employed as an intermediate in the manufacture of penicillin and various fluorocarbon refrigerants (Freon 22) and resins (Teflon), is apparently formed as a by-product of water chlorination[13,211], and may be produced in copious amounts during the bleaching of paper pulp according to Yung *et al.*[275]. A summary of source strengths and lifetimes for $CHCl_3$ and $CH_3Cl$, estimated using rate data measured by Howard and Evenson[107], with profiles for OH derived in the manner described by McElroy *et al.*[177], is given in Table 5.10. The table includes also comparable data for $CH_3I$. The lifetime for $CH_3I$ is set primarily by photolysis, and was obtained using cross sections measured by Porret and Goodeve[199].

**Table 5.10 Lifetimes and sources of halocarbons[275]**

| | $CHCl_3$ | $CH_3Cl$ | $CH_3I$ |
|---|---|---|---|
| Mixing ratio (v/v) ($10^{-12}$) | 20 | 500 | 1.2 |
| Major sink | $CHCl_3 + OH$ | $CH_3Cl + OH$ | $CH_3I + h\nu$ |
| Rate constant* | $k = 1.0 \times 10^{-12}$ $\exp(-630/T)$† | $k = 1.69 \times 10^{-12}$ $\exp(-1066/T)$‡ | $J = 6.0 \times 10^{-6}$ |
| Lifetime (s) | $1.0 \times 10^7$ | $2.8 \times 10^7$ | $1.7 \times 10^5$ |
| Source (if globally distributed) ($10^6$ metric ton $y^{-1}$) | 1.3 | 5.2 | 0.74 |
| Source (if spatially restricted, see text) ($10^6$ metric ton $y^{-1}$) | 0.32 | — | — |
| Global industrial source ($10^6$ metric ton $y^{-1}$) | See Ref. 275 | — | — |

*Bimolecular rate constants, $k$, and dissociation rate, $J$, are in units $cm^3\ s^{-1}$ and $s^{-1}$, respectively
†Howard and Evenson[107] (at 296 K) combined with an estimate of activation energies
‡Davis *et al.*[60]

Chlorine has been used extensively as a disinfectant for municipal water supplies and sewage since shortly after the turn of the century. It appears that the compound was employed first on a continuing basis at the filter plant of the City of Lincoln, England, in 1904, in an attempt to combat an epidemic of typhoid (Laubusch, 1962). It was introduced into the United States by G. A. Johnson in 1908, and was used initially at the Boonton Reservoir, which supplies the municipality of Jersey City, New Jersey[119,142]. Today more than 90% of all water treatment plants in the United States use chlorine, either alone or in conjunction with other agents. Approximately 4% of the world's chlorine production is taken up in water sanitation. A small fraction of the gas is transformed into chloroform and other haloforms during the chlorination process[13,211]. The haloform source is not large from the viewpoint of the present discussion, about $10^4$ metric tons per year. The production mechanism is of interest, however, since it would apply also during the bleaching of paper pulp. Commercial preparation of paper accounts for almost 20% of the world's industrial consumption of chlorine, similar to quantities of the gas used in the manufacture of plastics and various synthetic materials. Even a small yield of chlorinated organic compounds might have an important effect on the stratosphere, and could influence the chemistry of atmospheric ozone in a significant and possibly detectable manner.

According to Morris[182], the production of chloroform in water treatment systems begins with the reaction

$$Cl_2 + HCO_3^- \rightarrow HOCl + Cl^- + CO_2 \qquad (5.91)$$

followed by

$$HOCl \rightleftharpoons H^+ + OCl^- \qquad (5.92)$$

in which $Cl_2$ is hydrolysed to HOCl and $OCl^-$. It proceeds through a series of complicated reactions involving dissolved organic material, which may be written in the form

$$CH_3COR + 3HOCl \rightarrow CCl_3COR + 3H_2O \qquad (5.93)$$

followed by

$$CCl_3COR + H_2O \rightarrow CHCl_3 + RCOOH \qquad (5.94)$$

These and other reactions lead to the production of a series of a variety of halogenated species, and can account for most of the halogenated impurities measured by the U.S. Environmental Protection Agency (EPA) in their recent survey of the drinking waters of some 80 U.S. cities*. The EPA found evidence for 35 distinct halogenated species in the drinking waters of Miami, Florida, including $CHCl_3$, $CHCl_2Br$, $CHClBr_2$ and $CHBr_3$, and there are reasons to believe that a similarly rich suite of compounds should be formed during the bleaching of paper pulp.

One would expect that the bleaching process, which combines an abundant source of organic material with an ample supply of chlorine, should represent an ideal medium for synthesis of the various halocarbons of interest here. The production of haloforms in drinking water appears to be limited by the supply of dissolved organic material, a restriction which is most certainly absent in the paper bleaching process. A conversion efficiency as low as 8% could supply a global source of $CHCl_3$ as large as $4 \times 10^5$ tons $y^{-1}$, and would readily account for the source estimate implied by the data in Tables 5.9 and 5.10. By way of comparison, the yield of $CHCl_3$ in the treatment of

*Detection of halogenated methanes in the drinking waters of New Orleans led to a concern that the public water supplies of the United States might contain dangerous levels of various carcinogenic agents. The fears were fed by the well publicised physiological hazards thought to be associated with other chlorinated compounds, including DDT, dieldrin, polychlorinated biphenyls and vinyl chloride. The concerns led to 'The Safe Drinking Water Act', Public Law 93–523, which was signed into law by President Ford on December 16, 1974. The EPA survey was carried out in response to this law.

municipal water supplies, as inferred from the EPA data, is between 1 and 3%, according to Symons[242].

Decomposition of perchloroethylene, $Cl_2C{=}CCl_2$, could provide an additional source for atmospheric $CHCl_3$. The gas is used widely as a solvent, with annual release rates in excess of $10^6$ tons, according to the U.S. Tariff Commission. Howard and Evenson[107] found evidence that halogenated ethylenes react with OH with rates in excess of $10^{-12}$ cm$^3$ s$^{-1}$, to form OH addition compounds. The addition reaction should be faster than photolysis[247] and should be more rapid than possible competing reactions with $O_3$. The fate of the radical $CCl_2$—$CCl_2OH$ remains unclear. A number of experiments[55,56,99,161], however, suggest that dichloroacetyl chloride, $Cl_2CH$—CO—Cl, should be a major product. Photolysis of dichloroacetyl chloride could take place[30] at wavelengths as long as 3600 Å and could provide[275] a significant, though not easily calculable, source of $CHCl_3$.

The abundance of $CHCl_3$ listed in Table 5.9 would supply a source of stratospheric chlorine which could account for a reduction in ozone by *ca.* 0.1%. The reduction in ozone due to $CH_3Cl$ would be *ca.* 0.8% and there would be a similar loss due to $CCl_4$. As discussed in more detail below, there are reasons to believe that $CH_3Cl$ is produced primarily by certain classes of marine algae, although James Lovelock, as quoted in the June 19, 1975, issue of the *New Scientist*, speculates that there may be additional anthropogenic production due to burning vegetation. Lovelock points out that fire is the cheapest and most efficient means for land clearance. It is used extensively for this purpose by third world agriculture and Lovelock protests that those who would 'attack the aerosol industry, should also criticise the developing countries'. Carbon tetrachloride, as noted above, is produced in part by industrial activity, in part by natural processes. The natural source has not been identified as yet.

## 5.8 FUMIGANTS, GASOLINE ADDITIVES AND ALGAE

This section is concerned with the chemistry of atmospheric bromine. As was the case in the preceding discussion of chlorine, our interest is motivated by a concern that we should identify all possible sinks for atmospheric ozone. Bromine, like chlorine, can be an effective catalyst for recombination of odd oxygen. Recombination may proceed by reactions analogous to (5.83) + (5.84)

$$Br + O_3 \rightarrow BrO + O_2 \qquad (5.95)$$

followed by

$$BrO + O \rightarrow Br + O_2 \qquad (5.96)$$

although there may be additional recombination by (5.95), taken twice, followed by

$$BrO + BrO \rightarrow 2Br + O_2 \qquad (5.97)$$

The first path is equivalent to

$$(Br) + O + O_3 \rightarrow (Br) + 2O_2 \qquad (5.98)$$

The second, which could be effective at both low altitude and high latitude, corresponds to the net reaction

$$(2Br) + O_3 + O_3 \rightarrow (2Br) + 3O_2 \qquad (5.99)$$

Bromine is more effective than chlorine as a catalyst for recombination of ozone. Most of the free chlorine in the stratosphere is tied up as HCl, formed by reaction (5.86). The analogous reaction for bromine is endothermic, and Watson[254] was led to conclude that BrO might be the major form of gaseous bromine over an extensive height range. It seems more probable, however, as argued by Wofsy, McElroy and Yung[268], that HBr should dominate. The gas could be formed by the reactions

$$Br + HO_2 \rightarrow HBr + O_2 \qquad (5.100)$$

and

$$Br + H_2O_2 \rightarrow HBr + HO_2 \qquad (5.101)$$

and would be removed by

$$OH + HBr \rightarrow H_2O + Br \qquad (5.102)$$

with an additional contribution due to

$$h\nu + HBr \rightarrow H + Br \qquad (5.103)$$

The chemical model used by Wofsy *et al.*[268], taken for the most part from Watson[254], is included in Table 5.8.

Bromine is present in both gas and particulate phases in the troposphere, with the gas phase dominant by a factor which ranges between 4 and 20. The mixing ratio of gaseous bromine is *ca.* $1.5 \times 10^{-11}$ over Hawaii[183], and measurements near the South Pole suggest values[70] smaller by about a factor of 5. There are indications that the gaseous component is derived primarily from the volatilisation of bromine carried into the atmosphere by marine aerosols[68,69], with additional contributions due to methyl bromide[268] and particles released during the combustion of leaded gasoline[27,159,183]. Using available data for the gasoline source strength, approximately $1.8 \times 10^5$ tons $y^{-1}$ according to Klingman[137], and empirical values for the lifetime of atmospheric HBr, Wofsy *et al.*[268] concluded that marine aerosols should account for roughly 80% of the total global budget of atmospheric bromine, which they concluded should be about $10^6$ tons $y^{-1}$. The remainder is derived in roughly equal amounts from automobiles and from the decomposition of methyl bromide.

There are indications that the mixing ratio of gaseous forms of acidic bromine, most probably HBr, may approach $10^{-11}$ in today's stratosphere, similar to the data noted above for the troposphere[141]. The mixing ratio of acidic bromine appears to increase with increasing height between 15 and 19 km, and there is evidence for similar behaviour in the height distribution of the particulate form[62,141]. It is difficult to escape the conclusion that there must be a stratospheric source for particulate and gaseous acidic bromine. Decomposition of $CH_3Br$ offers the most plausible explanation. The gas would be removed primarily by

$$OH + CH_3Br \rightarrow H_2O + CH_2Br \qquad (5.104)$$

with additional loss at higher altitude due to photolysis. Methyl bromide has been detected recently in surface waters of the ocean and in Antarctic snow[205]. It is used extensively as an agricultural fumigant, and there are, almost certainly, natural sources, as discussed below.

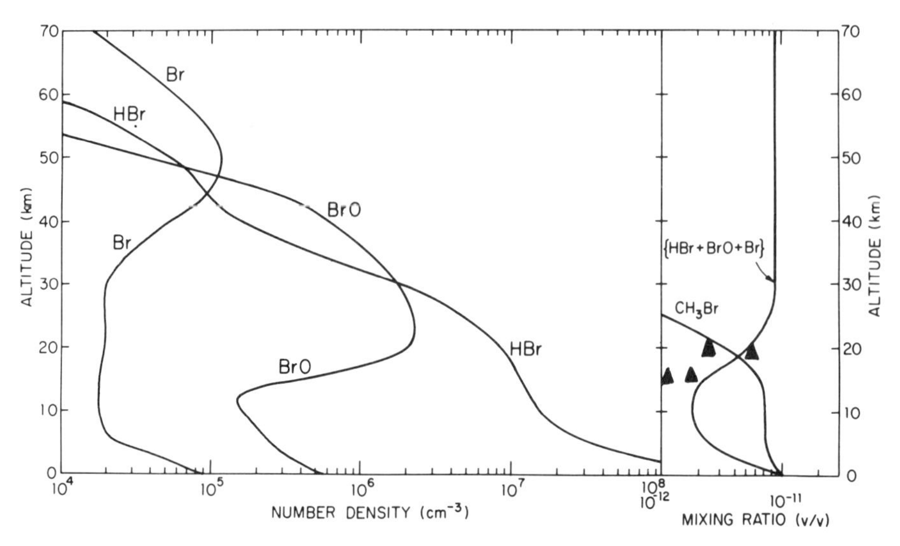


**Figure 5.19** Height profiles for Br, HBr, BrO and (Br + HBr + BrO); the data (▲) are from Ref. 141. Computations allow for transport of $N_2O$, $NO_x$, $CH_4$, CO, $O_3$, (HBr + Br + BrO) and $CH_3Br$, and accounted for complete $HO_x$ and $NO_x$ chemistry. The mean lifetime calculated for $CH_3Br$ (column concentration per flux) is 1.1 y (After Wofsy *et al.*[267], by courtesy of the American Association for the Advancement of Science)

Height profiles computed for the more abundant bromine species are shown in Figure 5.19. The computations were carried out with procedures similar to those used for $NO_x$ and chlorine, as described earlier. The mixing ratio of inorganic bromine was set equal to $10^{-11}$, consistent with Moyers and Duce[183] and Duce *et al.*[70]. The lifetime for inorganic bromide was taken equal to two weeks, based on the analysis presented by Moyers and Duce[183]. The surface flux of $CH_3Br$ was adjusted in order that the model should agree with the stratospheric data reported by Lazrus *et al.*[141]. In this manner, Wofsy *et al.*[268] derived a surface flux for $CH_3Br$ equal to $3 \times 10^6$ molecules $cm^{-2}$ $s^{-1}$, equivalent to an annual production of $7.7 \times 10^4$ tons $y^{-1}$. If their result is combined now with the data presented in Table 5.10, we may note that the methyl halides, $CH_3Cl$, $CH_3Br$ and $CH_3I$, are released to the atmosphere with concentrations, by weight, in the ratio 1 : 0.02 : 0.2. As emphasised by Wofsy *et al.*[268], these proportions are consistent with concentrations of chlorine, bromine and iodine found in certain seaweeds[106] It seems probable that marine algae should provide the major global source for $CH_3Cl$, $CH_3Br$ and $CH_3I$.

The source of $CH_3Br$ derived here, $7.7 \times 10^4$ tons $y^{-1}$, exceeds Klingman's[137] estimate for the total amount of $CH_3Br$ produced industrially in 1974 by about a factor of 4. Plonka[196] is of the opinion that only 25% of the industrial bromine is released to the atmosphere as a by-product of fumigation. Taken at face value, the analysis carried out by Wofsy *et al.*[268] would imply that the sea is the most probable source of atmospheric $CH_3Br$, and that it should be responsible for between 75 and 95% of the total concentration of the gas shown in Figure 5.19. The industrial contribution is growing rapidly, however, and the use of $CH_3Br$ in agriculture could play an important role in the atmospheric budget of the gas, in the not too distant future.

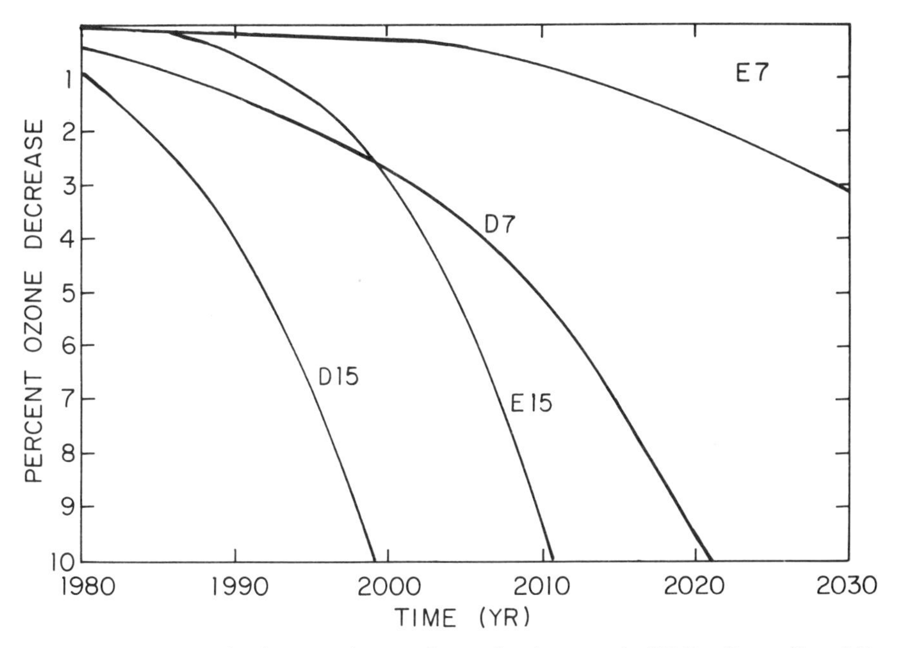


**Figure 5.20** Reduction in ozone due to release of anthropogenic $CH_3Br$. Curves D and E differ in the values assumed for the reaction of OH with $CH_3Br$: D, same rate as $CH_4$; E, preliminary measurement reported by Watson and Davis[255]. 7% and 15% annual increases are modelled by D7, E7, and D15, E15, respectively. The anthropogenic release was set equal to $10^4$ tons in 1974, approximately 50% of the total global production. The release rate estimated by Plonka[196] would imply a shift of the time axis by approximately 10 years for D7 and E7, 5 years for D15 and E15 (After Wofsy *et al.*[267], by courtesy of the American Association for the Advancement of Science)

Figure 5.20 shows several conceivable time profiles for the deficit in $O_3$ due to various possible models for the future agricultural consumption of $CH_3Br$. Curves D and E differ in the choice of rate constant for reaction (5.104). Curves E were derived with $k_{104}$ set equal to $2 \times 10^{-12}$ exp $(-1200/T)$, consistent with room-temperature measurements by Watson and Davis[255], and in agreement with high temperature data by Wilson[262]. Curves D were obtained with the rate constant for (5.104) set equal to the rate for the analogous reaction with $CH_4$, reaction (5.18). Production and use of $CH_3Br$ was assumed to increase at an annual rate of 7% per year in models D7 and E7, a value close to the historical trend. Curves D15 and E15 assumed a somewhat larger growth rate, approximately 15% per year. A growth rate of this magnitude might not be unreasonable, however. Cost factors currently restrict the application of methyl bromide to only a small number of high-value crops. One would expect that the agricultural use of methyl bromide should continue to grow, as food becomes more expensive and as the chemical industry succeeds in developing new and easier means for soil fumigation. It seems clear that an unconstrained growth in the application of methyl bromide could cause future problems for ozone. There is an urgent need for measurements of various bromine gases in the atmosphere, notably $CH_3Br$ and HBr, and there is a related need for laboratory studies of several key reactions, including (5.95), (5.96), (5.100) and (5.104).

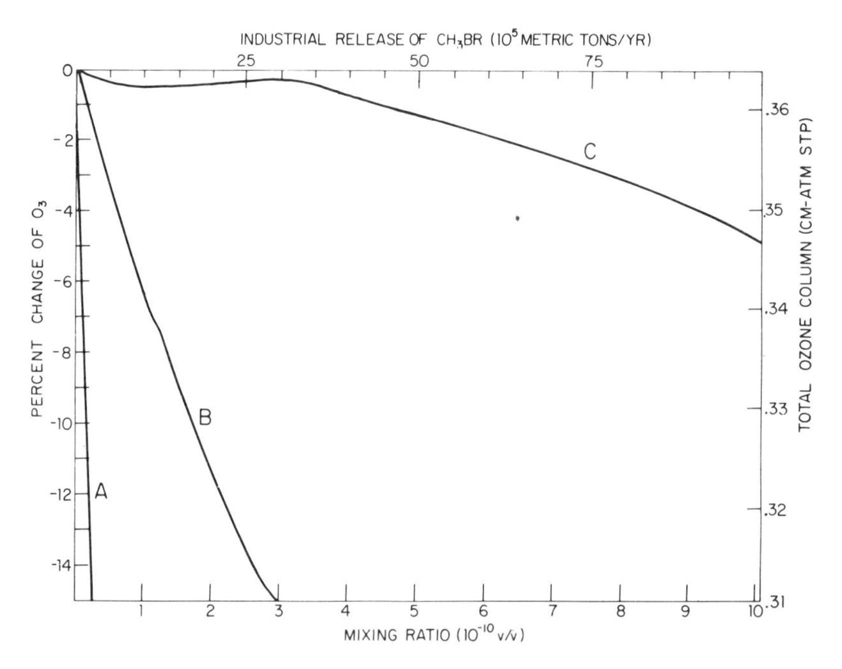


**Figure 5.21** Ozone reduction as a function of (HBr + Br + BrO). Different curves refer to different combinations of rate expressions as discussed in text. Curve B combines 'best estimates' for these rates. Curve A shows model where HBr formation does not occur. Curve C illustrates the effect of a choice designed to minimise the ozone perturbation. The upper axis shows the approximate industrial release rate of $CH_3Br$ which would produce the given stratospheric perturbation, using a rate for OH + $CH_3Br$ equal to $2 \times 10^{-12} \exp(-1200/T)$ (After Wofsy *et al.*[267], by courtesy of the American Association for the Advancement of Science)

Reductions in ozone, computed with various values for $k_{95}$ and $k_{100}$, are shown in Figure 5.21. The results shown here were derived subject to the constraint of a height-independent value for the mixing ratio of odd bromine, HBr + Br + BrO, and illustrate the manner in which one might expect the concentration of ozone to vary in response to a change in this parameter. The upper horizontal axis gives the chemical release rate required to produce a given perturbation in ozone. Curve B uses best estimates for the rate constants $k_{95}$ and $k_{100}$. Curve A shows how the results would change if BrO rather than HBr should turn out to be the major form of odd bromine (see, for example, Watson[254]). Curve C illustrates the effect of a choice for $k_{95}$ and $k_{100}$ designed to minimise the reduction in ozone. The results in Figures 5.19–5.21 imply a reduction in the contemporary concentration of $O_3$ of magnitude 0.3% due to bromine catalysis, of which approximately 0.2% may be attributed to $CH_3Br$ of marine origin, and 0.02% is associated with bromine contained in the exhaust streams of automobiles[268].

## 5.9 MARS

The Martian atmosphere is composed mainly of $CO_2$, with significant trace quantities of CO, $O_2$, $H_2O$, $O_3$ and most probably Ar, as noted earlier. Carbon dioxide is dissociated readily by sunlight at wavelengths below 2270 Å, and the central problem in Martian photochemistry is to account for the remarkable absence of the dissociation products, CO, O, $O_2$ and $O_3$.

The difficulties are severe, since $O_2$ may be formed efficiently by the reaction

$$O + O + CO_2 \rightarrow O_2 + CO_2 \qquad (5.105)$$

whereas the competing reaction

$$O + CO + CO_2 \rightarrow 2CO_2 \qquad (5.106)$$

fails to conserve electron spin for ground-state reactants and is, as a consequence, exceedingly slow at Martian temperatures[226,230,241], between about 120 and 250 K. The net, globally averaged, rate for dissociation of Martian $CO_2$ is[172] *ca.* $2 \times 10^{12}$ molecules $cm^{-2}\ s^{-1}$ and, in the absence of a suitably fast recombination process, the entire atmosphere could be transformed photochemically in a period as short as 3000 years. The concentrations of CO and $O_2$ now present in the atmosphere could be reproduced in as little as three and ten years, respectively.

**Table 5.11 Summary of present information on composition of the Martian atmosphere***

| *Constituent* | *Abundance*/cm atm |
|---|---|
| $CO_2$ | 7800 |
| CO | 5.6 |
| $O_2$ | 10.4 |
| $H_2O$ | ~3, variable |
| $H_2$ | ~0.4 |
| $O_3$ | $\sim 10^{-4}$ |
| $N_2$ | <400 |
| Ar + inert gases | <1560 |
| $SO_2$ | $<3 \times 10^{-3}$ |
| $N_2O$ | <200 |
| $CH_4$ | <10 |
| $C_2H_4$ | <2 |
| $C_2H_6$ | <1 |
| $NH_3$ | <2 |
| $NO_2$ | $<8 \times 10^{-4}$ |

*Data for $CO_2$–Ar are discussed in text. The upper limit for $NO_2$ is from Marshall[158]. Limits for other gases are from Kuiper[139]

This section is concerned primarily with the chemistry of the Martian atmosphere. Planetary dynamics is discussed extensively by Goody[86] and Gierasch[83], and more general reviews are given by Ingersoll and Leovy[117], Hunten[111] and McElroy[170]. Present information on composition is summarised in Table 5.11.

Various 'solutions' have been advanced over the past decade to resolve the so-called stability problem, i.e. to account for the survival of Martian $CO_2$. The simplest 'solution' held that $CO_2$ might not dissociate at wavelengths longer than 1000 Å. Dissociation at shorter wavelengths could be balanced by slow recombination through (5.106).

A second 'solution' invoked an intermediate $CO_3$ complex, formed by radiative recombination of $O(^1D)$ and $CO_2$. It was suggested that dissociation of $CO_2$ in the upper atmosphere might lead primarily to production of the metastable $O(^1D)$, which could be removed by

$$O(^1D) + CO_2 \rightarrow CO_3 + h\nu \quad (5.107)$$

followed by

$$CO_3 + CO \rightarrow 2CO_2 \quad (5.108)$$

Oxygen evolved in the lower atmosphere would be bound up initially as $O_2$, by (5.105), but could be recycled subsequently by

$$h\nu + O_2 \rightarrow O + O \quad (5.109)$$

followed by

$$O + O_2 + CO_2 \rightarrow O_3 + CO_2 \quad (5.110)$$

and

$$h\nu + O_3 \rightarrow O(^1D) + O_2 \quad (5.111)$$

with $O(^1D)$ removed by (5.107) and (5.108).

Neither of these 'solutions' stood the test of hard scrutiny by laboratory kineticists. The quantum yield of atomic oxygen, in photolysis of $O_2$, was measured as essentially unity, at a variety of wavelengths and pressures relevant to the Martian problem[38,77,229]. Quenching of $O(^1D)$ by $CO_2$ was found to yield $O(^3P)$ rather than $CO_3$, and previous claims for $CO_3$ in laboratory systems[40,132,248,252] were rejected on a variety of grounds by Clark[37].

It now seems clear that the solution to the stability problem must involve catalysis of O–CO recombination by trace quantities of H, OH and $HO_2$. In a recent paper, McElroy and Donahue[172] argued that recombination took place mainly in the lower atmosphere, through

$$O + HO_2 \rightarrow OH + O_2 \quad (5.112)$$

followed by

$$CO + OH \rightarrow CO_2 + H \quad (5.113)$$

and

$$H + O_2 + CO_2 \rightarrow HO_2 + CO_2 \quad (5.114)$$

with an additional contribution due to (5.113) followed by

$$H + O_3 \rightarrow OH + O_2 \quad (5.115)$$

and (5.110). Both sequences, (5.112) + (5.113) + (5.114) and (5.113) + (5.115) + (5.110), are equivalent to

$$(H) + CO + O \rightarrow (H) + CO_2 \quad (5.116)$$

Molecular oxygen was formed mainly by

$$O + OH \rightarrow O_2 + H \tag{5.117}$$

and removed by photolysis in the Herzberg continuum, (5.109), with additional loss[194] due to reaction (5.114) taken twice, followed by

$$HO_2 + HO_2 \rightarrow H_2O_2 + O_2 \tag{5.118}$$

$$h\nu + H_2O_2 \rightarrow 2OH \tag{5.119}$$

and 2 × (5.113), a sequence equivalent to

$$(2H) + 2CO + O_2 \rightarrow (2H) + 2CO_2 \tag{5.120}$$

Hydrogenous radicals were released by photolysis of $H_2O$, and by reaction of $O(^1D)$ with $H_2O$ and $H_2$. They were removed by

$$OH + HO_2 \rightarrow H_2O + O_2 \tag{5.121}$$

$$H + HO_2 \rightarrow H_2 + O_2 \tag{5.122}$$

and

$$H + HO_2 \rightarrow H_2O + O \tag{5.123}$$

A summary of the important chemistry, with best estimates for the relevant rate constants, is given in Table 5.12. Various source terms for CO and

**Table 5.12 Reactions and rate constants of relevance for Mars***

| *Reaction* | *Rate constant* |
|---|---|
| 1. $CO + O + CO_2 \rightarrow CO_2 + CO_2$ | $k_1 = 2 \times 10^{-37}$ |
| 2. $O + O + CO_2 \rightarrow O_2 + CO_2$ | $k_2 = 3 \times 10^{-33} (T/300)^{-2.9}$ |
| 3. $H + O_2 + CO_2 \rightarrow HO_2 + CO_2$ | $k_3 = 2 \times 10^{-31} (T/272)^{-1.3}$ |
| 4. $O + HO_2 \rightarrow OH + O_2$ | $k_4 = 7 \times 10^{-11}$ |
| 5. $CO + OH \rightarrow CO_2 + H$ | $k_5 = 9 \times 10^{-13} \exp(-500/T)$ |
| 6. $H + O_3 \rightarrow OH + O_2$ | $k_6 = 2.6 \times 10^{-11}$ |
| 7. $O + O_2 + CO_2 \rightarrow O_3 + CO_2$ | $k_7 = 1.4 \times 10^{-33} (T/300)^{-2.5}$ |
| 8. $O_3 + h\nu \rightarrow O_2 + O$ | $J_8 = 4.2 \times 10^{-3} s^{-1}$ |
| 9. $O + OH \rightarrow O_2 + H$ | $k_9 = 5 \times 10^{-11}$ |
| 10. $O_2 + h\nu \rightarrow O + O$ | $J_{10} = 5.8 \times 10^{-10} s^{-1}$ |
| 11. $CO + HO_2 \rightarrow CO_2 + OH$ | $k_{11} = <10^{-16}$ |
| 12. $HO_2 + HO_2 \rightarrow H_2O_2 + O_2$ | $k_{12} = 9.5 \times 10^{-12}$ |
| 13. $H_2O_2 + h\nu \rightarrow OH + OH$ | $J_{13} = 5.2 \times 10^{-5} s^{-1}$ |
| 14. $H + HO_2 \rightarrow H_2 + O_2$ | $k_{14} = 1 \times 10^{-11}$ |
| 15. $OH + HO_2 \rightarrow H_2O + O_2$ | $k_{15} = 2 \times 10^{-10}$ |

*For detailed references see McElroy and Donahue[172]. Rates for two-body reactions are given in units $cm^3$ $s^{-1}$, and rates for three-body reactions have units $cm^6$ $s^{-1}$

O, taken from McElroy and McConnell[173], are shown in Figure 5.22, and height profiles for O, $O_3$, H, OH and $HO_2$, calculated by McElroy and Donahue[172], are given in Figure 5.23. The relative importance of the various source and sink terms for O, CO and $O_2$ is illustrated in Figure 5.24.

Evidently recombination of CO and O takes placc primarily below 25 km, and reactions (5.112)–(5.114) play a major role. Molecular oxygen is formed

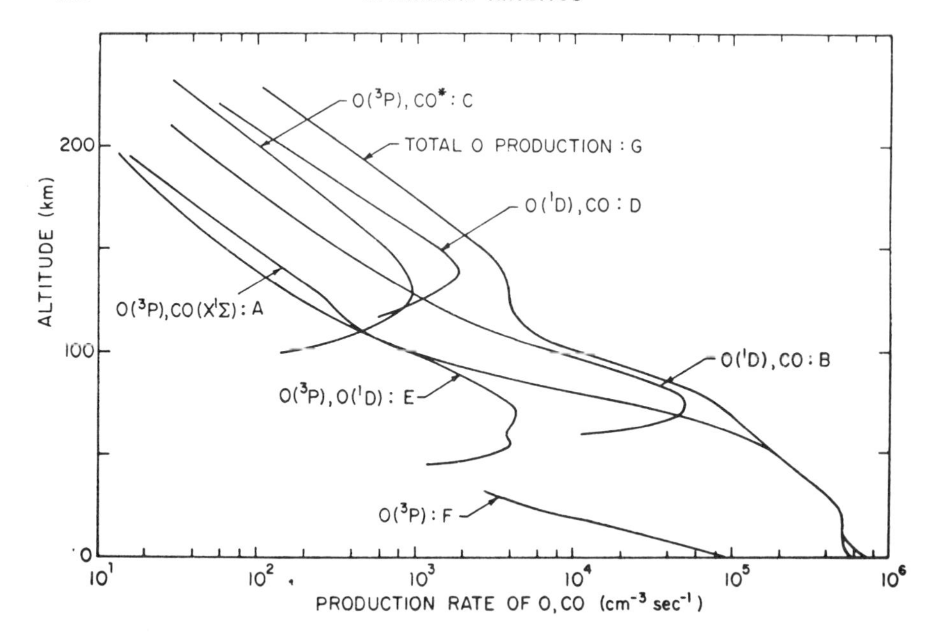


**Figure 5.22** Production rates of atomic oxygen and CO on Mars for a solar zenith angle of 60° and with solar fluxes reduced by a factor of 2 for day–night averaging. Curve A: $O(^3P)$ and $CO(^1\Sigma)$ from $CO_2$ by electron dissociation and photodissociation with 2270 $> \lambda >$ 1670 Å. Curve B: $O(^1D)$ and $CO(^1\Sigma)$ from $CO_2$ by photodissociation with 1670 $> \lambda >$ 1080 Å, assuming that the $O(^1S)$ produced for $\lambda <$ 1270 Å radiates to $O(^1D)$. Curve C: $O(^3P)$ and excited triplet states of CO from $CO_2$ by electron dissociation and by photolysis in the region $\lambda <$ 1080 Å. Curve D: $O(^1D)$ and CO from ionospheric reactions. Curve E: $O(^3P)$ and $O(^1D)$ from photolysis of $O_2$ for $\lambda <$ 1750 Å. Curve F: $O(^3P)$ from photolysis of $O_2$ in the Herzberg continuum. Curve G: Total O production rate (After McElroy and McConnell[173], by courtesy of the American Meteorological Society)

over a comparatively narrow height range, between 25 and 30 km, and reaction (5.117) is most important. The rate for photolysis of $H_2O$, derived with the concentrations for OH and $HO_2$, shown in Figure 5.23, and with the rate constant for (5.121) given by Hochanadel *et al.*[103], is $2.7 \times 10^9$ molecules $cm^{-2}\ s^{-1}$, and agrees well with an earlier estimate for this parameter[114] based on measured[150,219,220,246] concentrations of $H_2O$. The concentration of $O_3$ shown in Figure 5.24 corresponds to a vertical column density of $4 \times 10^{15}$ molecules $cm^{-2}\ s^{-1}$, smaller than the mean terrestrial abundance by a factor of *ca.* $10^3$.

The computations summarised by Figures 5.22–5.24 were constrained to give the escape route for H measured by Anderson and Hord[5], $1.8 \times 10^8$ atoms $cm^{-2}\ s^{-1}$. Escape of H from the Martian atmosphere proceeds through the classic mechanism outlined first by Jeans[118], and elaborated further by Chamberlain[31]. At temperatures which prevail in upper regions of the Martian atmosphere, *ca.* 350 K according to Stewart[234], significant numbers of H atoms have velocities in excess of 4.9 km $s^{-1}$, the minimum velocity required to escape the planet's gravitational field. These atoms can boil off, entering the interplanetary medium at rates which depend only on

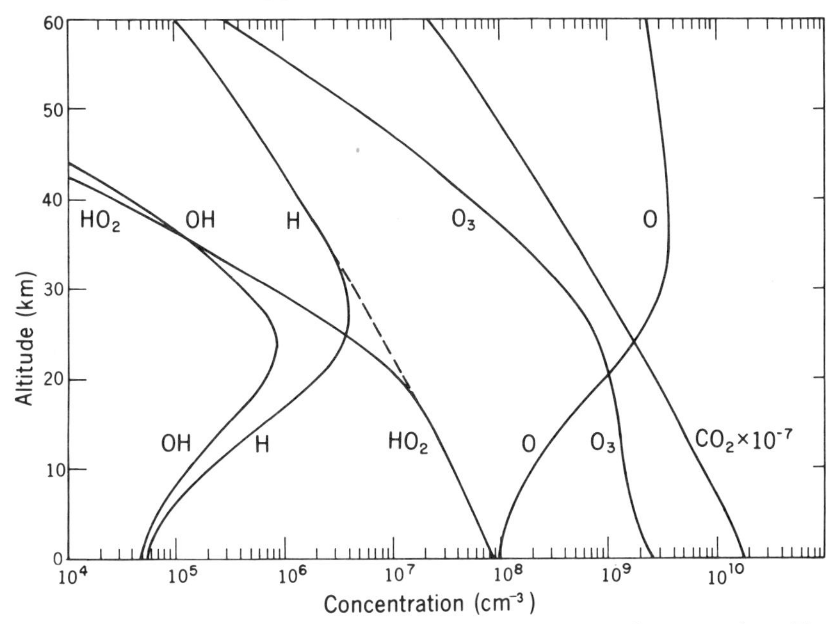


**Figure 5.23** Concentrations of principal constituents in the Martian atmosphere. The surface temperature is 220 K. The eddy diffusion coefficient is $15 \times 10^8$ cm$^2$ s$^{-1}$. The curve for odd hydrogen, shown as a broken line between the portions labelled $HO_2$ and H, is plotted for a density of odd hydrogen equal to $5 \times 10^{-10}$ times the $CO_2$ density. The $O_2$ and CO densities are, respectively $1.3 \times 10^{-3}$ and $8 \times 10^{-4}$ times the $CO_2$ density (After McElroy and Donahue[172], by courtesy of the American Association for the Advancement of Science)

temperature and on the density of atoms at the base of the exosphere*. The escaping H atoms are supplied by photolysis of $H_2O$. For every two H atoms which escape to space, one O atom is liberated, and it is clear that the escape process could provide a potent source of atmospheric $O_2$. It could account for the oxygen now present in the atmosphere in a time interval as short as $2 \times 10^5$ years. It could supply an abundance of oxygen similar to that in the Earth's atmosphere in as little as $3 \times 10^9$ years. We shall argue that the small abundance of oxygen in the contemporary atmosphere is a consequence of escape, which takes place by way of a non-thermal mechanism, driven by chemical reactions in the exosphere. It seems probable that the chemistry of the Martian atmosphere and the escape of H are controlled primarily by recombination of exospheric $O_2{}^+$, which proceeds mainly[19] by

$$O_2{}^+ + e^- \rightarrow O(^3P) + O(^1D) \qquad (5.124)$$

with an additional contribution due to

$$CO_2{}^+ + e^- \rightarrow CO + O \qquad (5.125)$$

*The exosphere is a region in which collision mean-free paths are sufficiently large that atoms may be assumed to describe simple ballistic orbits. The base of the exosphere, by convention, is taken at the level for which the collision mean-free path is equal to an atmospheric scale height. An atom moving up through the exobase with velocity equal to the escape velocity has a probability $1/e$ that it should escape to space without suffering further collisions.

The oxygen ions in (5.124) are formed by

$$h\nu + CO_2 \rightarrow CO_2^+ + e^- \tag{5.126}$$

and

$$h\nu + O \rightarrow O^+ + e^- \tag{5.127}$$

followed by

$$CO_2^+ + O \rightarrow CO + O_2^+ \tag{5.128}$$

and

$$O^+ + CO_2 \rightarrow O_2^+ + CO \tag{5.129}$$

The atoms in (5.124) are produced with an initial energy of 2.5 eV, significantly higher than the energy required for O escape, 1.99 eV. The oxygen atoms released in (5.125) can also escape and the net emission rate should be equal to approximately half the rate for ionisation of gas in the sunlit exosphere. McElroy[169] estimated an average rate for oxygen loss equal to $6 \times 10^7$ atoms $cm^{-2}$ $s^{-1}$, which could balance a mean rate for hydrogen escape of $1.2 \times 10^8$ atoms $cm^{-2}$ $s^{-1}$, in good agreement with the flux measured by Anderson and Hord[5].

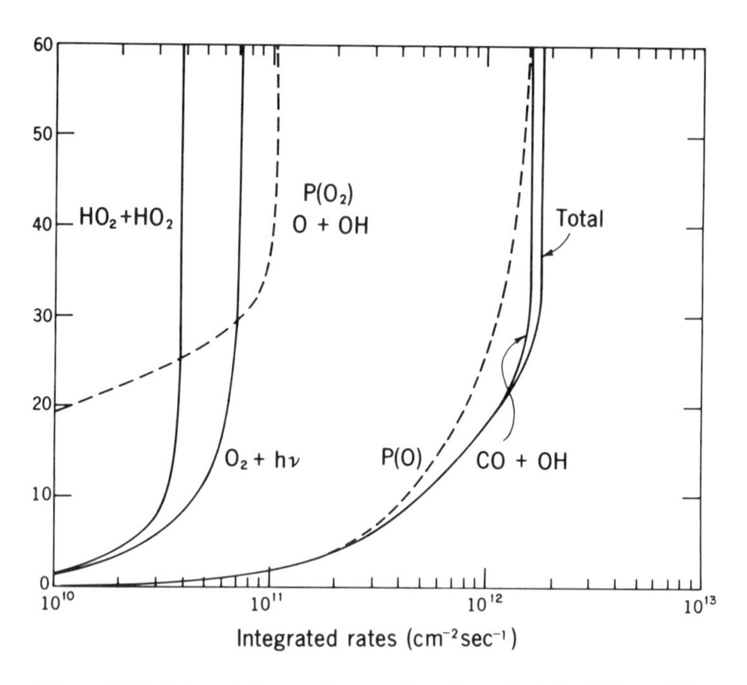


**Figure 5.24** Integrated rates for reactions important in $CO_2$ and $O_2$ formation and loss. The rates are integrated from the Martian surface to height $z$. The curve labelled P(O) is the integrated photolysis rate for $CO_2$. The contribution from $O + O + CO_2$ is only $2.3 \times 10^9$ $cm^2$ $s^{-1}$ and is not shown (After McElroy and Donahue[172], by courtesy of the American Association for the Advancement of Science)

The apparent balance between the escape rates for H and O is not accidental. It results from a complex interaction, involving the chemistry of the lower atmosphere. This interaction may be illustrated as follows*. Consider the consequence of an escape rate for H somewhat larger than the value which applies today. Excess hydrogen escape should lead to a net source for atmospheric $O_2$. The level of $O_2$ should rise, and the chemistry of the $CO_2$–$H_2O$ system should adjust accordingly. There should be a net decrease in the concentration of H at low altitudes, with an associated rise in the concentration of $HO_2$. Production of $H_2$, through (5.123), should decrease, with a consequent drop in the rate at which H atoms were supplied to the upper atmosphere. The process should continue, until the imbalance had been removed.

The system is similarly stable with respect to perturbations in the other direction. A decrease in the escape rate for H should lead to a change in the abundance of $O_2$, with a consequent rise in the production of $H_2$. The production of upper atmospheric H should increase, with appropriate change in the rate for escape of H, until the imbalance is removed as before. Molecular hydrogen acts as a powerful buffer in this cycle. It regulates the rate at which H atoms are released to the upper atmosphere. It ensures that, in an average sense, atoms should be supplied at precisely the rate required to balance escape of O. The buffer is relatively stable. The time constant associated with $H_2$ is *ca.* $10^3$ years, which may be compared with the much shorter times, of the order of 10 years, which apply for $O_2$ and CO, as discussed above.

It seems clear that exceedingly small concentrations of $H_2O$, H, OH and $HO_2$ can play a critical role in the chemistry of the Martian atmosphere. As we shall see below, catalytic chemistry appears to be implicated also for Venus.

## 5.10 VENUS

The atmosphere of Venus is composed mainly of $CO_2$, with trace quantities of $H_2O$, CO, HCl and HF. The abundance of CO is *ca.* $5 \times 10^{-5}$ that of $CO_2$, and could be produced photochemically in about 200 years. The abundance of $O_2$ is even smaller, less than $10^{-6}$ that of $CO_2$ according to Traub and Carleton[245]. We shall argue here that oxygen atoms produced above 85 km are removed by reactions which result in the formation of $O_2$, reactions (5.84), (5.105) and (5.117). Atoms produced below 85 km are removed by (5.112)–(5.114), with an additional contribution due to (5.110) followed by (5.83) and

$$ClO + CO \rightarrow Cl + CO_2 \tag{5.130}$$

with the latter equivalent to

$$(Cl) + O + CO \rightarrow (Cl) + CO_2 \tag{5.131}$$

Molecular oxygen formed above 85 km is transported downward, and

*The description here follows McElroy[169,170] and McElroy and Donahue[172]. The mechanism is amplified further by Liu and Donahue[152], who recently reported solutions to the coupled diffusion equations for CO, $O_2$, $H_2$, $H_2O_2$, odd hydrogen and oxygen.

removed by 2 × (5.114) + (5.118) + (5.119) + 2 × (5.113), equivalent to (5.120), with further loss associated with (5.114) followed by

$$Cl + HO_2 \rightarrow ClO + OH \quad (5.132)$$

and (5.113) + (5.126). The latter is equivalent to

$$(H) + (Cl) + O_2 + 2CO \rightarrow (H) + (Cl) + 2CO_2 \quad (5.133)$$

Free hydrogen is formed by

$$h\nu + HCl \rightarrow H + Cl \quad (5.134)$$

followed by

$$Cl + H_2 \rightarrow HCl + H \quad (5.135)$$

a sequence equivalent to

$$h\nu + H_2 \rightarrow H + H \quad (5.136)$$

Hydrogen chloride acts as a catalyst for the production of free hydrogen, a possibility noted first by McElroy[168] and elaborated further by Prinn[201].

Our interest here is directed primarily towards chemical processes which take place in the upper atmosphere, above the visible cloud deck. Venus is surrounded by an opaque blanket of whitish yellowish clouds which limit the penetration of visible solar radiation to altitudes near 70 km. According to recent ideas[198,202,225,266,271,273], these clouds are composed of partially hydrated $H_2SO_4$, and it seems probable that they are formed photochemically, with sulphur released by photolysis of COS [202]. Oxygen would be supplied most likely by photolysis of $CO_2$. Thus the chemistry of the upper atmosphere may be linked directly to the chemistry of the clouds. It is thought that the high temperatures, *ca.* 750 K, which prevail near the surface of Venus, result, at least in part, from a greenhouse effect regulated by the clouds. In this case the bulk chemistry of the lower atmosphere would be controlled, in an ultimate sense, by photolysis of $CO_2$ at wavelengths below 2000 Å, a remarkable situation of considerable scientific and perhaps practical significance.

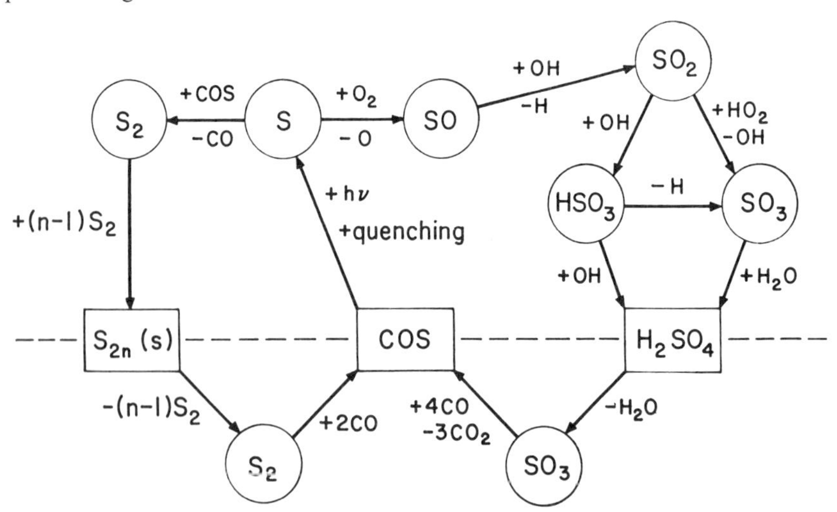


**Figure 5.25** Schematic representation of sulphur chemistry on Venus

Prinn's[203] model for the Venus cloud is illustrated schematically in Figure 5.25. Sulphur is formed by the reaction

$$h\nu + COS \rightarrow CO + S(^1D) \quad (5.137)$$

with COS released by high-temperature heterogeneous chemistry at the planetary surface[147]. The column rate for dissociation of COS is *ca.* $10^{13}$ molecules $cm^{-2}$ $s^{-1}$, and $H_2SO_4$ is formed by the reaction

$$S + O_2 \rightarrow SO + O \quad (5.138)$$

followed by

$$SO + OH \rightarrow SO_2 + H \quad (5.139)$$

$$SO_2 + HO_2 \rightarrow SO_3 + OH \quad (5.140)$$

and

$$SO_3 + H_2O \rightarrow H_2SO_4 \quad (5.141)$$

Sulphur may react also with COS to form $S_2$, either by

$$S(^1D) + COS \rightarrow S_2(^1\Delta) + CO \quad (5.142)$$

or by

$$S + COS \rightarrow S_2 + CO \quad (5.143)$$

and Prinn[203] points out that the Venus cloud may contain significant concentrations of elemental sulphur which could account for the dark markings observed in u.v. photographs of the planet[66,184,231]. The sulphuric acid droplets formed in (5.141) would diffuse slowly downward, and could be removed by thermal reactions with CO which should result in the formation of $CO_2$, COS and $H_2O$ at rates sufficient to balance the high altitude dissociation of these gases.

The major question in Prinn's scheme concerns the rate for reaction (5.138). For every two molecules of $H_2SO_4$ incorporated in the cloud we must supply three molecules of $O_2$, two molecules of COS, and two molecules of $H_2O$. The bulk reaction involved in the formation of the cloud may be written in the form

$$3O_2 + 2COS + 2H_2O \rightarrow 2H_2SO_4 + 2CO \quad (5.144)$$

The necessary source of free oxygen may be derived either by supply from above, or by *in situ* reactions such as

$$S(^1D) + CO_2 \rightarrow SO + CO \quad (5.145)$$

and

$$S + CO_2 \rightarrow SO + CO \quad (5.146)$$

These reactions were discussed briefly by Wofsy and Sze[266], who thought it unlikely that they could contribute significantly to the Venus oxygen problem. Further laboratory data would be desirable, either to confirm or deny the importance of (5.145) and (5.146).

The reaction scheme used by Sze and McElroy[243] to describe the chemistry of the upper atmosphere is summarised in Table 5.13. The rates for the various reactions included here are relatively well known, with several notable exceptions, as discussed below. Sze and McElroy[243], in their standard model,

**Table 5.13 Reactions and rate constants of relevance for Venus***

| *Reaction* | *Rate constant* |
|---|---|
| 1. $h\nu + CO_2 \rightarrow CO + O$ | $J_1 = 1 \times 10^{-6}$ |
| 2. $h\nu + O_2 \rightarrow O + O$ | $J_2 = 5 \times 10^{-6}$ |
| 3. $h\nu + O_3 \rightarrow O_2 + O$ | $J_3 = 8.8 \times 10^{-3}$ |
| 4. $h\nu + HCl \rightarrow H + Cl$ | $J_4 = 3.9 \times 10^{-6}$ |
| 5. $h\nu + H_2O \rightarrow H + OH$ | $J_5 = 9.2 \times 10^{-6}$ |
| 6. $h\nu + H_2O_2 \rightarrow 2OH$ | $J_6 = 1.1 \times 10^{-4}$ |
| 7. $h\nu + Cl_2 \rightarrow 2Cl$ | $J_7 = 2.7 \times 10^{-3}$ |
| 8. $CO + O + CO_2 \rightarrow 2CO_2$ | $k_1 = 2 \times 10^{-37}$ |
| 9. $O + O + CO_2 \rightarrow O_2 + CO_2$ | $k_2 = 3 \times 10^{-33}(T/300)^{-2.9}$ |
| 10. $H + O_2 + CO_2 \rightarrow HO_2 + CO_2$ | $k_3 = 2 \times 10^{-31}(T/273)^{-1.3}$ |
| 11. $O + HO_2 \rightarrow OH + O_2$ | $k_4 = 7 \times 10^{-11}$ |
| 12. $CO + OH \rightarrow CO_2 + H$ | $k_5 = 9 \times 10^{-13}\exp(-500/T)$ |
| 13. $H + O_3 \rightarrow OH + O_2$ | $k_6 = 2.6 \times 10^{-11}$ |
| 14. $O + O_2 + CO_2 \rightarrow O_3 + CO_2$ | $k_7 = 1.4 \times 10^{-33}(T/300)^{-2.5}$ |
| 15. $O + OH \rightarrow H + O_2$ | $k_8 = 5 \times 10^{-11}$ |
| 16. $HO_2 + HO_2 \rightarrow H_2O_2 + O_2$ | $k_9 = 9.5 \times 10^{-12}$ |
| 17. $H + HO_2 \rightarrow H_2 + O_2$ | $k_{10} = 1 \times 10^{-11}$ |
| 18. $OH + HO_2 \rightarrow H_2O + O_2$ | $k_{11} = 2 \times 10^{-10}$ |
| 19. $H + HO_2 \rightarrow 2OH$ | $k_{12} = 3 \times 10^{-11}$ |
| 20. $H + H + CO_2 \rightarrow H_2 + CO_2$ | $k_{13} = 2.6 \times 10^{-32}$ |
| 21. $Cl + H_2 \rightarrow HCl + H$ | $k_{14} = 8 \times 10^{-11}\exp(-2480/T)$ |
| 22. $Cl + Cl + CO_2 \rightarrow Cl_2 + CO_2$ | $k_{15} = 2.7 \times 10^{-32}$ |
| 23. $Cl + O_3 \rightarrow ClO + O_2$ | $k_{16} = 1.8 \times 10^{-11}$ |
| 24. $Cl + O_2 + CO_2 \rightarrow ClOO + CO_2$ | $k_{17} = 1.7 \times 10^{-33}$ |
| 25. $Cl + ClOO \rightarrow Cl_2 + O_2$ | $k_{18} = 1.5 \times 10^{-10}$ |
| 26. $Cl + ClOO \rightarrow ClO + ClO$ | $k_{19} = 1.44 \times 10^{-12}$ |
| 27. $O + ClO \rightarrow O_2 + Cl$ | $k_{20} = 5.3 \times 10^{-11}$ |
| 28. $ClO + H_2 \rightarrow$ product | $k_{12} < 5 \times 10^{-16}$ |
| 29. $ClO + O_3 \rightarrow ClOO + O_2$ | $k_{22} < 5 \times 10^{-15}$ |
| 30. $ClO + ClO \rightarrow Cl_2 + O_2$ | $k_{23} = 2.3 \times 10^{-14}$ |
| 31. $ClO + ClO \rightarrow Cl + ClOO$ | $k_{24} = 1.2 \times 10^{-12}\exp(-1260/T)$ |
| 32. $ClO + ClO + CO_2 \rightarrow Cl_2 + O_2 + CO_2$ | $k_{25} = 1 \times 10^{-31}$ |
| 33. $O + Cl_2 \rightarrow ClO + Cl$ | $k_{26} = 9.3 \times 10^{-21}\exp(-1560/T)$ |
| 34. $OH + HCl \rightarrow OH + Cl$ | $k_{27} = 2 \times 10^{-13}\exp(-310/T)$ |
| 35. $O + HCl \rightarrow OH + Cl$ | $k_{28} = 1.75 \times 10^{-12}\exp(-2260/T)$ |
| 36. $ClOO + CO_2 \rightarrow Cl + O_2 + CO_2$ | $k_{29} = 10^{-15}$ |
| 37. $ClO + CO \rightarrow Cl + CO_2$ | $k_{30} = 1.7 \times 10^{-15}$ |
| 38. $ClOO + CO \rightarrow ClO + CO_2$ | $k_{31} \approx 10^{-15}$ |
| 39. $Cl + HO_2 \rightarrow HCl + O_2$ | $k_{32} = 10^{-13}$ |
| 40. $Cl + HO_2 \rightarrow ClO + OH$ | $k_{33} = 10^{-13}$ |
| 41. $H + HCl \rightarrow H_2 + Cl$ | $k_{34} = 1 \times 10^{-11}\exp(-1605/T)$ |
| 42. $OH + H_2 \rightarrow H_2O + H$ | $k_{35} = 6.7 \times 10^{-12}\exp(-2010/T)$ |
| 43. $O + H_2 \rightarrow OH + H$ | $k_{36} = 5.1 \times 10^{-11}\exp(-5100/T)$ |
| 44. $OH + O_3 \rightarrow H + O_2$ | $k_{37} = 1.3 \times 10^{-12}\exp(-950/T)$ |
| 45. $HO_2 + O_3 \rightarrow OH + 2O_2$ | $k_{38} = 1.3 \times 10^{-12}\exp(-820/T)$ |
| 46. $O + O_3 \rightarrow 2O_2$ | $k_{39} = 1.3 \times 10^{-11}\exp(-2140/T)$ |
| 47. $CO_2^+ + H_2 \rightarrow HCO_2^+ + H$ | $k_{40} = 1 \times 10^{-9}$ |

*For detailed references see Sze and McElroy[243]. Rates for two-body reactions are given in units $cm^3\ s^{-1}$, and rates for three-body reactions have units $cm^6\ s^{-1}$

assumed that the rates for (5.132), and for the analogous reaction leading to formation of HCl

$$Cl + HO_2 \rightarrow HCl + O_2 \qquad (5.147)$$

were comparable, and they adopted temperature-independent rates in both cases equal to $10^{-13}$ $cm^3$ $s^{-1}$. A rapid rate for (5.147) would inhibit the catalytic scheme for recombination of CO and $O_2$, and would permit a

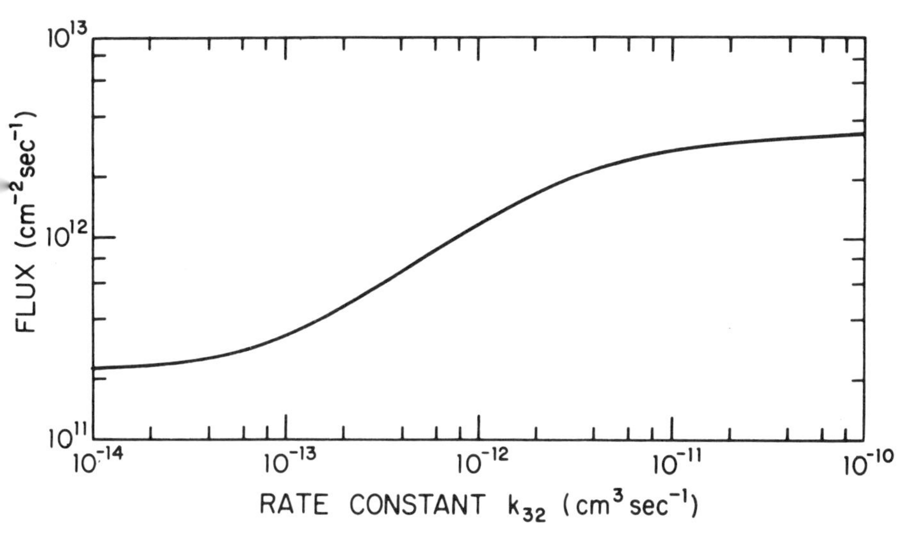


**Figure 5.26** Downward flux of $O_2$ at 62 km as a function of the rate constant for the reaction: $Cl + HO_2 \rightarrow HCl + O_2$ (After Sze and McElroy[243], by courtesy of Pergamon Press)

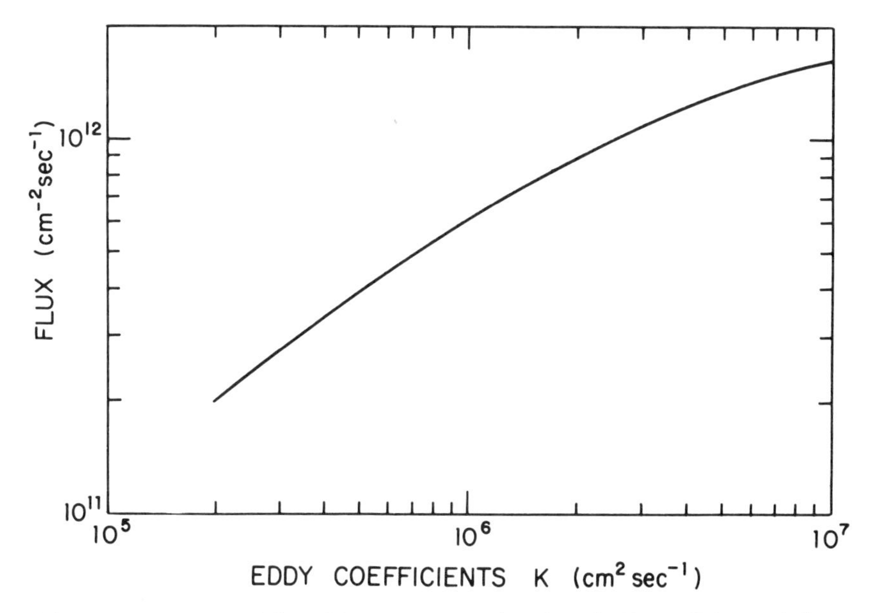


**Figure 5.27** Downward flux of $O_2$ at 62 km as a function of eddy coefficient $K$ (After Sze and McElroy[243], by courtesy of Pergamon Press)

relatively larger flux of $O_2$ to penetrate to low altitudes, where it could participate in the cloud chemistry summarised by (5.137)–(5.146). Sze and McElroy[243] explored the manner in which their results might be affected by a change in the rate for (5.132). The results of this exercise are summarised in Figure 5.26.

The low altitude flux of $O_2$ is a sensitive function also of the value adopted for the eddy mixing coefficient $K$, as illustrated by Figure 5.27. The variation in the height profile for $O_2$, as a function of $K$ with the standard chemical model in Table 5.13, is shown in Figure 5.28, while profiles for O, $O_2$, $O_3$,

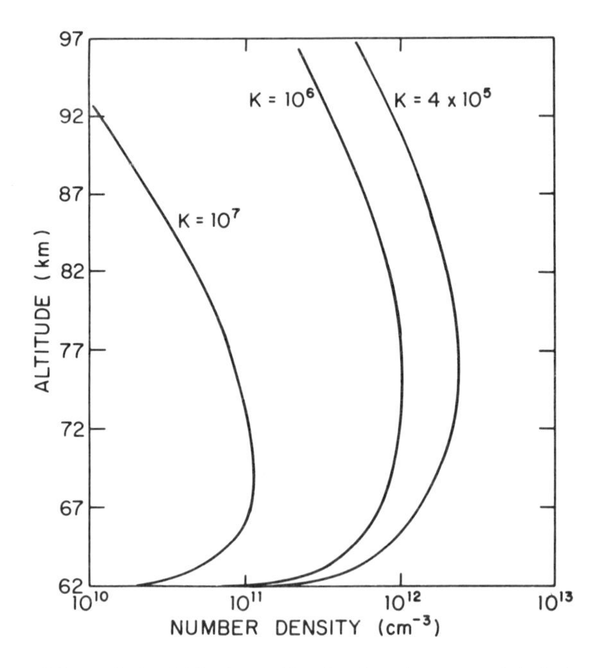


**Figure 5.28** Altitude profiles for $O_2$. Curves are labelled by different values assumed for the eddy coefficient $K$, where $K$ is taken independent of altitude. The mixing ratio of $O_2$ is set equal to zero at 62 km (After Sze and McElroy[243], by courtesy of Pergamon Press)

H, OH, $HO_2$, $H_2O_2$, HCl, Cl, ClO, ClOO and $Cl_2$ are shown in Figures 5.29 and 5.30. The results in Figures 5.29 and 5.30 were obtained with

$$K(z) = 6.5 \times 10^{14}[n(z)]^{-\frac{1}{2}} \quad z \leqslant z^* \tag{5.148}$$

$$= K^* \quad z \geqslant z^*$$

where

$$K(z) = K^* \tag{5.149}$$

and $n(z)$ indicates the total molecular density at $z$. The functional form denoted by (5.148) + (5.149) should be appropriate if mixing is due primarily to upward propagation of internal gravity waves[149,264]. The quantity $K^*$

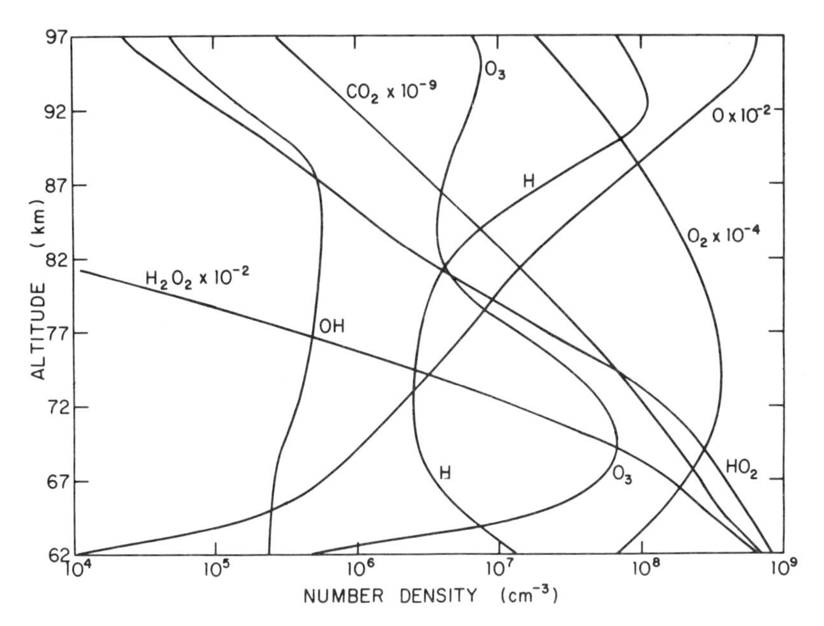


**Figure 5.29** Number densities of O, $O_2$, $O_3$, H, OH, $HO_2$, $H_2O_2$ and $CO_2$ (After Sze and McElroy[243], by courtesy of Pergamon Press)

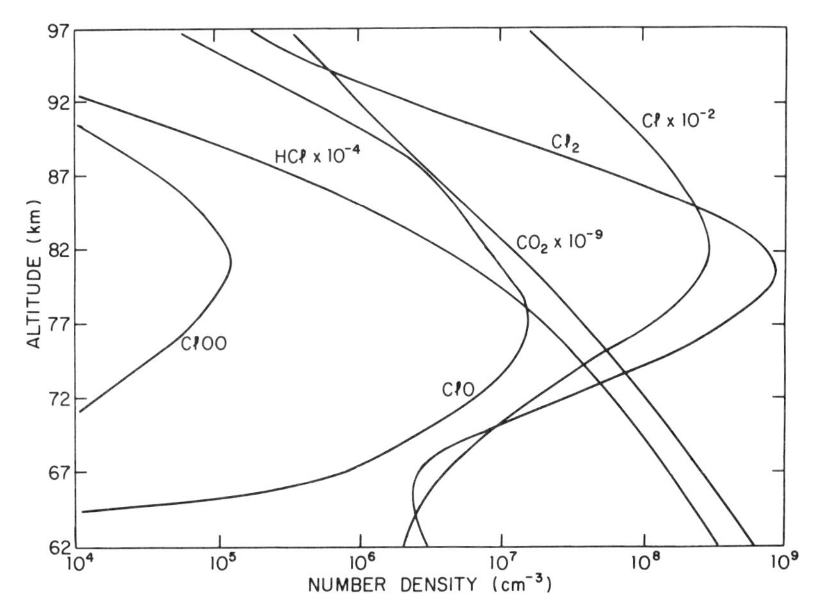


**Figure 5.30** Number densities of HCl, Cl, ClO, ClOO, $Cl_2$ and $CO_2$ (After Sze and McElroy[243], by courtesy of Pergamon Press)

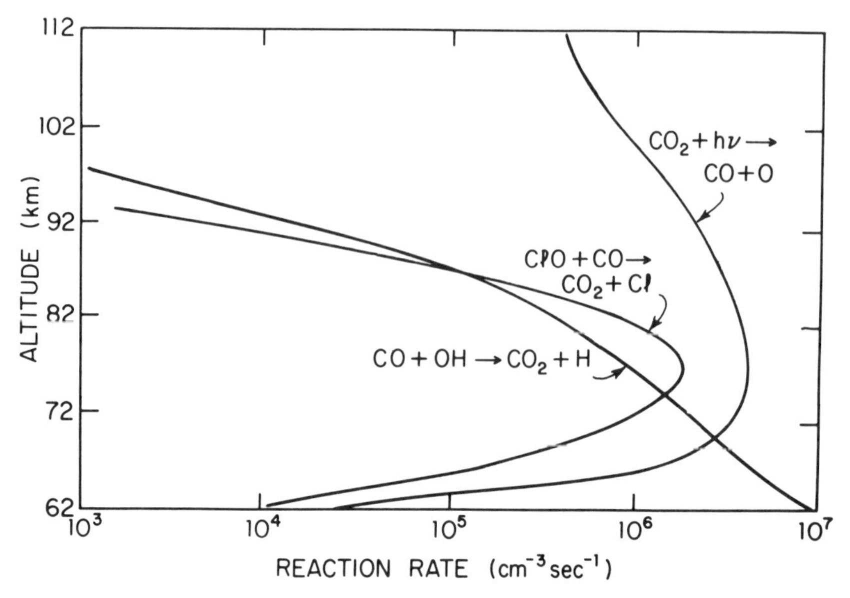


**Figure 5.31** Reaction rates for key reactions leading to production ($CO_2 + h\nu$) and removal ($CO + OH$ and $CO + ClO$) of CO (After Sze and McElroy[243], by courtesy of Pergamon Press)

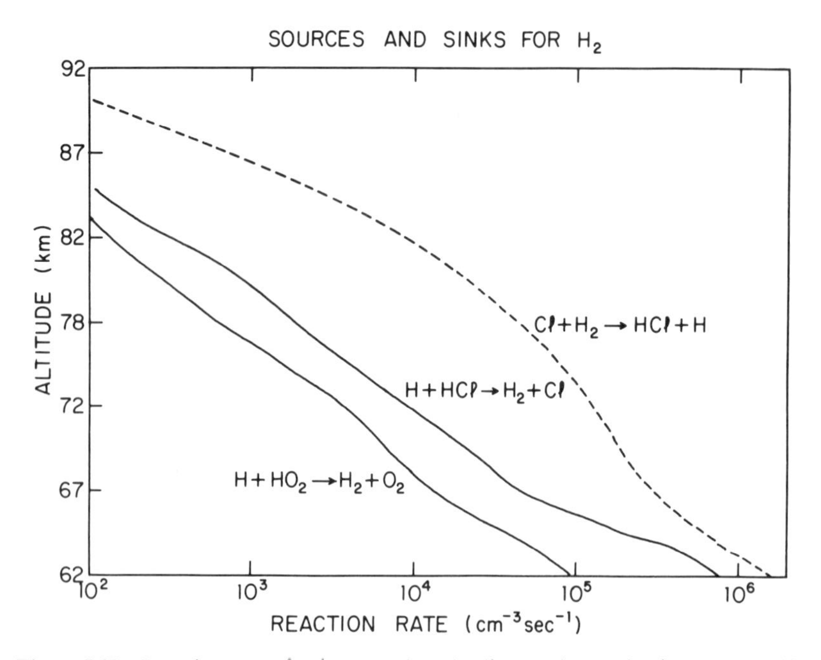


**Figure 5.32** Reaction rates for key reactions leading to the production ($H + HCl$ and $H + HO_2$) and removal ($Cl + H_2$) of $H_2$ (After Sze and McElroy[243], by courtesy of Pergamon Press)

was set equal to $6 \times 10^6$ cm$^2$ s$^{-1}$ for the computations summarised by Figures 5.29–5.35.

The results in Figures 5.29 and 5.30 were obtained by numerical solution of a set of coupled equations analogous to (5.79) + (5.80). They allowed for chemistry and transport of odd hydrogen ($H + OH + HO_2 + H_2O_2$), odd

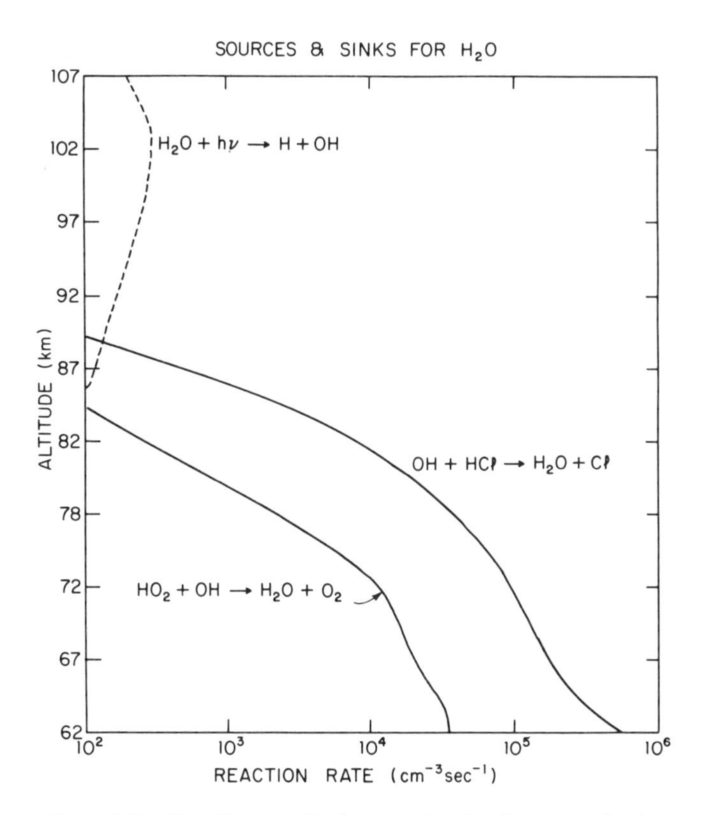


**Figure 5.33** Reaction rates for key reactions leading to production ($OH + HCl$ and $HO_2 + OH$) and removal ($H_2O + h\nu$) of $H_2O$ (After Sze and McElroy[243], by courtesy of Pergamon Press)

oxygen ($O + O_3$), free chlorine ($Cl + Cl_2 + ClO + ClOO$), $H_2$, $H_2O$, HCl, CO and $O_2$. Production and loss rates for CO as a function of altitude, with the standard chemical model and with $K^*$ set equal to $6 \times 10^6$ cm$^2$ s$^{-1}$, are shown in Figure 5.31, while similar data are presented in Figures 5.32–5.34 for $H_2$, $H_2O$ and HCl. Figure 5.35 gives a summary of various reactions which involve formation and destruction of an O—O bond. Molecular hydrogen is formed mainly by (5.35) and removed by (5.135). Reaction (5.87) is a major source for stratospheric $H_2O$, and an important sink for HCl. Reaction (5.84) is the predominant source for $O_2$, and the O—O bond is destroyed by (5.119), with additional contributions due to (5.132) and (5.138).

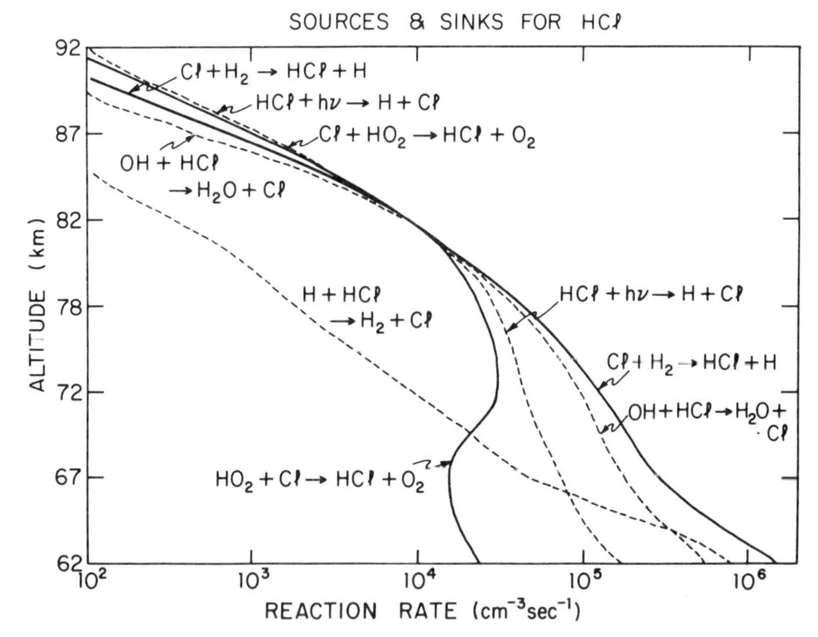


**Figure 5.34** Reaction rates for key reactions leading to production (Cl + $H_2$ and HO + Cl) and to removal (OH + HCl, HCl + *hν* and H + HCl) (After Sze and McElroy[243], by courtesy of Pergamon Press)

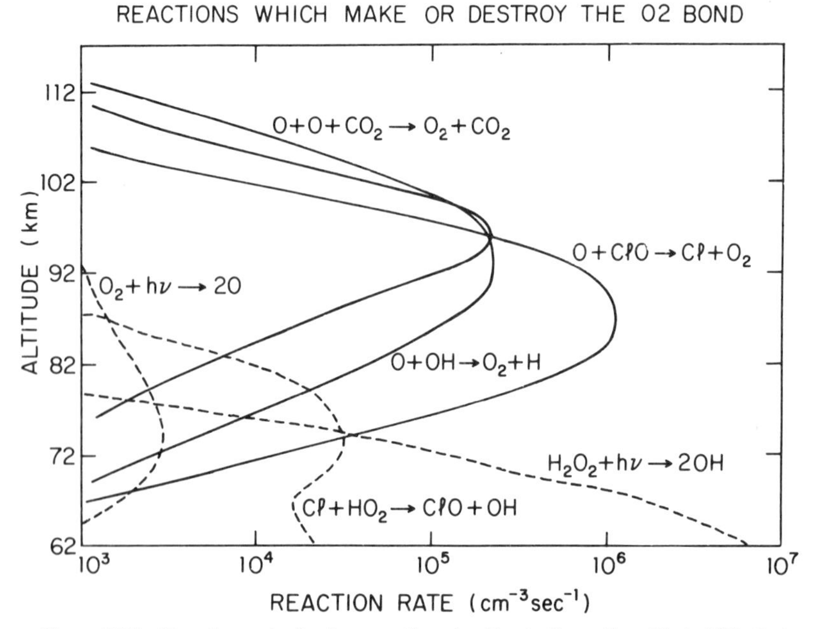


**Figure 5.35** Reaction rates for key reactions leading to formation (O + ClO, O + OH and O + O + CO)$_2$ and to the destruction of the O—O bond ($HO_2$ + $HO_2$, $O_2$ + *hν* and Cl + $HO_2$) (After Sze and McElroy[243], by courtesy of Pergamon Press)

## 5.11 JUPITER

Jupiter is the largest planet in the solar system, with a total mass equal to approximately 318 times that of the Earth. The planet has an equatorial radius[109] of 71 600 km and a mean density of 1.314 g $cm^{-3}$. Its density is such that hydrogen and helium must be major atmospheric constituents, and most models for Jupiter implicitly assume that the planet is composed of material of essentially solar composition.

A summary of the best current information on composition is given in Table 5.14. With the exception of the result for helium, the table is based on analyses of data obtained with Earth-based telescopic equipment. The results for $H_2$, $CH_4$, $NH_3$ and $CH_3D$ were derived from measurements of reflected sunlight, which dominates Jupiter's spectrum at wavelengths less than 5 μm. The results for $C_2H_5$, $C_2H_2$ and $PH_3$ were obtained from studies of the emission spectrum at longer wavelengths. Molecular hydrogen was detected first by Kiess *et al.*[136]. Helium was observed only recently, by Judge and Carlson[127], with instrumentation on the deep-space probe Pioneer 10, and the results for $CH_4$, $NH_3$, $CH_3D$, $C_2H_6$ and $C_2H_4$ given in Table 5.14 were taken from Belton[14], Owen[191], Beer *et al.*[12] and Ridgway[207].

**Table 5.14 Composition of Jupiter's atmosphere**

| *Gas* | *Abundance*$^a$ | *Mixing ratio*$^b$ | *Wavelength*$^c$ |
|---|---|---|---|
| $H_2$ | 67 | 1 | 0.82 |
| He | | 0.2 | 0.06 |
| $CH_4$ | 45 | $\sim 10^{-3}$ | 1.1 |
| $NH_3$ | 13 | $\sim 10^{-4}$ | 1.1 |
| $CH_3D$ | 2.6 | $\sim 10^{-7}$ | 4.6 |
| $C_2H_6$ | $10^{-4}$ | $4 \times 10^{-3d}$ | 12 |
| $C_2H_2$ | $2 \times 10^{-6}$ | $8 \times 10^{-5d}$ | 13 |
| $PH_3$ | $^e$ | $^e$ | |
| $H_2O$ | $^f$ | $^f$ | 5 |

$^a$ The abundance of $H_2$ is expressed in km atm, while the abundance of $CH_3D$ is given in cm atm. Abundances for all other species are in m atm
$^b$ Mixing ratio by volume with respect to $H_2$
$^c$ Approximate wavelength (μm) at which the gas was detected
$^d$ These abundances and ratios are uncertain
$^e$ Observed by S. T. Ridgway, but abundance not determined as yet
$^f$ Positive detection reported by R. Treffers *et al.*

The abundances quoted here refer to concentrations of individual species above the visible cloud deck. Jupiter appears to be surrounded by at least two distinct cloud layers. The upper layer is optically thin and diffuse, composed most probably of frozen ammonia. The lower layer, according to thermochemical studies by Lewis[145,146], may be formed from a dense blanket of $NH_4SH$, and there may be additional cloud layers at lower altitudes containing water, aqueous ammonia, ammonia chloride and silicates. For recent discussions, emphasising the physical and chemical properties of Jupiter's atmosphere, the reader is referred to McElroy[167,171], Danielson and Tomasko[57], Axel[9], Hunt[110] and Newburn and Gulkis[185]. The discussion in the remaining part of this section is restricted to chemical processes in the upper atmosphere, and is patterned after extensive theoretical studies by Strobel[236–239].

Dissociation of $CH_4$ is energetically possible at wavelengths below 2700 Å, but proceeds to a significant extent only at wavelengths shorter than 1600 Å. Approximately 70% of the total $CH_4$ photolysis takes place at Lyman $\alpha$. The primary paths are

$$h\nu + CH_4 \rightarrow {}^1CH_2 + H_2 \quad (5.150)$$

and

$$h\nu + CH_4 \rightarrow CH + H + H_2 \quad (5.151)$$

for which the branching ratios are 92% and 8%, respectively[240].

The radical $^1CH_2$ is removed primarily by

$$^1CH_2 + H_2 \rightarrow CH_3 + H \quad (5.152)$$

and CH is lost either by

$$CH + CH_3 \rightarrow C_2H_4 + H \quad (5.153)$$

or by

$$CH + H_2 + M \rightarrow CH_3 + M \quad (5.154)$$

with (5.153) predominant in regions of the atmosphere where the density is less than $10^{15}$ cm$^{-3}$. The radical $CH_3$ is removed by

$$CH_3 + H + M \rightarrow CH_4 + M \quad (5.155)$$

and by

$$CH_3 + CH_3 + M \rightarrow C_2H_6 + M \quad (5.156)$$

Reaction (5.153) provides the major source for $C_2H_4$. Ethane is formed mainly by (5.156) and is relatively stable in the upper atmosphere of Jupiter, in contrast to ethylene which may be photolysed to give acetylene

$$h\nu + C_2H_4 \rightarrow C_2H_2 + H_2 \quad (5.157)$$

which in turn may polymerise, by reactions initiated by

$$h\nu + C_2H_2 \rightarrow C_2H_2^* \quad (5.158)$$

followed by

$$C_2H_2^* + C_2H_2 \quad C_4H_4 \quad (5.159)$$

The most abundant hydrocarbons, according to Strobel[239], in order of decreasing importance are $CH_4$, $C_2H_6$, $C_2H_2$, $C_2H_4$ and $CH_3$. Ethane, acetylene and hydrogen, formed in Jupiter's mesosphere and stratosphere, are transported downward and converted ultimately to $CH_4$ at the high temperatures and pressures which prevail below the visible cloud deck. The downward flux of $C_2H_6$, $C_2H_2$ and $H_2$ is balanced by a corresponding upward flux of $CH_4$.

The upper atmosphere contains significant concentrations of $NH_3$. Ammonia may be photolysed below 2300 Å, through

$$h\nu + NH_3 \rightarrow NH_2 + H \quad (5.160)$$

which may be followed either by

$$NH_2 + NH_2 + M \rightarrow N_2H_4 + M \quad (5.161)$$

or by

$$NH_2 + H + M \rightarrow NH_3 + M \quad (5.162)$$

In either case there should be significant production of $N_2H_4$, and Strobel[237] made the interesting suggestion that hydrazine formed in this manner could condense in Jupiter's stratosphere and might contribute to the high-altitude haze discussed by Danielson and Tomasko[47], Axel[9] and Hunt[110].

Photochemical studies of Jupiter are in their infancy as yet. Pioneers 10 and 11 flew by Jupiter in December 1973 and December 1974, respectively, with a battery of instruments designed primarily to map the particles and fields environment of the planet. Plans are now at an advanced stage for more intensive studies, from a pair of three-axis stabilised Mariner spacecraft to be launched in 1977 and 1978. The Mariner missions are expected to carry out an extensive programme of measurements, designed not only to advance our appreciation for Jupiter, but to further also our understanding of the major Jovian satellites. Jupiter is surrounded by a suite of 13 satellites, at least two of which, Io and Ganymede, are known to have thin atmospheric envelopes. Some indirect evidence suggests that Io may be covered with a layer of ammonia ice containing trace quantities of sodium, potassium and calcium[175,177]. Photolysis of $NH_3$ followed by escape of H could lead to a significant source of $N_2$ on Io. Likewise Ganymede appears to be covered with a layer of water ice. Photolysis of $H_2O$ could provide a significant source of $O_2$ in this case, and one might expect Ganymede to retain a thin atmosphere of $O_2$, with important trace quantities of $O_3$, O, H and $H_2$. Escape of oxygen from Ganymede might be regulated by a non-thermal mechanism[274] similar to that discussed earlier for Mars, with energetic atoms formed not only by (5.124), but with a significant contribution expected also due to photolysis, reaction (5.42).

## 5.12 CONCLUDING REMARKS

The topics in this review have ranged far and wide, from Ganymede to Venus, from discussions of primitive Earth to models for the possible perturbation of the present atmosphere by aerosol sprays, agricultural fumigants, and nitrogenous fertilisers. It has become clear in recent years that we cannot expect to develop a detailed appreciation for the Earth in isolation. Earth is a planet and we can hope to understand it fully only by viewing it in context.

Jupiter contains a relatively complete mix of the gases which must have been present in the primitive nebula. Its atmosphere is turbulent, with the turbulence driven by a powerful internal energy source. The planet emits somewhere between 2.0 and 2.5 times the amount of energy it absorbs from the sun[8,33]. Chemical processes in the upper atmosphere lead to synthesis of complex organic molecules, which are removed, however, rapidly by downward transport and consumed ultimately by high-temperature reactions below the clouds.

Venus receives less energy from the sun than does Earth, a curious consequence of its relatively high visual albedo. Despite this deficiency, the planet manages to maintain a surface temperature of 750 K. We suggested that the high surface temperature of Venus was a result, at least in part, of a greenhouse mechanism, in which thermal radiation from the lower atmosphere was trapped by gas and particles, whose optical properties allowed relatively deep penetration of a small amount of visible solar radiation. Could such a greenhouse mechanism occur for Earth? Is it possible that the increased concentrations of $CO_2$, $CFCl_3$, $CF_2Cl_2$, $CCl_4$, and other elements associated with modern technological society, could generate a terrestrial greenhouse of Venusian proportions? The terrestrial atmosphere contains within itself the seeds of an instability. A small rise in surface temperature could lead to an increase in the rate at which water evaporates from the ocean. An increase in the concentration of atmospheric $H_2O$ could lead to a more efficient greenhouse, which could raise the surface temperature, and so on. We cannot provide an unambiguous answer to this puzzle. We have learned only recently to pose the question.

Mars has abundant sources of water, bound chemically to its surface rocks. Could Mars in times past have enjoyed a more temperate climate? Is it possible that much of this water might have been released at one time to the atmosphere, that conditions might have been more favourable for the origin and evolution of life? We cannot say for certain. There is evidence, however, that large parts of the Martian surface were covered once with running liquid water. We know that violent dust storms spring up from time to time, enveloping the entire planet, and raising the mean surface temperature by as much as 50 K in some regions. Mars shows a variety of geologic features, a range of elevations which extend from the tip of Olympicus Mons, the largest volcano known to man, to the depths of Valles Marineris, a remarkable canyon, more than 5000 km long and no less than 6 km deep. What happened to the massive quantities of gas and particulate material which must have vented to the atmosphere during the last explosion of Olympicus? What was the fate of the vast quantity of solid matter which must have been moved during the formation of Valles Marineris? Our ignorance is such that we cannot answer these questions, any more than we can provide answers to the host of similar questions which arise when we attempt to describe the early history of Earth.

What was the complexion of the Earth prior to the origin and evolution of life? We argued here that the origin of life was inevitable, given the conditions which prevailed in the primitive atmosphere and ocean. If life were to terminate now, we would argue that it should begin again after a brief interlude, a lapse of a few billion years, assuming of course that the sun should continue to provide its beneficient supply of radiant energy. Much of the current atmosphere has been influenced and worked over by life, a concept elaborated imaginatively in recent years by Lovelock and Margulis[155]. It is difficult to define in non-controversial terms the nature of an abiological Earth.

The variety of threats to ozone pose serious questions for environmental science. The fertiliser problem is particularly challenging. Wasteful use of chemical fertiliser may set in motion a chain of chemical and biological processes which could lead ultimately to a rise in the level of atmospheric $N_2O$, with a consequent drop in the concentration of $O_3$. A change in the acidity of rain could have a similar effect, and there are reasons to believe that the perturbation in $N_2O$ may have already begun[86,222]. The rise in $N_2O$ observed during the sixties may be attributed to the general growth in industrial activity which took place during the early part of this century. We might expect the trend in $N_2O$ to stabilise, or to reverse, as recent environmental legislation takes effect. The major impact of chemical fertiliser should arise in the future, and we should prepare now to develop measures appropriate to minimise the ultimate environmental impact. The task is urgent, and, as noted earlier, it will require an unusual mixture of scientific, legal, political and economic skills.

## Acknowledgements

This research was supported in part by the Atmospheric Sciences Division of the National Science Foundation under Grant GA33990X, and by the National Aeronautics and Space Administration under contracts NAS-1-10492 and NASA NSG 2031 to Harvard University. I am indebted particularly to my colleagues S. C. Wofsy and Y. L. Yung who helped mould many of the ideas in this paper, to J. Elkins for his invaluable assistance, to H. D. Holland who protected me from several potential pitfalls in the discussion of the cycles for oxygen and nitrogen and to H. Brooks who stimulated much of the discussion of the possible role of fertiliser. Particular thanks are due to D. Herschbach for monumental patience and editorial understanding The work would not have seen the light of day without the patient cooperation of L. Karanewsky who suffered through many changes of style, handwriting and temperament, during the various iterations which preceded the final form of this manuscript.

## References

1. Albers, E. A., Hoyermann, K., Wagner, H. G. G. and Wolfrum, J. (1971). *Thirteenth International Symposium on Combustion*, 81 (Pittsburgh, Pa.: Combustion Institute)
2. Alexander, M. (1961). *Introduction to Soil Microbiology* (New York: Wiley)
3. Anderson, J. G. (1971). *J. Geophys. Res.*, **76,** 7820
4. Anderson, J. G. (1975). *Geophys. Res. Lett.*, **2,** 231
5. Anderson, D. E. and Hord, C. W. (1971). *J. Geophys. Res.*, **76,** 6666
6. Angel, J. K. and Korshover, J. (1973). *Mon. Wea. Rev.*, **101,** 426
7. Angström, A. and Högberg, L. (1952). *Tellus*, **4,** 271
8. Aumann, H. H., Gillespie, C. M. and Low, F. J. (1969). *Astrophys. J.*, **157,** L69
9. Axel, L. (1972). *Astrophys. J.*, **173,** 451
10. Bates, D. R. and Hays, P. B. (1967). *Planet. Spa. Sci.*, **15,** 189
11. Bates, D. R. and Nicolet, M. (1950). *J. Geophys. Res.*, **55,** 301
12. Beer, R. and Taylor, F. W. (1973). *Astrophys. J.*, **179,** 309
13. Bellar, T. A., Lichtenberg, J. J. and Kroner, R. C. (1974). *J. Amer. Water Works Assoc.*, **66,** 703
14. Belton, M. J. S. (1969). *Astrophys. J.*, **157,** 469
15. Bemand, P. P., Clyne, M. A. and Watson, R. T. (1973). *J. Chem. Soc. Faraday Trans. I*, **69,** 1356
16. Benson, S. W., Cruickshank, F. R. and Shaw, R. (1969). *Int. J. Chem. Kin.*, **1,** 29
17. Berkner, L. V. and Marshall, L. C. (1965). *The Origin and Evolution of Atmospheres and Oceans*, Chap. 6 (P. J. Brancazio and A. G. W. Cameron, editors) (New York: Wiley)
18. Biedenkapp, D., Hartshorn, L. G. and Bair, E. J. (1970). *Chem. Phys. Lett.*, **5,** 379
19. Biondi, M. A. (1969). *Can. J. Chem.*, **47,** 1711
20. Blake, D. and Lindzen, R. (1973). *Mon. Wea. Rev.*, **101,** 783

21. Bolin, B. and Eriksson, E. (1959). *The Atmosphere and Sea in Motion*, 130 (New York: The Rockefeller Institute Press)
22. Brasseur, G. and Nicolet, M. (1973). *Planet. Space Sci.*, **21,** 939
23. Broderick, A. J., Moreley English, J. and Forney, A. K. (1973). AIAA paper 73-508
24. Broecker, W. S. and Peng, T. H. (1974). *Tellus*, **26,** 21
25. Buat-Menard, P. (1970). *Thesis*, University of Paris
26. Buat-Menard, P. and Chesselet, R. (1971). *C.R. Acad. Sci. Paris*, **272,** 1330
27. Cadle, R. D. (1972). *Trans. Amer. Geophys. U.*, **53,** 812
28. Callear, A. B. and Pilling, M. J. (1970). *Trans. Faraday Soc.*, **66,** 1886
29. Campbell, I. M. and Thrush, B. A. (1967). *Proc. Roy. Soc. (London)*, **296,** 222
30. Capey, W. D., Majer, J. R. and Robb, J. C. (1968). *J. Chem. Soc. B*, 447
31. Chamberlain, J. W. (1963). *Planet. Space Sci.*, **11,** 901
32. Chapman, N. (1930). *Mem. Royal Meteor. Soc.*, **3,** 103
33. Chase, S. C., Ruiz, R. D., Munch, G., Neugebaner, X., Schroeder, M. and Trafton, L. M. (1974). *Science*, **183,** 315
34. Chesselet, R., Morelli, J. and Buat-Menard, P. (1972). *The Changing Chemistry of the Oceans*, 93 (New York: Wiley Interscience)
35. Chet, I. (1975). Personal communication
36. Cicerone, R. J., Stolarski, R. S. and Walters, S. (1974). *Science*, **185,** 1165
37. Clark, I. D. (1971). *J. Atmos. Sci.*, **28,** 847
38. Clark, I. D. and Noxon, J. F. (1970). *J. Geophys. Res.*, **75,** 7307
39. Clarke, F. W. (1924). *Bull. U.S. Geol. Surv.*, **770,** 841
40. Clere, M. and Reiffsteck, A. (1968). *J. Chem. Phys.*, **48,** 2799
41. Cline, J. D. (1973). *Thesis*, University of California, Los Angeles
42. Cloud, P. (1972). *Amer. J. Sci.*, **272,** 537
43. Clyne, M. A. A. and Coxon, J. A. (1966). *Trans. Faraday Soc.*, **62,** 2175
44. Clyne, M. A. A. and Walker, R. F. (1973). *J. Chem. Soc. Faraday Trans. I*, **69,** 1547
45. Clyne, M. A. A. and Watson, R. T. (1974). *J. Chem. Soc. Faraday Trans. I*, **70,** 2250
46. Clyne, M. A. A. and White, I. F. (1971). *Trans. Faraday Soc.*, **67,** 2068
47. Codispoti, L. A. (1972). *Thesis*, University of Washington
48. Colegrove, F. D., Johnson, F. S. and Hanson, W. B. (1966). *J. Geophys. Res.*, **71,** 2227
49. Craig, H. (1957). *Tellus*, **9,** 1
50. Crutzen, P. J. (1970). *Quart. J. Royal Meteor. Soc.*, **96,** 320
51. Crutzen, P. J. (1974). *Can. J. Chem.*, **52,** 1569
52. Crutzen, P. J., Isaksen, I. S. A. and Reid, G. C. (1975). *Science*, **189,** 457
53. Cunnold, D. M., Alyea, F. N., Phillips, N. A. and Prinn, R. G. (1974). *IAMAP/IAPSO Conference*, Melbourne, Australia, Jan. 14–25
54. Cunnold, D. M., Alyea, F., Phillips, N. and Prinn, R. (1975). *J. Atmos. Sci.* **32,** 170
55. Dahlberg, J. A. (1969). *Acta Chem. Scand.*, **23,** 3081
56. Dahlberg, J. A. and Kihlman, I. B. (1970). *Acta Chem. Scand.*, **24,** 644
57. Danielson, R. E. and Tomasko, M. G. (1969). *J. Atmos. Sci.*, **26,** 889
58. Davidson, N. (1951). *J. Amer. Chem. Soc.*, **73,** 468
59. Davis, D. D., Payne, W. A. and Stief, L. J. (1973). *Science*, **179,** 280
60. Davis, D. D., Watson, R. T., *et al.* (1975). Personal communication
61. Day, M. J., Stamp, D. V., Thompson, K. and Dixon-Lewis, G. (1971). *13th Symp. Combustion*, 705 (Pittsburgh, Pa.: The Combustion Institute)
62. Delany, A. C., Shedlovsky, J. P. and Pollack, W. H. (1974). *J. Geophys. Res.*, **79,** 5646
63. Delwiche, C. C. (1970). *The Biosphere* (San Francisco: W. H. Freeman)
64. Dickinson, R. E. (1973). *J. Geophys. Res.*, **78,** 4451
65. Dixon-Lewis, G., Wilson, W. E. and Westenberg, A. (1966). *J. Chem. Phys.*, **44,** 2877
66. Dollfus, A. (1968). *The Atmospheres of Venus and Mars* (J. C. Brand and M. B. McElroy, editors) (New York: Gordon and Breach)
67. Duce, R. A. (1969). *J. Geophys. Res.*, **74,** 4597
68. Duce, R. A., Wasson, J. T., Winchester, J. W. and Burns, F. (1963). *J. Geophys. Res.*, **68,** 3943
69. Duce, R. A., Winchester, J. W. and Van Nahl, T. W. (1965). *J. Geophys. Res.*, **70,** 1775
70. Duce, R. A., Zoller, W. H. and Moyers, J. L. (1973). *J. Geophys. Res.*, **78,** 7802
71. Durie, R. A. and Ramsey, P. A. (1958). *Can. J. Phys.*, **36,** 35
72. Emery, K. O., Orr, W. L. and Rittenberg, S. C. (1955). *Nutrient budgets in the ocean*, 299 (Los Angeles: University of Southern California Press)

73. Eriksson, E. (1952). *Tellus*, **4,** 215
74. Eriksson, E. (1960). *Tellus*, **11,** 63
75. Evans, W. F. J., Hunten, D. M., Llewellyn, E. J. and Vallance Jones, A. (1968). *J. Geophys. Res.*, **73,** 2885
76. Farmer, C. B. (1974). *Can. J. Chem.*, **52,** 1544
77. Felder, W., Morrow, W. and Young, R. A. *J. Geophys. Res.*, **75,** 7311
78. Foley, H. M. and Ruderman, M. A. (1973). *J. Geophys. Res.*, **78,** 4441
79. Fudali, R. F. (1965). *Geochim. Cosmochim. Acta*, **29,** 1063
80. Galante, T. J. and Gislason, E. A. (1973). *Chem. Phys. Lett.*, **18,** 231
81. Gauthier, M. and Snelling, D. R. (1971). *J. Chem. Phys.*, **54,** 4317
82. Gibson, G. E. and Bayliss, N. S. (1933). *Phys. Rev.*, **44,** 188
83. Gierasch, P. J. (1970). *Earth Extraterrestrial Sci.*, **1,** 171
84. Glueckauf, E. (1951). *Compendium Meteorology*, 3 (T. F. Malone, editor) (New York: American Meteorological Society)
85. Godiksen, P. H., Fairhall, A. W. and Reed, R. J. (1968). *J. Geophys. Res.*, **73,** 4461
86. Goody, R. M. (1969). *Planet. Space Sci.*, **17,** 1319
87. Gordon, S., Mulac, W. and Nangia, P. (1971). *J. Phys. Chem.*, **75,** 2087
88. Green, J. (1959). *Geol. Soc. Amer. Bull.*, **70,** 1127
89. Greenberg, R. I. and Heicklen, J. (1970). *Int. J. Chem. Kinetics*, **2,** 185
90. Griener, N. R. (1968). *J. Phys. Chem.*, **72,** 406
91. Greiner, N. R. (1969). *J. Chem. Phys.*, **51,** 5049
92. Griggs, M. (1968). *J. Chem. Phys.*, **49,** 857
93. Grobecker, A. J. (1974). *Acta Astronaut.*, **1,** 179
94. Hahn, J. (1974). *Tellus*, **26,** 160
95. Hall, T. C., Jr. and Blacet, F. E. (1952). *J. Chem. Phys.*, **20,** 1745
96. Hamilton, P. B., Shug, A. L. and Wilson, P. W. (1959). *Proc. Nat. Acad. Sci. U.S.A*, **43,** 297
97. Hampson, J. (1974). *Nature*, **250,** 189
98. Hardy, R. W. F. and Havelka, U. P. (1975). *Science*, **188,** 633
99. Haszeldine, R. N. and Nyman, F. (1959). *J. Amer. Chem. Soc.*, **81,** 387
100. Hering, W. S. and Borden, T. R. (1967). *Environ. Res. Papers*, Air Force Cambridge Res. Labs.
101. Hilsenrath, E., Seiden, L. and Goodman, P. (1969). *J. Geophys. Res.*, **74,** 6837
102. Hippler, H. and Troe, J. (1973). *Chem. Phys. Lett.*, **19,** 607
103. Hochanadel, C. J., Ghorley, J. A. and Ogren, P. J. (1972). *J. Chem. Phys.*, **56,** 4426
104. Holland, H. D. (1973). *Proc. Symp. Hydrogeochemistry and Biogeochemistry*, Vol. 1, 68 (Washington: The Clarke Co.)
105. Holland, H. D. (1975). Personal communication
106. Hoppe, H. A. (1969). *Marine Algae, A survey of Research and Utilization* (T. Levring, H. A. Hoppe and O. Schmid, editors) (New York: Cram, DeGruyter)
107. Howard, C. J. and Evenson, K. M. (1975). Personal communication
108. Howard, P. H. and Hanchett, A. (1975). *Science*, **189,** 217
109. Hubbard, W. B. and Van Flandern, T. C. (1972). *Astron. J.*, **77,** 65
110. Hunt, G. E. (1972). *Paper, AAS Meeting*, Kona, Hawaii
111. Hunten, D. M. (1971). *Space Sci. Rev.*, **12,** 539
112. Hunten, D. M. (1973). *J. Atmos. Sci.*, **30,** 726
113. Hunten, D. M. (1975). *4th CIAP Conference*, Cambridge, Mass.
114. Hunten, D. M. and McElroy, M. B. (1970). *J. Geophys. Res.*, **75,** 31
115. Hutchinson, G. E. (1944). *Amer. Sci.*, **32,** 178
116. Hutchinson, G. E. (1954). *The Earth as a Planet* (G. P. Kuiper, editor) (Chicago: University of Chicago Press)
117. Ingersoll, A. P. and Leovy, C. B. (1971). *Ann. Rev. Astron. Ap.*, **9,** 147
118. Jeans, H. H. (1925). *The Dynamical Theory of Gases* (London and New York: Cambridge University Press)
120. Johnson, F. S., Purcell, J. D., Tousey, R. and Watanabe, K. (1952). *J. Geophys. Res.*, **57,** 157
121. Johnston, H. S. (1968). *NSRDS Nat. Bur. Stand. Report*, No. 20
122. Johnston, H. S. (1971). *Science*, **173,** 517

123. Johnston, H. S., Whitten, G. and Birks, J. (1973). *J. Geophys. Res.*, **78,** 6107
124. Johnston, H. S., Kattenhorn, D. and Whitten, G. (1975). Lawrence Livermore Laboratory, LBL 3548
125. Jones, M. J. (1971). *Tellus*, **23,** 459
126. Jones, E. J. and Wulf, O. R. (1937). *J. Chem. Phys.*, **5,** 873
127. Judge, D. L. and Carlson, R. W. (1974). *Science*, **183,** 317
128. Huie, R. E., Herron, J. T. and Davis, D. D. (1972). *J. Phys. Chem.*, **76,** 2653
129. Junge, C. E. (1956). *Tellus*, **8,** 127
130. Junge, C. E. (1957). *Tellus*, **9,** 528
131. Junge, C. E. (1963). *Air Chemistry and Radioactivity* (New York: Academic Press)
132. Katakis, D. and Taube, H. (1962). *J. Chem. Phys.*, **36,** 416
133. Kaufman, F. (1964). *Ann. Geophys.*, **20,** 106
134. Kaufman, F. (1969). *Can. J. Chem.*, **47,** 1917
135. Kaufman, F. (1974). Paper presented at Summer Advanced Study Inst., Physics and Chemistry of Atmospheres, University of Liège, Belgium, July 29–August 9
136. Kiess, C. C., Corliss, C. H. and Kiess, H. K. (1960). *Astrophys. J.*, **132,** 221
137. Klingman, C. L. (1974). *Mineral Yearbook* (Washington: U.S. Dept. of the Interior)
138. Krueger, A. J., Heath, D. F. and Mateer, C. L. (1972). *NASA TM-X-66108, NTIS-N73-12379*
139. Kuiper, G. P. (1952). *The Atmospheres of the Earth and Planets* (G. P. Kuiper, editor) (Chicago: The University of Chicago Press)
140. Kurylo, M. J. (1972). *J. Phys. Chem.*, **76,** 3518
141. Lazrus, A., Gandrud, B., Woodard, R. and Sedlacek, W. (1975). Paper presented at CIAP Conference, Cambridge, Mass.
142. Leal, J. L. (1909). *Proc. Amer. Water Works Assn.*, 100
143. Leopold, L. B., Wolman, M. G. and Miller, J. P. (1964). *Fluvial Processes in Geomorphology* (San Francisco: W. H. Freeman)
144. Lettau, H. (1951). *Compendium of Meteorology*, 320 (T. F. Malone, editor) (Baltimore: Waverly
145. Lewis, J. S. (1969). *Icarus*, **10,** 365
146. Lewis, J. S. (1969). *Icarus*, **10,** 393
147. Lewis, J. S. (1970). *Earth Planet. Sci. Lett.*, **10,** 73
148. Lewis, J. S. (1974). *Sci. Amer.*, **230,** 50
149. Lindzen, R. S. (1971). *Mesospheric Models and Related Experiments* (Dordrecht, Netherlands: D. Reidel)
150. Little, S. J. (1971). *Planetary Atmospheres*, 241 (C. Sagan, T. C. Owen and H. I. Smith, editors) (Dordrecht, Netherlands: D. Reidel)
151. Liu, S. C. and Donahue, T. M. (1974). *J. Atmos. Sci.*, **31,** 1118
152. Liu, S. C. and Donahue, T. M. (1975). *Icarus*, to be published
153. Lovelock, J. E. (1974). *Nature*, **252,** 292
154. Lovelock, J. E. (1975). Personal communication
155. Lovelock, J. E. and Margulis, L. (1974). *Tellus*, **26,** 2
156. Lovelock, J. E., Maggs, R. J. and Wade, R. J. (1973). *Nature*, **241,** 194
157. Lowenstein, M. (1971). *J. Chem. Phys.*, **54,** 2282
158. Marshall, J. V. (1964). *Commun. Lunar Planet. Lab.*, **2,** 167
159. Martens, C. S., Wesolowski, J. J., Kaufer, R. and John, W. (1973). *Atmos. Environ.*, **7,** 905
160. Mason, B. (1966). *Principles of Geochemistry* (New York: Wiley)
161. Mathias, E., Sanhueza, E., Hisatsune, I. C. and Heicklen, J. (1974). *Can. J. Chem.*, **52,** 3852
162. McCarthy, R. L. (1975). Personal communication
163. McConnell, J. C. (1973). *J. Geophys. Res.*, **78,** 7812
164. McConnell, J. C. (1974). *Can. J. Chem.*, **52,** 1625
165. McConnell, J. C. and McElroy, M. B. (1973). *J. Atmos. Sci.*, **30,** 1465
167. McElroy, M. B. (1969). *J. Atmos. Sci.*, **26,** 5
168. McElroy, M. B. (1970). Lectures presented at the University of Colorado, Summer School on Planetary Atmospheres, July
169. McElroy, M. B. (1972). *Science*, **175,** 443
170. McElroy, M. B. (1973). *Adv. At. Mol. Phys.*, **9,** 323

171. McElroy, M. B. (1975). *Atmospheres of Earth and the Planets* (B. M. McCormac, editor) (Dordrecht, Netherlands: D. Reidel)
172. McElroy, M. B. and Donahue, T. M. (1972). *Science*, **177,** 986
173. McElroy, M. B. and McConnell, J. C. (1971). *J. Atmos. Sci.*, **28,** 879
174. McElroy, M. B. and Wofsy, S. C. (1974). Testimony presented to the Environmental Protection Agency, Boston, Mass., November 14
175. McElroy, M. B. and Yung, Y. L. (1975). *Ap. J.*, **196,** 227
176. McElroy, M. B., Sze, N. D. and Yung, Y. L. (1973). *J. Atmos. Sci.*, **30,** 1437
177. McElroy, M. B., Wofsy, S. C., Penner, J. E. and McConnell, J. C. (1974). *J. Atmos. Sci.*, **31,** 287
178. Midgely, T. and Henne, A. L. (1930). *Ind. Eng. Chem.*, **22,** 542
179. Mitchell, R. (1975). Personal communication
180. Molina, M. J. and Rowland, F. S. (1974). *Nature*, **249,** 810
181. Molina, M. J. and Rowland, F. S. (1974). *Geophys. Res. Lett.*, **1,** 309
182. Morris, J. C. (1975). Draft report to Office of Research and Development, EPA (RD 683), Washington
183. Moyers, J. L. and Duce, R. A. (1972). *J. Geophys. Res.*, **77,** 5229
184. Murray, B., Belton, M., Danielson, G. *et al.* (1974). *Science*, **187,** 1307
185. Newburn, R. L. and Gulkis, S. (1973). *Space Sci. Rev.*, **14,** 179
186. Newell, R. E. (1964). *Pure Appl. Geophys.*, **58,** 145
187. Nicholas, J. E. and Norrish, R. G. W. (1968). *Proc. Roy. Soc.* (*London*), **A307,** 391
188. Nicolet, M. (1965). *J. Geophys. Res.*, **65,** 679
189. Norrish, R. G. W. and Neville, G. H. J. (1934). *J. Chem. Soc.*, 1864
190. Noxon, J. F. (1970). *J. Chem. Phys.*, **52,** 1852
191. Owen, T. (1969). *Icarus*, **10,** 355
192. Paraskevopoulous, G. and Cvetanovic, R. J. (1969). *J. Amer. Chem. Soc.*, **91,** 7572
193. Paraskevopoulous, G. and Cvetanovic, R. J. (1971). *Chem. Phys. Lett.*, **9,** 603
194. Parkinson, T. M. and Hunten, D. M. (1972). *J. Atmos. Sci.*, **29,** 1380
195. Piper, D. Z. and Codispoti, L. A. (1975). *Science*, **188,** 15
196. Plonka, J. H. (1975). Personal communication
197. Poldervaart, A. (1955). *Crust of the Earth* (A. Poldervaart, editor), *Geol. Soc. Amer. Special Paper*, **62,** 119
198. Pollack, J. B. *et al.* (1973). Paper presented at AMS–DPS meeting, March
199. Porret, D. and Goodeve, C. F. (1938). *Proc. Roy. Soc.* (*London*), **A165,** 31
200. Prabhakara, C., Allison, L. G., Conrath, B. J. and Steranka, J. (1971). NASA-TN-D-6443
201. Prinn, R. G. (1971). *J. Atmos. Sci.*, **29,** 1058
202. Prinn, R. G. (1973). *Science*, **182,** 1132
203. Prinn, R. G. (1975). *J. Atmos. Sci.*, **32,** 1237
204. Ramanathan, K. G. (1975). *Science*, **196,** 50
205. Rasmussen, R. A. (1975). Personal communication
206. Richards, F. A. (1965). *Chemical Oceanography*, Chap. 6 (New York: Academic Press)
207. Ridgway, S. T. (1974). *Astrophys. J.*, **187,** L41
208. Robinson, E. and Robbins, R. C. (1968). Final Report, Project PR-6755, Stanford Research Institute
209. Romand, J. and Vodar, B. (1948). *C.R. Acad. Sci.* (*Paris*), **226,** 238
210. Ronov, A. B. (1959). *Geochemistry*, **5,** 510
211. Rook, J. J. (1974). *Water Treatment Exam.*, **23,** 234
212. Rowland, F. S. and Molina, M. J. (1975). *Rev. Geophys. Space Phys.*, **13,** 1
213. Rubey, W. W. (1951). *Bull. Geol. Soc. Amer.*, **62,** 1111
214. Ruderman, M. A. and Chamberlain, J. W. (1975). *Planet. Space Sci.*, **23,** 247
215. Rutten, M. G. (1970). *Space Life Sci.*, **2,** 5
217. Schiff, H. I. (1974). *Can. J. Chem.*, **52,** 1536
218. Schofield, K. (1967). *Planet. Space Sci.*, **15,** 643
219. Schorn, R. A. (1971). *Planetary Atmospheres*, 223 (C. Sagan, T. C. Owen, and H. J. Smith, editors) (Dordrecht, Netherlands: D. Reidel)
220. Schorn, R. A., Spinrad, H., Moore, R. C., Smith, H. J. and Giver, L. P. (1967). *Ap. J.*, **147,** 743
221. Schott, G. and Davidson, N. (1958). *J. Amer. Chem. Soc.*, **80,** 1841

222. Schutz, K., Junge, C., Beck, R. and Albrecht, B. *J.* (1970). *Geophys. Res.*, **75,** 2230
223. (1974). *Science and Technology of Aerosol Packaging* (J. J. Sciarra and L. Stoller, editors) (New York: Wiley)
224. Scott, P. M. and Cvetanovic, R. J. (1971). *J. Chem. Phys.*, **54,** 1440
225. Sill, G. T. and Carm, O. (1972). *Commun. Lunar Planet. Lab.*, **9,** 191
226. Simonaitis, R. and Heicklen, J. (1972). *J. Chem. Phys.*, **56,** 2004
227. Simpson, S. A. (1961). *Science and Space* (L. V. Berkner and H. Odishaw, editors) (New York: McGraw-Hill)
228. Skirrow, G. (1965). *Chemical Oceanography*, Vol. 1 (J. P. Riley and G. Skirrow, editors) (London, New York: Academic Press)
229. Slanger, T. G. and Black, G. (1971). *J. Chem. Phys.*, **55,** 2164
230. Slanger, T. G., Wood, B. J. and Black, G. (1972). *J. Chem. Phys.*, **57,** 233
231. Smith, B. A. (1967). *Science*, **158,** 114
233. Stedman, D. H., Steffenson, D. and Niki, H. (1970). *Chem. Phys. Lett.*, **7,** 173
234. Stewart, A. I. (1972). *J. Geophys. Res.*, **77,** 54
235. Stolarski, R. S. and Cicerone, R. J. (1974). *Can. J. Chem.*, **52,** 1610
236. Strobel, D. F. (1969). *J. Atmos. Sci.*, **26,** 906
237. Strobel, D. F. (1973). *J. Atmos. Sci.*, **30,** 489
238. Strobel, D. F. (1973). *J. Atmos. Sci.*, **30,** 1205
239. Strobel, D. F. (1974). *Astrophys. J.*, **192,** L47
240. Strobel, D. F. (1975). *Atmospheres of Earth and the Planets* (B. M. McCormack, editor) (Dordrecht, Netherlands: D. Reidel)
241. Stuhl, F. and Niki, H. (1971). *J. Chem. Phys.*, **55,** 3943
242. Symons, J. M. (1975). Interim report to Congress, Water Supply Research Lab., National Env. Res. Center, Cincinnati, Ohio
243. Sze, N. D. and McElroy, M. B. (1975). *Planet. Space Sci.*, **23,** 763
244. Takas, G. A. and Glass, G. P. (1973). *J. Phys. Chem.*, **77,** 1948
245. Traub, W. A. and Carleton, N. P. (1973). Paper presented at American Astronomical Society Third Annual Meeting, Tucson, Arizona
246. Tull, R. G. (1971). *Planetary Atmospheres*, 237 (C. Sagan, T. C. Owen and H. J. Smith, editors) (Dordrecht, Netherlands: D. Reidel)
247. Tyerman, W. J. R. (1969). *Trans. Faraday Soc.*, **65,** 2948
248. Ung, A. Y. and Schiff, H. I. (1966). *Can. J. Chem.*, **44,** 1981
249. Vaccaro, R. F. (1965). *Chemical Oceanography*, Vol. 1 (J. P. Riley and G. Skirrow, editors) (London, New York: Academic Press)
250. Van Valen, L. (1971). *Science*, **171,** 439
251. Walker, J. C. G. (1974). *Amer. J. Sci.*, **274,** 193
252. Warneck, P. (1964). *J. Chem. Phys.*, **41,** 3435
253. Warneck, P. (1972). *J. Geophys. Res.*, **77,** 6589
254. Watson, R. T. (1975). *CIAP Monograph*, Vol. 1, in press
255. Watson, R. T. and Davis, D. D. (1975). Meeting of the American Chemical Society, Philadelphia, Pa., April 9
256. Wayne, L. G. and Yost, D. M. (1951). *J. Chem. Phys.*, **19,** 41
257. Westenberg, A. A. and DeHaas, N. (1972). *J. Chem. Phys.*, **57,** 5375
258. Westenberg, A. A., Roscoe, J. M. and DeHaas, N. (1970). *Chem. Phys. Lett.*, **7,** 597
259. Widman, R. P. and DeGraff, B. A. (1973). *J. Phys. Chem.*, **77,** 1325
260. Wilkniss, P. E., Lamontagne, R. A., Larson, R. E., Swinnerton, J. W., Dickson, C. R. and Thompson, T. (1973). *Nature*, **245,** 45
261. Willis, W. H. and Sturgis, M. B. (1944). *Soil Sci. Soc. Amer. Proc.*, **9,** 106
262. Wilson, W. E. (1965). *10th Comb. Symp.*, 47
263. Wofsy, S. C. (1974). *Proc. Third Conf. on Climatic Impact Assessment Program* (Washington: Dept. of Transportation)
264. Wofsy, S. C. and McElroy, M. B. (1973). *J. Geophys. Res.*, **78,** 2619
265. Wofsy, S. C. and McElroy, M. B. (1974). *Can. J. Chem.*, **52,** 1582
266. Wofsy, S. C. and Sze, N. D. (1975). *Atmospheres of Earth and the Planets* (B. M. McCormack, editor) (Dordrecht, Netherlands: D. Reidel)
267. Wofsy, S. C., McElroy, M. B. and Sze, N. D. (1975). *Science*, **187,** 535
268. Wofsy, S. C., McElroy, M. B. and Yung, Y. L. (1975). *Geophys. Res. Lett.*, **2,** 215
269. Wofsy, S. C., McConnell, J. C. and McElroy, M. B. (1972). *J. Geophys. Res.*, **77,** 4477
270. Wong, E. L. and Belles, F. E. (1971). *NASA Technical Note*, TN D-6495

271. Young, A. T. (1973). *Icarus*, **18,** 564
272. Young, L. D. G. and Young, A. T. (1973). *Astrophys. J.*, **179,** L39
273. Young, R. A., Black, G. and Slanger, T. G. (1968). *J. Chem. Phys.*, **49,** 4758
274. Yung, Y. L. and McElroy, M. B. (1975). To be published
275. Yung, Y. L., McElroy, M. B. and Wofsy, S. C. (1975). *Geophys. Res. Lett.*, in press
276. Anderson, J. G. and Kaufman, F. (1973). *Chem. Phys. Lett.*, **19,** 483
277. Baulch, D. L., Drysdale, D. D. and Horne, D. G. (1973). *Chem. Kin. Data Survey V* (Washington: Nat. Bur. Stand)
278. Buchanan, R. E. and Fulmer, E. K. (1928). *Physiology and Biochemistry of Bacteria*, Vol. 1 (Baltimore: Williams and Wilkins)
279. Clark, I. P. and Wayne, R. P. (1969). *Proc. Roy. Soc. (London)*, **A314,** 111
280. Davidson, J. A., Sadowski, C. M., Schiff, H. I., Streit, G. E., Howard, C. J., Jennings, D. A. and Schmeltekopf, A. L. (1976). *J. Chem. Phys.*, **64,** 57
281. Davis, D. D., Wong, W. and Schiff, R. (1974). *J. Phys. Chem.*, **78,** 463
282. DeMore, W. B. (1973). *Science*, **180,** 735
283. DeMore, W. B. and Tschuikow-Roux, E. (1974). *J. Phys. Chem.*, **78,** 1447
284. Graham, R. A. and Johnston, H. S. (1974). *J. Chem. Phys.*, **60,** 4628
285. Hack, W., Hoyermann, K. and Wagner, H. G. (1975). *Int. J. Chem. Kin. Symp.*, **1,** 329
286. Johnston, H. S. and Graham, R. (1973). *J. Phys. Chem.*, **77,** 62
287. Johnston, H. S. and Selwyn, G. (1975). *Geophys. Res. Lett.*, **2,** 549
288. Johnston, H. S., Chang, S. G. and Whitten, G. (1975). *J. Phys. Chem.*, **78,** 1
289. McCrumb, J. L. and Kaufman, F. (1972). *J. Chem. Phys.*, **57,** 1270
290. McElroy, M. B., Elkins, J. W., Wofsy, S. C. and Yung, Y. L. (1976). *Rev. Geophys. Space Phys.*, in press
291. Margitan, J. J., Kaufman, F. and Anderson, J. G. (1975). *Int. J. Chem. Kin. Symp.*, **1,** 281
292. Payne, W. A., Stief, L. J. and Davis, D. D. (1973). *J. Amer. Chem. Soc.*, **95,** 7614
293. Streit, G. E., Howard, C. J., Schmeltekopf, A. L., Davidson, J. A. and Schiff, H. I. (1976). *J. Chem. Phys.*, in press
294. Tsang, W. (1973). *Int. J. Chem. Kin.*, **5,** 947
295. Zahniser, M. S., Kaufman, F. and Anderson, J. G. (1974). *Chem. Phys. Lett.*, **27,** 507

# 6
# Chemiluminescence in the Liquid Phase: Electron Transfer

**L. R. FAULKNER**
University of Illinois

## 6.1 BASIC ASPECTS OF CHEMICAL EXCITATION

In the 1870s, it was first realised that light emission could accompany condensed-phase chemical reactions occurring outside a biological environment. A large body of knowledge has since been built up by many investigators[1]. The precursors required for emission from a host of classically important bioluminescent and chemiluminescent systems have apparently been identified, and many have been traced through often complex reaction sequences to final products. In a large number of cases the emitting substances and the luminescing states are known. Understanding in these areas seems relatively secure and has recently been reviewed in several articles[1].

However, an aspect which continues to elude definition for almost every relatively complex chemiluminescent system is the nature of the process which actually introduces electronic excitation. The intermediates whose reactions govern the excitation are quite transient and are therefore very difficult to characterise. Nonetheless, one can infer some probable features of chemical excitation simply by considering the conditions for which a reaction path yielding excited products could compete with paths yielding species in the ground electronic states. After all, some processes such as the firefly reactions have excitation yields near unity. If a chemical transformation proceeds preferentially or even significantly to an excited product, one must seek means by which the ground-state pathway is inhibited. Obviously, the conservation of spin or orbital symmetry could be involved. However, a more general rationale is based on the concept of chemical excitation as a kinetic manifestation of the Franck–Condon principle, which is itself a recognition that a molecule usually cannot accept energy instantaneously by becoming mechanically excited. Nuclear positions and momenta can be altered only over a significant time. Consequently, a reaction path to ground-state products, which involves the largest possible mechanical energy change, would be inhibited if the chemical transformation were very fast and energetic. Thus one suspects that the immediate precursors to chemical excitation often are energy-rich species which undergo very exoergic reactions involving little or no immediate mechanical restructuring.

Much recent literature has focused on two kinds of chemical excitation that possess just this sort of mechanistic simplicity, *viz.* the unimolecular decomposition of dioxetanes and certain very energetic electron transfer processes[2,3]. Both of these chemiluminescence mechanisms have been discovered and characterised only in the past decade. In contrast to many older systems, however, the dioxetane and electron transfer systems are chemically uncomplicated; hence they enable one to concentrate experimentally on the actual act of excitation. Much has been learned about the requirements for chemical excitation in these processes, and they are now provoking attention as candidates for the key steps in more complex chemiluminescent and bioluminescent systems.

Dioxetane chemiluminescence is reviewed in Chapter 7 by Wilson. Electron transfer chemiluminescence is treated here in a complementary way, with emphasis on what is and is not known about the nature of the excitation process.

## 6.2 THE PHENOMENON OF REDOX EXCITATION

### 6.2.1 Introduction

A prototype for the bimolecular excitation model that we consider in this section is the reaction between the anion and cation radicals derived from 9,10-diphenylanthracene (DPA). Such a process can be conveniently carried out in an aprotic solvent such as *N,N*-dimethylformamide (DMF) or acetonitrile, and the reaction products are clearly two DPA molecules. Accompanying the chemical change is a blue luminescence, which has a spectral distribution identical to DPA fluorescence; thus excitation to the first excited singlet state is a consequence of the reaction.

The key elements in the rationale for this observation are the size of the free energy change driving the electron transfer and the timescale over which the actual transfer takes place. For the reaction to ground state products

$$DPA^{-\cdot} + DPA^{+\cdot} \rightarrow 2DPA \tag{6.1}$$

one can calculate from electrochemical standard potentials that $\Delta G^\circ = -3.2$ eV. The time in which this large reaction energy must be accommodated is not known precisely. It is known, however, that the redox process occurs at each encounter[5–7]. Because the reactants here are oppositely charged, and virtually no structural modification follows the transfer, it is reasonable to postulate a transfer duration in the Franck–Condon range. Under such a circumstance, a substantial likelihood exists that the reaction energy will be converted to electronic excitation if an excited state is energetically accessible. In the DPA(+)/DPA(−) reaction, excitation to the first excited singlet requires *ca.* 3.0 eV; hence direct redox excitation to the emitting state is possible and a particularly simple mechanism for emission emerges. The transformation can be represented schematically by the following molecular orbital diagram:

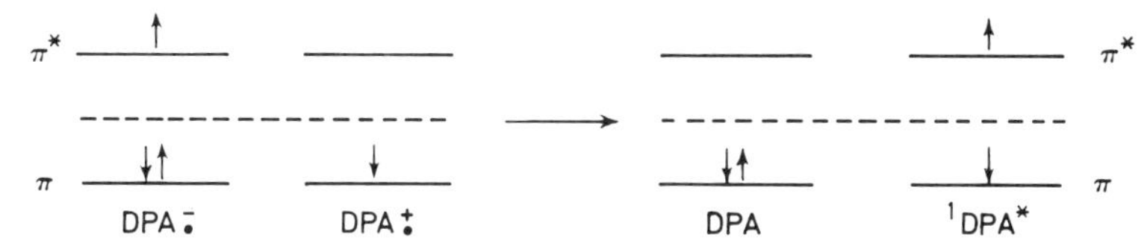


By transferring electrons between corresponding π* or π orbitals, rather than from π* on DPA(−) to π on DPA(+), the system can produce the emitting state directly and can greatly reduce the amount of energy which must be taken up mechanically. However, we shall find below that such simplicity will not always apply and that more complicated schemes for producing the emitting singlet will be required. Nevertheless, we will rely almost constantly on the redox process to yield excitation at some level (singlet or triplet) in each chemiluminescent system.

In the seven years immediately following the first reports of radical ion chemiluminescence[9–12], almost all work was dedicated to the elucidation of mechanisms. By 1972 the qualitative features of these systems seemed largely understood; hence efforts shifted toward the quantitative questions surrounding the efficiencies of excited state generation and light production. Several reviews covering literature through 1968 exist[2]. Among them, the third by Hercules[2g] gives an especially interesting account of the early history. Recent months have seen specialised articles covering efficiency measurements[13], theoretical principles[14], radical ion reactions in non-polar media[15–17], the role of exciplexes in redox chemiluminescence[15–18], and experimental methods[19]. This chapter aims at a comprehensive summary and analysis; hence it presents a collection and examination of the evidence supporting mechanistic conclusions and an interpretation of the existing quantitative data. Brief introductions to the range of chemical and spectroscopic behaviour and to experimental methods are necessary preludes.

### 6.2.2 The range of chemical possibilities

Chemiluminescence from the DPA(+)/DPA(−) reaction is but one example of emission from radical ion annihilations. The property is virtually general whenever an excited state of a luminescent product molecule is energetically accessible. The reactant radicals may be derived from the same precursor or from different ones, for example:

$$R^{-\cdot} + R^{+\cdot} \rightarrow {}^1R^* + R \qquad (6.2a)$$

$$R^{-\cdot} + TMPD^{+\cdot} \rightarrow {}^1R^* + TMPD \qquad (6.2b)$$

$$BQ^{-\cdot} + R^{+\cdot} \rightarrow {}^1R^* + BQ \qquad (6.2c)$$

Here, R, BQ and TMPD represent rubrene, *p*-benzoquinone, and *N*,*N*,*N*,'*N*'-tetramethyl-*p*-phenylenediamine, respectively. Note also that, in the mixed systems, excitation may reside on either ion precursor. All three processes yield emission from rubrene's first excited singlet state, $^1R^*$, but one should not regard (6.2) as describing elementary reactions (see Section 6.4). Though all the foregoing examples involve emission from aromatic hydrocarbons, many cases of luminescence from heterocyclic systems also exist. To date, emission has been recorded for several hundred radical ion reactions, and large tabulations of these observations appear in several sources[2,20–24]

Because experimentation with radical ion annihilations has dominated redox chemiluminescence, the entire field has often been defined in terms of these charge recombination processes, but numerous examples of redox excitation outside this limited chemical sphere render such a definition incomplete. An important recent development is the observation of luminescence from certain metal chelates, such as tris(bipyridyl)ruthenium(II)[25–29]. The phosphorescent charge-transfer state of this complex ion is apparently populated directly in the reaction[27,28]

$$Ru(bipy)_3{}^{3+} + Ru(bipy)_3{}^{+} \rightarrow [Ru(bipy)_3{}^{2+}]^* + Ru(bipy)_3{}^{2+} \qquad (6.3)$$

Martin, *et al.*[29] have recorded the same emission as a companion to

$$Ru(bipy)_3{}^{3+} + e^-{}_{aq} \rightarrow [Ru(bipy)_3{}^{2+}]^* \qquad (6.4)$$

where $e^-{}_{aq}$ represents a radiolytically produced aquated electron. Mayeda and Bard have reported the generation of singlet ($^1\Delta_g$) oxygen by oxidation of superoxide ion with ferricenium ion in acetonitrile[30]. In addition, there are many chemiluminescent oxidations of aromatic radical anions by such species as benzoyl peroxide, $HgCl_2$, $Pb(OAc)_4$ and $Cl_2$[2d,10,31,32]. The possible roles of electron transfer excitation in more complex chemiluminescent processes, such as electrochemical preannihilation emission[33–40], the faradaic reductions of some alkyl halides[41], and the oxidations of luminol and certain linear hydrazides[43–45], have also been considered in the literature. Redox excitation is therefore not restricted to any special class of molecule. Its basic requirement is the existence of a facile, energetic electron transfer embedded somewhere in the overall chemical reaction. Radical ion annihilations have received experimental emphasis because their interpretation is usually simpler; the whole reaction often comprises the one elementary step of interest.

### 6.2.3 Spectroscopic complications

The cases we have examined above are spectroscopically simple in that the light is emitted by and can be identified with only one of the products. Many, probably most, systems involve emission from more than one molecular species. For example, in the reaction of the thianthrene cation with the anion radical derived from 2,5-diphenyl-1,3,4-oxadiazole (PPD), either product may become excited, and a dual emission is seen[46]. In other cases one may observe emission from a radical annihilation product together with one or more longer wavelength bands that cannot be ascribed to either product individually. These usually arise from one of three mechanisms: (1) emission from a product of a side reaction or formation of an emitting exciplex either (2) subsequent to electron transfer or (3) directly from the radical ions.

The prime example of the first case is the long-wavelength luminescence accompanying the anthracene anion–cation reaction in DMF. Separate studies have shown that most of this light arises from 9-anthranol produced from decay of anthracene(+)[47,48]. Excitation of anthranol apparently comes about by energy transfer from singlet excited anthracene which is produced in the primary process.

Chemiluminescence from the pyrene(−)/TMPD(+) reaction is the classic example of case (2)[49,50]. The long-wavelength emission here is easily identified with the well-characterised pyrene excimer $^1Py_2^*$, which arises from the first excited pyrene singlet by association with a ground state molecule[51–53]

$$^1Py^* + Py \rightarrow {}^1Py_2^* \tag{6.5}$$

or from two triplets via the annihilation reaction[52,53]

$$^3Py^* + {}^3Py^* \rightarrow {}^1Py_2 \tag{6.6}$$

The excimer emission results from the radiative dissociation

$$^1Py_2^* \rightarrow 2Py + h\nu \tag{6.7}$$

Heteroexcimers, which are excited state complexes in which the partners differ[15,20a,54–57] may also be formed by an association process similar to (6.5) or by a mixed triplet–triplet annihilation. They ordinarily have charge-transfer character[57]; hence they arise in systems for which excited state donor–acceptor interactions are possible. Again, emission results from radiative dissociation.

Working with tetrahydrofuran (THF), 2-methyltetrahydrofuran (MTHF), and 1,2-dimethoxyethane (DME) solutions, Weller and Zachariasse found heteroexcimer emission from several reactions of TMPD(+), including that with naphthalene(—)[20a,58]. They showed it to arise partially by mixed annihilation of triplets created in the redox step[20a].

$$^3\text{TMPD}^* + {}^3(\text{Naphthalene})^* \rightarrow (\text{TMPD}^+ \text{ Naphthalene}^-) \quad (6.8)$$

A more important result of their extensive experiments was the discovery that heteroexcimers are often formed directly from the radical ions in non-polar media[15,17,20a,59]. Consider the reaction in THF between the anthracene (An) anion and the cation radical derived from tri-*p*-tolylamine (TPTA):

$$\text{An}^{-\cdot} + \text{TPTA}^{+\cdot} \rightarrow (\text{TPTA}^{+}\cdot\text{An}^{-}\cdot) \quad (6.9)$$

In one sense the heteroexcimer can be regarded as a specially structured contact ion pair; thus an excited state capable of radiative relaxation is created by association of the reactants without full electron transfer. The heteroexcimer component dominates emission in this particular case, but its contribution depends dramatically upon solvent polarity, apparently because more polar media destabilise the heteroexcimer with respect to the ions.

In general, heteroexcimer emission is a much less obvious feature of redox chemiluminescence from polar media such as acetonitrile or DMF, and until very recently there was no evidence for direct formation of these exciplexes from the reactants. This situation has led to a heavy emphasis on solvent polarity in the mechanistic discussion below. In recent publications, Bard and his co-workers have reported observations indicating that some heteroexcimers actually do form directly in high dielectric media[18]. The categories used here for exposition thus should not be regarded as rigid.

## 6.3 EXPERIMENTAL TECHNIQUES

### 6.3.1 Approaches

Redox excitation requires energetic electron transfer reactions; hence every experiment involves potent oxidants and reductants. The difficulties encountered in studying these processes are often related to the generation and preservation of such reactive substances prior to and during the experiment.

A few compounds, such as TMPD and TPTA, will yield stable perchlorate salts of their cation radicals[60,61]. One can therefore carry out comparatively long experiments by bulk reaction of these cations with anions generated in solvents such as THF and DME by alkali metal reduction[20a,62,63]. This method is well suited to studies of temperature effects, and it is especially useful for work with non-polar media. It has the additional advantage over the electrochemical alternatives that a base electrolyte is not required to render the reaction mixture conducting.

The faradaic methods, which involve *in situ* electrochemical generation of reactant species from parent compounds, are often complementary to the bulk reaction approach. Several such techniques offer precise experimental control and provide a continuously variable timescale ranging from hours to microseconds. This last feature provides flexibility to deal with short-lived reactants and allows a wide range of time-dependent stimuli. Most workers have, accordingly, made extensive use of these methods. They have been reviewed in detail recently[19], but we shall briefly introduce them here because they are not widely understood.

### 6.3.2 The control of electrochemical dynamics

The interconversion at an electrochemical interface between, say, a hydrocarbon and its radical anion can be described by opposing transfer reactions

$$R + e^- \underset{k_o}{\overset{k_r}{\rightleftharpoons}} R^{-\cdot} \tag{6.10}$$

The heterogeneous rate constants $k_r$ and $k_o$ depend exponentially upon the boundary potential difference between the electrode and solution phases. Changes in this boundary potential can be enforced or observed by coupling the working electrode to a reference electrode whose boundary potential is fixed by maintaining the participants in its electrode reaction at constant activity. The saturated calomel electrode (SCE) is a familiar example. The parameters $k_o$ and $k_r$ are equal when the working electrode potential (*vs.* the reference) is equal to the thermodynamic standard potential $E°$. At more negative potentials $k_r$ dominates $k_o$; hence $R^-$ will dominate R at the electrode surface. The opposite situation applies at potentials more positive than $E°$. For very large rate constants it can be shown that the surface concentrations are always reconciled (neglecting activity coefficients) to the potential by

$$E = E° + \frac{RT}{nF} \ln \left( \frac{C_R}{C_{R^-}} \right)_{\text{surface}} \tag{6.11}$$

If the surface concentrations are rendered different from the bulk values by forcing the electrode potential away from equilibrium, mass transport processes will tend to eliminate the concentration gradients. Maintenance of equation (6.11) therefore demands net conversion of R into $R^-$ or *vice versa*, and a current will flow at the working electrode.

Electroneutrality requires that the charge passed at the interface be compensated elsewhere in the solution phase; hence the working electrode must be linked to a counter electrode. Whatever current flows in reduction (or oxidation) at the working electrode must flow in oxidation (or reduction) at the counter electrode.

This three-electrode system is manipulated electronically by a potentiostatic device. It senses the potential difference between the working and reference electrodes and, through a feedback loop, forces through the counter and working electrodes whatever current is required to maintain the sensed potential equal to a programmed value. An electrochemically inactive supporting electrolyte, such as 0.1 M tetra-n-butylammonium perchlorate

(TBAP), is required to render electronic potential control and precise treatment of mass transport tractable.

### 6.3.3 Cyclic voltammetry

Candidate materials are often characterised by a method termed cyclic voltammetry[64–66], which involves the application of a triangular potential waveform to a small working electrode (e.g. a 0.05 cm$^2$ Pt disc) immersed in an unstirred solution. Rubrene's voltammetric response is displayed as an

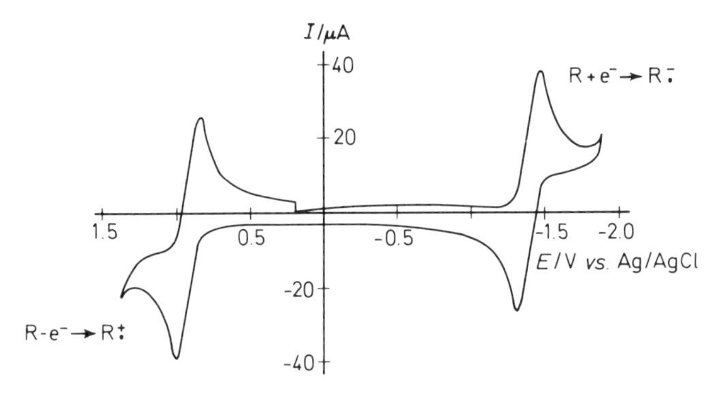


**Figure 6.1** Cyclic voltammetric curve. 0.593 mM rubrene in benzonitrile with 0.1 M TBAP. Potential scan begins at + 0.2 V

example in Figure 6.1. As a starting point, a potential is chosen at which no electrode reaction occurs. As the potential is scanned negatively, one eventually reaches the region in which rubrene is reduced. A cathodic (positive by convention) current flows as the result of the interplay between (6.11) and Fick's laws of diffusion. If, as in this case, the electrode reaction is chemically reversible and features very fast heterogeneous kinetics ('reversible system'), then reduction begins in the region of $E^\circ$ $(\mathrm{R}/\mathrm{R}^-)$ and a peak is recorded at a slightly more extreme potential. Once the peak has been passed, the surface concentration of R is virtually zero and $\mathrm{R}^-$ predominates.

This reduction product is examined for stability by a subsequent reversal in scan direction. As the potential moves positively past $E^\circ$ $(\mathrm{R}/\mathrm{R}^-)$, relation (6.11) demands renewed dominance of R at the electrode surface; hence reoxidation of the $\mathrm{R}^-$ takes place and an anodic (negative) current is recorded.

The positive scan eventually reaches the region in which rubrene is oxidised. An anodic peak therefore appears as $\mathrm{R}^+$ is created, and a second reversal causes a cathodic peak resulting from the reversion of $\mathrm{R}^+$ to R as the potential tends toward the starting point.

Nicholson and Shain have shown that the forward peak potential for a reversible reduction is given at 25 °C by[64]

$$E_p(R/R^-\cdot) = E^\circ(R/R^-\cdot) + \frac{0.0591}{n}\log\left(\frac{D_{R^-}}{D_R}\right)^{\frac{1}{2}} - \frac{0.0285}{n}\ \text{(volts)}$$

$$E_p(R/R^-\cdot) = E_{\frac{1}{2}}(R/R^-\cdot) - \frac{0.0285}{n}\ \text{(volts)} \tag{6.12}$$

Similar relations hold for oxidations. Because the diffusion coefficients $D_R$ and $D_{R-}$ do not differ widely, $E_{\frac{1}{2}}(R/R^-)$, which is the polarographic half-wave potential, is virtually the same as $E^\circ(R/R^-)$. The two dynamic parameters $E_{\frac{1}{2}}(R/R^-)$ and $E_p(R/R^-)$ can therefore be regarded as revealing thermodynamic information, and they have been useful in defining reaction energetics for redox excitation.

### 6.3.4 Electrochemical methods for chemiluminescence

Two parallel approaches have been employed in the electrogeneration of chemiluminescence (ECL). Perhaps the most straightforward has been the use of two separate generating electrodes for simultaneous creation of reactants, which are then brought together convectively for reaction. Since electrogeneration can occur continuously, a steady reaction rate can be attained. The alternate approach involves sequential generation of reactants at a single electrode; hence it features transients in the reaction rate.

Maloy, Prater and Bard implemented the steady-state method in a most elegant fashion by using a rotating ring–disc combination[49,67]. Figure 6.2 shows that this device has two carefully machined, concentric Pt electrodes. Both are working electrodes, and a dual potentiostat must be employed. When the assembly is rotated rapidly (10–100 r.p.s.) the surrounding medium follows a well-defined convective transport pattern, which has been described by Levich and others[68,69]. The solution moves axially to the vicinity of the electrode, where it receives a radial component and moves outward. The disc and ring are set at potentials which separately yield the two reactants; hence they converge for reaction just below the face of the ring. Although a steady luminescence is recorded, changes in rotation rate allow a variable timescale for the study.

The sequential approach has been implemented most commonly in a multiple step experiment. One deals with a planar Pt microelectrode immersed in an unstirred, deaerated solution of the ion precursors. The initial potential is set at a value at which no electrode reaction occurs, and the experiment begins when the electrode potential is stepped to a value which yields the first reactant. After a time $t_f$, which may range from 10 μs to 10 s, a second step moves the potential to a value which produces the second reactant. In this transition an infinitesimal amount of the initial reactant is destroyed on the electrode surface, but as the second becomes established there the first reactant is sealed from further destruction at the interface. This event therefore creates an isolated zone of the first reactant, which tends to diffuse back toward the electrode, where its concentration is zero. As the second reactant, formed on the electrode, diffuses outward, the two meet in a very thin zone near the surface and annihilate. The accompanying pulse of light has a shape which depends on mass transport dynamics and mechanistic parameters, and it has been subjected to theoretical analysis by Feldberg[70–72] and by Bezman and Faulkner[73,74].

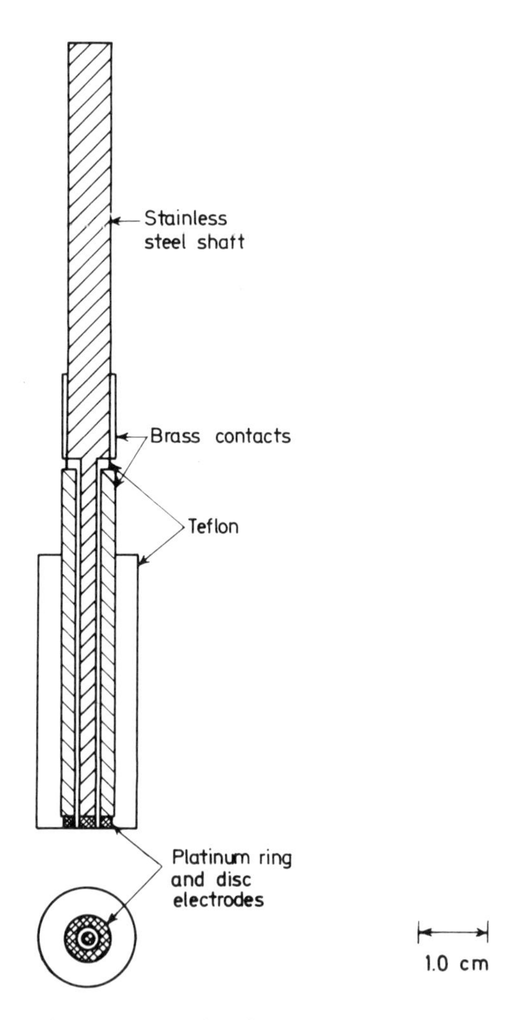


**Figure 6.2** A rotating ring–disc electrode (RRDE). Solution contacts only the faces of the electrodes and the Teflon mantle (Adapted from Maloy *et al.*[67], by courtesy of the American Chemical Society)

A number of studies have involved trains of pulses for creating pseudo steady-state light levels for spectral examination. These are easily created by rectangular potential waveforms from which alternate periods of oxidation and reduction result. One may view the technique as a multiple cycle variation of the single pulse method described above.

## 6.4 MECHANISMS FOR EXCITATION AND CHEMILUMINESCENCE

### 6.4.1 Basic energy calculations

An allusion has already been made to the utility of reaction energetics as a guide for mechanistic interpretation. Analyses of available energy have therefore appeared occasionally since the earliest work[2b,20a,58,75,76]. Briefly stated, the goal of such an analysis is to determine whether the energy released from a redox process is sufficient to leave one of the product molecules in an excited state. In a first approach, one could answer this question by defining the sign of $\Delta G°$ for the reaction,

$$R^+ + R^- \rightarrow R^* + R \tag{6.13}$$

The calculation involves a straightforward comparison of electrochemical and spectroscopic information, which can be implemented via the following cycle:

$$\begin{array}{ccc} (R^+ + R^-)_{is} & \xrightarrow{\Delta G°_{ex}} & (R^* + R)_{es} \\ \Delta G°_g \downarrow & & \uparrow \Delta G°_{resolv} \\ (R + R)_{gs} & \xrightarrow{\Delta G°_{abs}} & (R^* + R)_{gs} \end{array} \tag{6.14}$$

The subscripts 'is', 'gs' and 'es' refer here to solvation states characteristic of the ions, the ground states, and the excited state–ground state pair, respectively. One may regard the symbol $(R^+ + R^-)$ as a solvent-shared ion pair with an ionic spacing of the average reaction distance. Thus the free energy of redox excitation is

$$\Delta G°_{ex} = \Delta G°_g + \Delta G°_{abs} + \Delta G_{resolv} \tag{6.15}$$

The quantity $\Delta G°_{abs}$ concerns spectroscopic absorption. Since entropy effects and $PV$ work are negligible, this free energy may be equated with the 0,0 transition energy, $\Delta E^*$. The energy for electron transfer to ground state products approximates

$$\Delta G°_g = E°(R/R^-) - E°(R^+/R) + e^2/\varepsilon a \quad \text{(in eV)} \tag{6.16}$$

where the last term is merely an estimate for the coulombic stabilisation energy of an ion pair with interionic distance $a$ in a medium with dielectric constant $\varepsilon$. Thus, the criterion for energy sufficiency in redox excitation may be stated

$$E°(R^+/R) - E°(R/R^-) - e^2/\varepsilon a - \Delta G°_{resolv} \geqslant \Delta E^* \tag{6.17}$$

Hoytink[14,75] and Weller and Zachariasse[20a,58] have suggested that the value of $a$ lies between 5 and 10 Å for full electron transfer processes. With acetonitrile or DMF ($\varepsilon \approx 37$), the coulombic term therefore adopts a value near 0.05 eV, which is small enough to neglect, as most workers have implicitly done for these media. For benzonitrile ($\varepsilon = 26$) and the ether solvents ($\varepsilon \approx 7$), it takes on the more significant values of 0.05–0.10 and 0.2–0.4 eV, respectively. The resolvation energy is more difficult to ascertain, so it is most convenient to neglect it. This approach is probably realistic for aromatic hydrocarbons and other molecules for which small spectroscopic Stokes shifts are encountered.

Several authors, mindful of the timescale over which single redox events occur, have felt uncomfortable with $\Delta G^\circ_{ex}$ as an indicator of excitation thresholds. In their view, the entropy change associated with (6.13) probably would have little effect on the likelihood of excitation because it arises almost entirely from solvation changes which probably could not be manifested in a Franck–Condon transition. If so, one more properly would examine the change in internal energy $\Delta E^\circ_{ex}$, which is about the same as $\Delta H^\circ_{ex}$. Thus $\Delta G^\circ_{ex}$ would be added to $T\Delta S^\circ_{ex} \approx T\Delta S^\circ_{g}$. For hydrocarbon systems in fairly polar media, $T\Delta S^\circ_{g} = 0.1 \pm 0.1$ eV[75,77,78], so the difference between $\Delta H^\circ_{ex}$ and $\Delta G^\circ_{ex}$ is hardly significant in such a circumstance. Larger differences between the two approaches would arise from less ideal situations, viz. (a) reactions in non-polar media with high ion populations; (b) reactions involving ionic products [e.g. chemiluminescence from $Ru(bipy)_3{}^{2+}$]; or (c) reactions in which the nuclear configuration of the ions differs significantly from that of the thermalised excited state–ground state pair. In each such case the entropy change is probably larger than that characterising hydrocarbon systems in polar media. In addition, there may even be components of the internal energy that could not be manifested on a Franck–Condon timescale.

Despite the conceptual distinctions between these two approaches, there presently is no experimental basis for favouring one over the other. Knowledge about the behaviour of chemiluminescent systems in the threshold region is just too scarce. Moreover, the kinetic considerations outlined in Section 6.5.2 suggest that sharp thresholds might not really exist. One must therefore avoid overinterpretation of results from basic energy calculations, but they certainly seem valid as rough indices of excited state accessibility in redox reactions.

Table 6.1 summarises the energy balance, as $\Delta H^\circ_{ex}$, for several important chemiluminescent systems. Unless otherwise noted, emission from each system comes only from the first excited singlet of the ion precursor listed in the third column. The enthalpies for excitation of this precursor which are listed in the last two columns have been calculated under the assumption that $T\Delta S^\circ_{ex} = 0.1 \pm 0.1$ eV.

Only a few reactions shown here have negative or near zero $\Delta H^\circ_{ex}$. These are termed 'energy sufficient', because a single redox event yields virtually all the energy required for excitation to the emitting state. Direct population of that singlet state is therefore a possible accompaniment to electron transfer, and luminescence may arise by the so-called S-route:

$$R^+ + R^- \rightarrow {}^1R^* + R \qquad (6.18a)$$

$$^1R^* \rightarrow R + h\nu \qquad (6.18b)$$

Evidence exists for the actual operation of this scheme in several systems (see Section 6.4.2).

Most reactions are energy deficient, in that a single emitted photon contains substantially more energy than is provided by a single electron transfer. The S-route is therefore not a viable mechanism for emission, and a more

**Table 6.1 Energy balance for simple cases of redox excitation**

| *Reactants*[a] | | *Emitter (from first excited singlet)* | $\Delta H^\circ_{ex}$ *for emitter excitation*[b]/eV | | *Ref.* |
|---|---|---|---|---|---|
| *Oxidant* | *Reductant* | | *First excited singlet* | *Lowest triplet* | |
| Thianthrene(+) | PPD(−) | Thianthrene, PPD[c] | −0.5[d], +0.3[e] | −0.7, −0.1[e] | 46 |
| DPA(+) | DPA(−) | DPA | −0.1 | −1.3 | 76 |
| Rubrene(+) | Rubrene(−) | Rubrene | 0.0 | −1.1[g] | 76 |
| TPP(+) | TPP(−) | TPP | +0.1 | −0.9[f] | 76 |
| 10-MP(+) | DPA(−) | DPA | +0.3 | −0.9 | 80 |
| 10-MP(+) | Fluoranthene(−) | Fluoranthene[h] | +0.6 | −0.1 | 76 |
| 10-MP(+) | Rubrene(−) | Rubrene | +0.8 | −0.9[g] | 80 |
| DMPD(+) | DPA(−) | DPA | +0.8 | −0.4 | 80 |
| TMPD(+) | Rubrene(−) | Rubrene | +0.8 | −0.3[g] | 76 |
| DMPD(+) | Fluoranthene(−) | None | +0.9 | +0.2 | 80 |
| TMPD(+) | DPA(−) | DPA | +1.0 | −0.2 | 76 |
| Rubrene(+) | *p*-Benzoquinone(−) | Rubrene | +1.0 | −0.1[g] | 76 |
| TMPD(+) | Anthracene(−) | Anthracene | +1.1 | −0.3 | 76 |
| TMPD(+) | TPP(−) | TPP | +1.1 | +0.1[f] | 76 |
| TMPD(+) | Fluoranthene(−) | None | +1.2 | +0.5 | 80 |

[a] Reactions in DMF unless otherwise noted. DPA = 9,10-diphenylanthracene; PPD = 2,5-diphenyl-1,3,4-oxadiazole; TPP = 1,3,6,8-tetraphenylpyrene; 10-MP = 10-methylphenothiazine; DMPD = *N,N*-dimethyl-*p*-phenylenediamine; TMPD = *N,N,N′,N′*-tetramethyl-*p*-phenylenediamine
[b] Calculated from information supplied in Ref. at right
[c] Reaction in $CH_3CN$. Only *ca.* 1% of emission is due to PPD
[d] Large apparent Stokes shift makes this figure uncertain. Singlet level derived from fluorescence data
[e] Data for thianthrene excitation listed first
[f] Triplet energy taken as that for pyrene
[g] Triplet energy taken as that for tetracene
[h] 10-MP emission is difficult to distinguish from that of fluoranthene, so emitter identification is ambiguous

complicated scheme must be invoked. From mechanistic studies extending through 1972, a picture involving triplet intermediates has evolved. This 'T-route' may be written

$$R^+ + R^- \rightarrow {}^3R^* + R \quad (6.19a)$$

$$^3R^* + {}^3R^* \rightarrow {}^1R^* + R \quad (6.19b)$$

$$^1R^* \rightarrow R + h\nu \quad (6.19c)$$

In this mechanism, triplet–triplet annihilation (6.19b) provides the pooling of energy from two separate redox events. Of course, ancillary reactions such as singlet and triplet quenching must be included for a complete description.

### 6.4.2 Mechanistic aspects common to all media

Direct evidence in support of the conclusions stated above came from four sources: (1) studies of magnetic effects on chemiluminescence; (2) a detailed examination of the excimer/monomer emission ratio in systems displaying excimer emission; (3) analysis of luminescence transients accompanying the step experiment; and (4) interception of triplet intermediates from certain energy deficient systems. This historical sequence is a tortuous route for exposition of the evidence; hence we shall discuss first (4) and (2), which concern results from specific chemical systems, then (1) and (3), which have yielded more general findings.

#### *6.4.2.1 Triplet interception*

An obvious experiment to confirm or deny the intermediacy of triplet states in the luminescence scheme for energy deficient systems is to observe the effect of an otherwise inert triplet quencher on the emission spectrum or intensity. Difficulty in finding a suitable donor–acceptor pair for triplet–triplet energy transfer unfortunately delayed positive results for some time. The problem lies in the strong correlation between a molecule's lowest triplet level and its standard potentials for electrode processes. Within a series of compounds of similar nature, the lower the triplet lies, the easier electrochemical conversion usually comes about. Thus one is taxed to find an interceptor with a suitably low triplet energy and with sufficiently extreme electrochemical transformations that its homogeneous and heterogeneous electron transfer processes will not interfere with the reaction to be probed. Success can be achieved if the ion precursors have a rather different electronic nature from that of the interceptor, and useful results were reported by Freed and Faulkner for oxidations of the fluoranthene anion by cations derived from aromatic amines, particularly 10-methylphenothiazine (10-MP)[80].

The fluoranthene(−)/10-MP(+) reaction in DMF yields the broad emission spectrum displayed in Figure 6.3a. Structure is slight, but it suggests

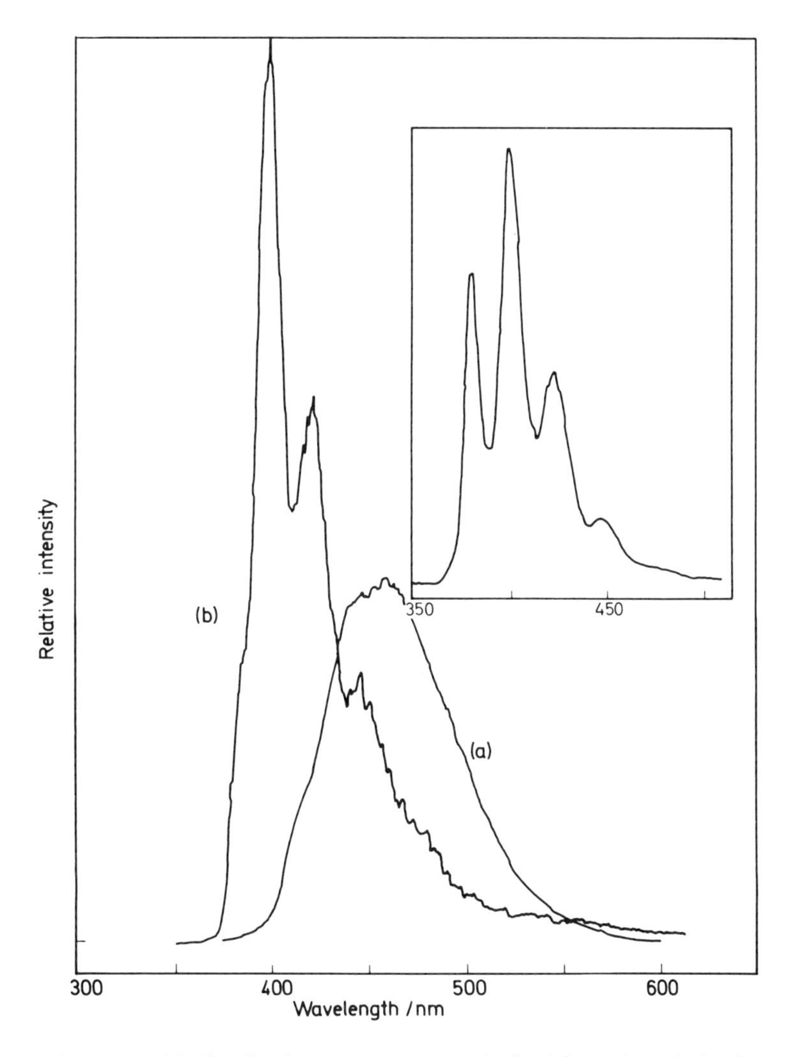


**Figure 6.3** (a) Chemiluminescence spectrum obtained from electrolysis of a DMF solution containing 1 mM fluoranthene and 1 mM 10-MP. Alternating steps at −1.75 V and + 0.88 V *vs.* SCE were used. (b) Chemiluminescence spectrum obtained under the same conditions with 1 mM added anthracene. Inset shows anthracene fluorescence spectrum for a $10^{-5}$ M DMF solution. Reabsorption reduces the 0,0 intensity in (b) (Adapted from Freed and Faulkner[80], by courtesy of the American Chemical Society)

fluoranthene's first excited singlet as the emitting species. However, fluoranthene and 10-MP have very similar fluorescence spectra, so one cannot rule out a chemiluminescence component from 10-MP. In any circumstance, Table 6.1 shows the system to be markedly energy deficient for emitter production, but reference to Table 6.2 shows that either product triplet is probably accessible.

**Table 6.2 Electrochemical and spectroscopic data for interception experiments**

| *Compound*, R | $E_p(R/R^+)$[b] | $-E_p(R/R^-)$[b] | $E(^1R^*)$[c]/eV | $E(^3R^*)$/eV |
|---|---|---|---|---|
| (a) Reductant precursor: | | | | |
| Fluoranthene | NO[d] | 1.76 | 3.0 | 2.3[e] |
| (b) Oxidant precursors: | | | | |
| TMPD | 0.24 | NR[d] | 3.5 | 2.7[f] |
| DMPD | 0.40 | NR[d] | 3.5 | 2.7[g] |
| 10-PP | 0.83 | NR[d] | 3.4 | 2.4[h] |
| 10-MP | 0.82 | NR[d] | 3.4 | 2.4[h] |
| (c) Interceptors: | | | | |
| Pyrene | 1.29 | 2.15 | 3.4 | 2.1[e] |
| Anthracene | 1.36 | 2.00 | 3.2 | 1.8[i] |
| *trans*-Stilbene | NO[d] | 2.16 | 3.8 | 2.1[i] |
| Naphthalene | NO[d] | 2.55 | 3.9 | 2.6[i] |

[a] Abbreviations are defined in note (*a*) to Table 6.1. Data taken from Refs. 76 and 80
[b] Cyclic voltammetric peak potentials for DMF solutions at ~200 mV $s^{-1}$ scan rate. In V *vs.* SCE
[c] From fluorescence spectra in DMF
[d] NO = not oxidised; NR = not reduced
[e] Clar, E. and Zander, M. (1956). *Chem. Ber.*, **89**, 749
[f] Skelly, D. W. and Hammill, W. H. (1965). *J. Chem. Phys.*, **43**, 3497
[g] Estimated from TMPD data
[h] Estimated from value for phenothiazine. Lhoste, J. M. and Merceille, J. B. (1968). *J. Chem. Phys.*, **65**, 1889
[i] McGlynn, S. P., Padhye, M.R. and Kasha, M. (1955). *J. Chem. Phys.*, **23**, 593
[j] Hammond, G. S. *et al.* (1964). *J. Amer. Chem. Soc.*, **86**, 3197

Table 6.2 collects electrochemical and spectroscopic data for the compounds used in the interception experiments[80]. One can see that fluoranthene is more easily reduced and 10-MP is more easily oxidised than all the potential interceptors. The reactant ions for the redox process at hand can therefore be electrogenerated from DMF solutions of fluoranthene and 10-MP without interference from any of these other substances, should they be added.

The spectroscopic studies featured light generation by continuous cycling of a working electrode between $-1.70$ and $+0.90$ V *vs.* SCE. Solutions contained 1 mM levels of the ion precursors and the candidate interceptor in DMF with 0.1 M TBAP. Addition of anthracene had the dramatic effect of transforming the broad fluoranthene-like emission to anthracene's charateristic structured spectrum, as shown in Figure 6.3b. Addition of pyrene had a similar effect, whereas *trans*-stilbene quenched chemiluminescence entirely. Naphthalene had no effect.

Each molecule which altered the emission has a lower triplet than fluoranthene (or 10-MP), whereas naphthalene does not. All the observations may be explained by invoking interception of the triplet intermediates yielding normal chemiluminescence. With anthracene and pyrene, the interception produces the triplet states of those molecules, which may themselves undergo triplet–triplet annihilation and yield their own fluorescence as a result. On the other hand, the stilbene triplet is too short-lived to participate in annihilation (because it undergoes *cis–trans* isomerisation[81]; Section 6.5.1), so one expects the observed total quenching in that case. Similar results have also been recorded for reactions between cation radicals derived from 10-phenylphenothiazine (10-PP) and TPTA[82–84].

Related studies showed that no emission accompanied the oxidation of fluoranthene(−) by TMPD(+) or by the *N,N*-dimethyl-*p*-phenylenediamine cation[62,80]. Since these reactions yield just slightly less energy (1.9 and 2.1 eV, respectively) than is required for population of the fluoranthene triplet at 2.3 eV, the negative results tend to confirm the view that access to that state is a prerequisite for emission.

These parallel experiments leave little doubt that triplets, probably those of fluoranthene[80,82], are required intermediates for chemiluminescence from the systems at hand. In spectroscopic studies, fluoranthene has been shown to undergo triplet–triplet annihilation[85]; hence the evidence strongly supports the T-route as the mechanism for chemiluminescence in the present cases and, by inference, in other energy-deficient aromatic systems as well.

#### 6.4.2.2 *Analysis of excimer emission*

Long-wavelength excimer bands in chemiluminescence have drawn special attention since Chandross, Longworth and Visco suggested that they might sometimes arise directly from a homomolecular ion annihilation such as that producing pairs of 9,10-dimethylanthracene (DMA) molecules[86]. In addition, they held out possibilities for testing the importance of triplet–triplet annihilation in the emission mechanism. This opportunity seemed to extend from detailed studies conducted by Birks, Parker and Stevens into excimer bands in prompt and delayed fluorescence[52,53,87]. An excimer, $A_2{}^*$, of a molecule A may arise in prompt fluorescence only through the optically excited singlet

$$^1A^* + A \rightleftharpoons A_2^* \tag{6.20}$$

On the other hand, triplet–triplet annihilation, which yields P-type delayed fluorescence, apparently has parallel pathways to $^1A^*$ and $A_2^*$.

$$^3A^* + {^3A^*} \begin{cases} \rightarrow {^1A^*} + A \\ \quad\quad \downarrow\uparrow \\ \rightarrow {^1A_2^*} \end{cases} \tag{6.21}$$

There is widespread agreement on this point, even though controversy remains over the existence of a common intermediate for the two paths.

Parker and Hatchard[88] and Birks[89] have shown, however, that one can disregard some mechanistic detail and treat the system through the scheme:

$$\xrightarrow{R_M} {^1A^*} \underset{k_d}{\overset{k_a[A]}{\rightleftharpoons}} A_2^* \xleftarrow{R_D} \qquad 1/\tau_M \downarrow \qquad \downarrow 1/\tau_D \tag{6.22}$$

Here, $R_M$ and $R_D$ are independent rates of formation of excited monomer and dimer, and $\tau_M^{-1}$ and $\tau_D^{-1}$ represent the sums of the pseudo-first-order rate constants for all processes other than those depicted, which dispose of $^1A^*$ and $A_2^*$. Of course, the latter quantities contain the two rate constants for fluorescence, $k_f$ and $k'_f$, respectively. In a system at steady state, these relations imply that the integrated dimer-to-monomer emission ratio, $\phi_D/\phi_M$, is given by

$$\frac{\phi_D}{\phi_M} = \frac{k'_f[A_2^*]}{k_f[A^*]} = \left[\frac{k'_f\tau_D}{k_f(1 + \alpha\tau_D k_d)}\right]\left\{\frac{\alpha}{\tau_M} + (1 + \alpha)k_a[A]\right\} \tag{6.23}$$

The quantity $\alpha$ is $R_D/R_M$.

Note that $\phi_D/\phi_M$ is always a linear function of [A], and when $\alpha = 0$, as in prompt fluorescence, one expects a zero intercept and a slope of $k'_f k_a \tau_D/k_f$. Whenever the dimer is produced independently, a non-zero intercept and a slope no less than for $\alpha = 0$ should be observed. Prompt and delayed fluorescence emissions from several molecules conform to these respective patterns[52,53,88,89]; hence the results for delayed fluorescence have been cited as evidence for direct excimer generation in triplet–triplet annihilation.

Parker and Short were the first to recognise the utility of scheme (6.22) for treating excimer bands in redox chemiluminescence[90]. Because the origins of $^1A^*$ and $A_2^*$ are not specified, relation (6.23) can hold without alteration. These authors studied the dimer/monomer emission ratio for the 9,10-dimethylanthracene anion–cation reaction in DMF with 0.1 M tetraethylammonium perchlorate as supporting electrolyte; luminescence was generated as a pulse train by alternating-step electrolysis. It is not entirely clear that the steady state kinetics implied by (6.23) pertain to this situation, but the practice has been used and defended for the analysis of the dynamically similar single-pulse experiments[73]. The linear plots reported for $\phi_D/\phi_M$ *vs.* DMA concentration (a) displayed larger slopes than those for prompt fluorescence, but (b) the intercepts were substantially larger than those found for delayed fluorescence[90,91]. These observations were taken as evidence for direct production of the excited dimer in the electrogenerated chemiluminescence (ECL), but Parker and Short ruled out triplet–triplet annihilation as the main source of excited dimers on the basis of observation (b). Since the redox process is virtually energy sufficient [about the same as DPA(+)/DPA(−), Table 6.1], there is no compelling reason to invoke triplet intermediates, and one can view the excimer emission as resulting from

$$DMA^{+\cdot} + DMA^{-\cdot} \rightarrow (DMA)_2^* \qquad (6.24)$$

as suggested first by Chandross *et al.*[86]. Werner, Chang and Hercules, who later observed a much weaker excimer emission from considerably more dilute DMA solutions, challenged this interpretation and suggested that the longer-wavelength component observed earlier could be due to side reactions of DMA(+), participation of dimeric cations ($DMA_2^{+}$ or $DMA_2^{2+}$), or a concentration-dependent T-route component[48]. This issue has not yet been settled.

In a later study of pyrene(−)/TMPD(+), Maloy and Bard also relied on scheme (6.22)[50]. In this case a process like (6.24) is clearly ruled out, and energy calculations reveal a marked energy deficiency. Nonetheless, the

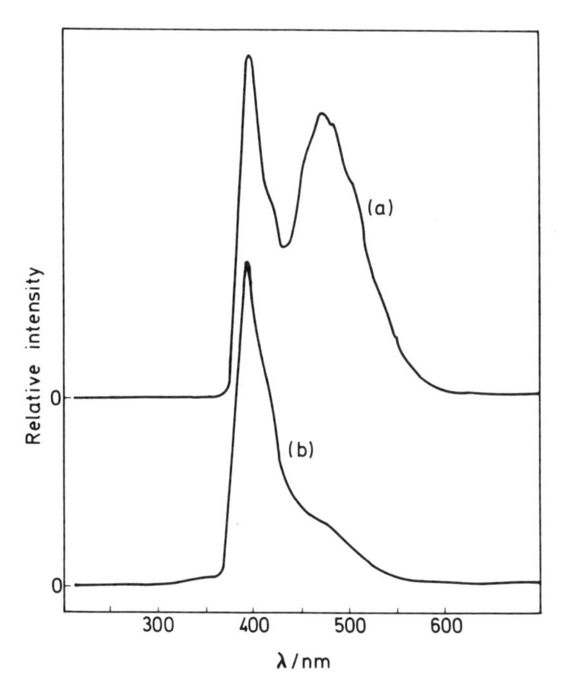


**Figure 6.4** (a) Chemiluminescence from the pyrene(−)/TMPD(+) reaction in DMF. Electrolysis involved solutions which were 1 mM in TMPD and 5 mM in pyrene. (b) Fluorescence from the same solution as excited by 350 nm radiation. The pyrene excimer band shows a maximum at *ca.* 470 nm (Adapted from Maloy and Bard[50], by courtesy of the American Chemical Society)

chemiluminescence spectrum, as earlier workers had noted[49,62], displays both a strong monomer component and a sizable excimer emission. Figure 6.4 shows that the latter is much more significant in ECL than in prompt fluorescence. Since Maloy and Bard generated their luminescence from DMF solutions at steady state with a rotating ring–disc, relation (6.23) should hold strictly. Their results for $\phi_D/\phi_M$ as a function of pyrene concentration, which are shown in Figure 6.5, clearly support independent production of the

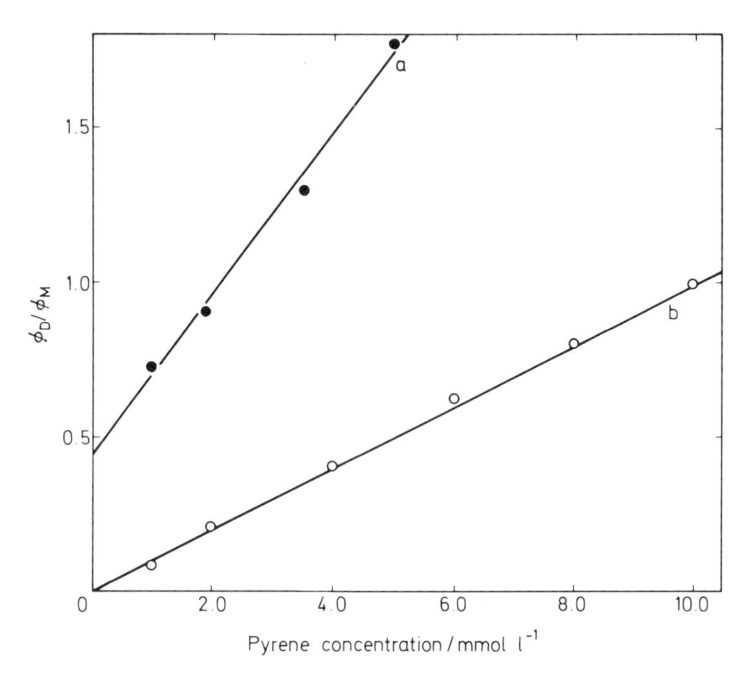


**Figure 6.5** Excimer-to-monomer emission ratios for pyrene in DMF. (a) ECL from pyrene(−)/(TMPD(+). (b) Prompt fluorescence (Adapted from Maloy and Bard[50], by courtesy of the American Chemical Society)

excited dimer and monomer species. The means by which these states are populated cannot be identified, but the triplet–triplet annihilation implicit in the T-route is an attractive possibility.

In a complementary study, Weller and Zachariasse examined the temperature dependence of the DMA excimer band arising from the energy-deficient TMPD(+)/DMA(−) reaction in THF[92]. They employed bulk reaction of $2 \times 10^{-5}$–$2 \times 10^{-4}$ M solutions of the anion with $TMPD^{+\cdot}$ $ClO_4^-$. The perchlorate is not very soluble in THF, so it is introduced as the solid, but the actual redox process seems to involve dissolved cations[20a]. In any case, the DMA ground-state species was always at a much lower concentration than that required (∼10 mM) to induce excimer emission via (6.20). Figure 6.6 displays the excimer-to-monomer emission ratio *vs.* temperature for chemiluminescence and for delayed fluorescence from DMA dissolved in ethanol and DMF. The existence of a maximum in $\phi_D/\phi_M$ is a general

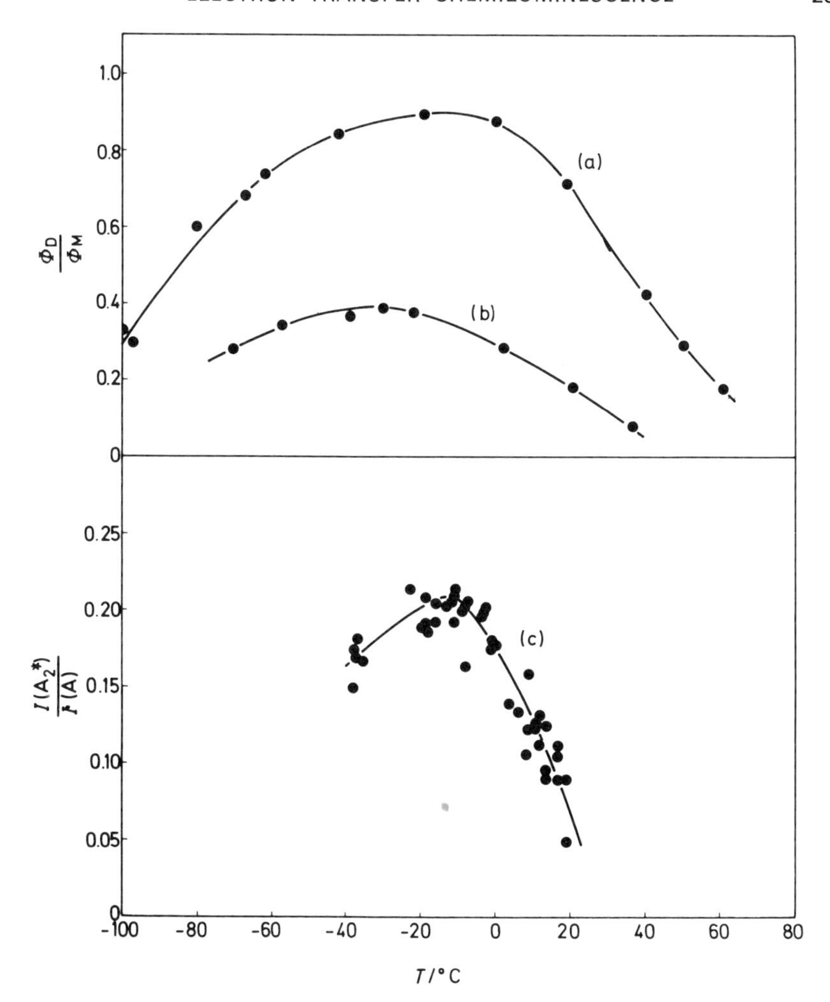


**Figure 6.6** Excimer-to-monomer emission ratios for 9,10-dimethylanthracene (DMA) luminescence. (a) Delayed fluorescence of $5.5 \times 10^{-5}$ M DMA sensitised by 313 nm excitation of 2.4 mM naphthalene in ethanol. (b) Delayed fluorescence from the same system in DMF. (c) Chemiluminescence from the DMA(−)/TMPD(+) reaction in THF (Adapted from Parker and Joyce[91] and Weller and Zachariasse[92], by courtesy of the Chemical Society and North Holland Publishing Company)

feature of excimer formation via triplet–triplet annihilation, though its magnitude and position in the 0 to −60 °C range varies with the emitter and the medium[52,53,91]. The decline with higher temperatures is ascribed to thermal dissociation of the excimer, whereas that at lower temperatures apparently reflects selective inhibition, through increased viscosity, of the excimer channel in (6.21) with respect to that producing excited monomer. The latter seems to operate via a longer-range interaction which is comparatively insensitive to solvent viscosity. From the similar temperature dependence of $\phi_D/\phi_M$ for the chemiluminescence and for DMA delayed fluorescence, Weller and Zachariasse suggested that the former arises via the T-route.

### 6.4.2.3 *Magnetic field effects*

The idea that the effect of a magnetic field on chemiluminescent systems might yield mechanistic information was derived from work reported by Johnson *et al.* concerning field effects on delayed fluorescence from crystalline anthracene[93–95]. These authors found the luminescence intensity from a single crystal to depend in a complex fashion on the strength of an external field. The size of the deviation from the zero-field intensity was markedly anisotropic, but a common functional form for intensity *vs.* field strength

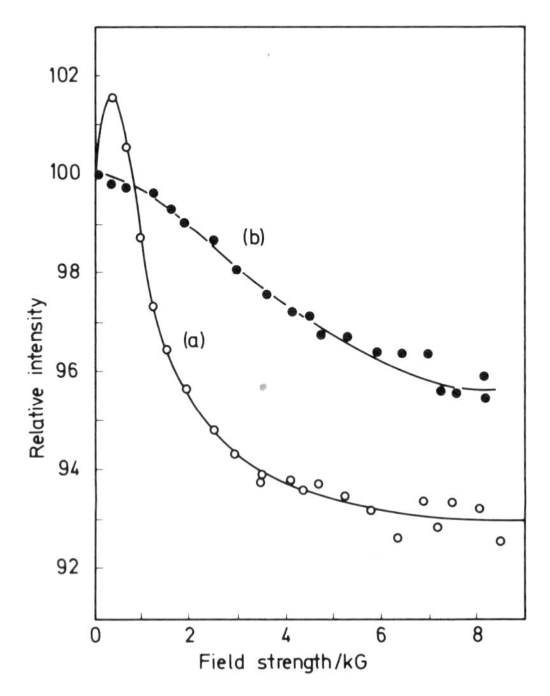


**Figure 6.7** Magnetic effects on room temperature delayed fluorescence from anthracene. (a) Polycrystalline powder. (b) DMF solution, $8 \times 10^{-5}$ M; excitation at 366 nm (Adapted from Faulkner and Bard[107], by courtesy of the American Chemical Society)

applied to all orientations. Figure 6.7a, which presents the effect for a powder sample of anthracene, shows the typical pattern. Using pulsed field methods, Johnson *et al.* traced the effect to a field-dependent rate constant for the triplet exciton annihilation

$$T_1 + T_1 \xrightarrow{k(H)} S_1 + S_0 \qquad (6.25)$$

Merrifield has presented a theory[96,97], extended by others[98–102], which successfully accounts for these observations. The essence of it is that an angular momentum selection rule exists for the conversion of the reactant pair of triplets into the product singlet pair. Since each triplet may be in one of three sublevels, the reactant pair state is one of nine possibilities, each of which can be formed with equal probability. This pair state is a complex combination of singlet, triplet and quintet components, but Merrifield postulates that the probability for conversion to the pure singlet is proportional to the singlet amplitude of the pair. The macroscopic rate constant therefore reflects the average transmission probability for the nine pair states. Since the triplet eigenfunctions change with magnetic field strength, the singlet amplitude distribution depends on the field strength, and a field-dependent macroscopic rate constant is forecast. Merrifield's theory correctly predicts the shape of the dependence, and it has been used to extract some quantitative information from the magnitude of the effect.

The theory also suggests that a field-dependent rate constant should characterise any bimolecular process featuring a change in the reactant pair's spin character, if it is not externally constrained by factors such as diffusion. Magnetic effects were consequently discovered and treated theoretically for such transformations as the fission of tetracene excitons into two triplets[103–106], the fluid solution triplet–triplet annihilation process[107,108], and quenching of triplets by doublet and triplet species (e.g. radical ions and

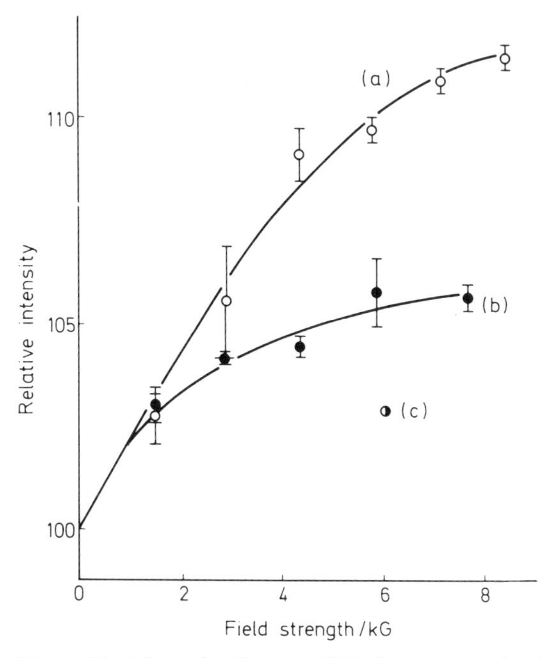


**Figure 6.8** Magnetic effects on ECL from rubrene(+)/rubrene(−). (a) In DMF. (b) In benzonitrile. (c) In THF. All solutions contained 0.1 M tetra-n-butylammonium perchlorate (Adapted from Tachikawa[8], by courtesy of the author)

**Table 6.3 Magnetic field effects**

| Reactants[a] Oxidant | Reductant | Solvent[b] | $\Delta H^{\circ}_{ex}$ /eV[c] | Field effect /%[d] | Field strength /kG | Ref. |
|---|---|---|---|---|---|---|
| Thianthrene(+), [1.0] | PPD(−), [0.95] | AN | − 0.5[g] | − 0.4 ± 0.3 | 6.15 | 46 |
| DPA(+), [11.0] | DPA(−), [11.0] | DMF[e] | − 0.1 | − 0.3 ± 0.8 | 7.5 | 76 |
| TPP(+), [0.15] | TPP(−), [0.15] | DMF | + 0.1 | + 0.2 ± 0.2 | 6.15 | 76 |
| Rubrene(+), [0.5] | Rubrene(+), [0.5] | THF[f] | − 0.1 | + 2.9 | 6.0 | 116[a] |
| [2.0] | [2.0] | BZN | + 0.1 | + 6 ± 1 | 6.0 | 116[a] |
| [0.1] | [0.1] | AN | + 0.1 | + 7.8 ± 0.2 | 6.0 | 116[a] |
| [0.94] | [0.94] | DMF | 0.0 | + 8.5, + 7.8[h] | 5.0 | 118 |
| [1.0] | [1.0] | DMF | 0.0 | + 9.5 ± 0.2 | 5.7 | 76 |
| 10-MP(+), [2.4] | Fluoranthene(−), [3.0] | DMF | + 0.6 | + 7.0 ± 0.4 | 6.15 | 76 |
| TMPD(+), [1.5] | Rubrene(−), [1.0] | DMF | + 0.8 | + 28.2 ± 0.2 | 5.7 | 76 |
| TMPD(+), [0.5] | Tetracene(−), [0.5] | DMF | + 0.9 | + 15.7 | 6.15 | 114 |
| TMPD(+), [14.6] | DPA(−), [10.5] | DMF[e] | + 1.0 | + 5.6 ± 0.3 | 6.0 | 76 |
| TMPD(+), [0.45] | DNA(−), [0.27] | DMF | + 1.0 | + 18.5 ± 0.4 | 5.7 | 8 |
| Rubrene(+), [1.0] | *p*-Benzoquinone(−), [2.4] | DMF | + 1.0 | + 15.7 ± 0.2 | 5.7 | 76 |
| TMPD(+), [14.4] | Anthracene(−) [10.5] | DMF[e] | + 1.1 | + 16 | 6.25 | 76 |
| TMPD(+), [0.23] | TPP(−), [0.15] | DMF | + 1.1 | + 26.0 ± 0.2 | 6.15 | 76 |
| TPTA(+), [1.0] | DMA(−), [1.0] | THF | + 0.3 | + 12.4 | 6.15 | 116[b] |
| [1.0] | [1.0] | DMF | + 0.4 | + 7.0 | 7.5 | 116[b] |
| TMPD(+), [1.0] | Pyrene(−), [1.0] | DMF | + 1.3 | + 17 ± 1, + 19 ± 1[i] | 6.15 | 116[b] |
| TPTA(+), [1.0] | 9-MA(+), [1.0] | THF | + 0.4 | + 8.5, + 5.6[k] | 6.15 | 116[b] |

[a] DNA = 9,10-di-α-naphthylanthracene; DMA = 9,10-dimethylanthracene; 9-MA = 9-methylanthracene. Other abbreviations are defined to note (*a*) in Table 6.1. Reactant precursor concentrations [mM] are bracketed
[b] AN = acetonitrile with 0.1 M tetra-n-butylammonium perchlorate (TBAP); THF = tetrahydrofuran with 0.2 M TBAP; DMF = *N*,*N*-dimethylformamide with 0.1 M TBAP
[c] For excitation to emitter's first excited singlet state. See Table 6.1 and cited reference for parameters needed in this calculation
[d] Percentage change in intensity from zero-field value
[e] 0.2 M TBAP
[f] 0.1 M TBAP
[g] See note (*d*) to Table 6.1
[h] First entry pertains to initial cathodic step. Second is for initial anodic step
[i] Depends on TBAP concentration. See Section 6.2.3.3
[j] Entries for monomer and excimer emissions, respectively
[k] Entries for 415 nm (9-MA) and 520 nm (heteroexcimer?) emissions, respectively

oxygen)[100,101,109–112]. Only the latter two categories are important for our discussion below. We should note in advance that the prediction of theory and the result of optical luminescence experiments is that both processes for fluid media are almost always inhibited, with respect to their zero-field rates, by an impressed field. A theoretical exception to this rule may apply to triplet–triplet interactions involving unusually large charge-transfer splitting of pair states, but no experimental evidence has been recorded for it[100,101]. Typical results for anthracene triplet annihilation are shown in Figure 6.7b.

The possibility that different behaviour patterns might be observed for energy-deficient and energy-sufficient chemiluminescent systems led Faulkner and Bard to begin their studies of magnetic effects in 1968[113]. The T- and S-routes were hypothetical schemes for these cases at the time, and the triplet annihilation step postulated for the former suggested that energy-deficient cases might exhibit a field dependence that would not be observed for the energy-sufficient ones. Their initial work was extended by Tachikawa and Bard[8,13,46,76,114–116] and, later, by Santhanam[117,118]. Accounts of experimental detail are available[8,19,76]; hence we shall mention here only that ECL intensity was measured in each case by the heights of pulses created in alternating step electrolysis. Field effects on these intensities were commonly observed, but, surprisingly, they always took the form of monotonic enhancements over the zero field emission rates like those shown in Figure 6.8. An enhancement appears ordinarily to approach a saturation level; hence the comprehensive summary of reported data in Table 6.3 lists only the enhancement for a single field strength near 6 kG. Note that, apart from the direction of the effect, the original anticipation was borne out: every markedly energy-deficient system shows an enhancement, whereas each energy-sufficient or marginally sufficient reaction does not. The rubrene homomolecular ion annihilation is an exception we shall examine separately below.

The chemiluminescence mechanism for energy-deficient systems apparently involves field-influenced steps which are absent from the energy sufficient cases. Paramagnetic species probably participate in those steps, but triplet–triplet annihilation cannot be the sole source of the magnetic effect because that process is impeded by the field whereas the ECL intensity is increased. Several other possibilities must therefore be considered. Among them, in principle, are diffusion of charged species and the electrode processes used to generate the ions. However, these processes have been ruled out by Tachikawa's examination of field effects on the electrochemistry of TMPD, TPP and perylene[8]. The field failed to influence voltammetric or step behaviour. By the Merrifield treatment[96–102], one could also suspect the ion annihilation as the source of the ECL enhancements, but it is difficult to explain the field-independence of the energy sufficient systems. However, a completely consistent rationale has been developed within the framework of the T-route by considering field effects on the quenching of triplet intermediates by paramagnetic species. Each chemiluminescent reaction considered here takes place in a zone which is co-occupied by radical ions and residual or electrogenerated oxygen. As effective quenchers, they would partially, even largely, control the lifetimes of triplet states and, consequently, the emission rate from a T-route system.

Thus an inhibition of the quenching, as predicted by theory, would give rise to longer lifetimes and a greater opportunity for triplet–triplet annihilation. Since the intensity of luminescence arising from the latter process depends on the square of the triplet lifetime (whenever triplet annihilation does not influence the lifetime), the increase in lifetime enforced by the field will ordinarily override the decline in the rate constant for triplet annihilation. One should therefore observe a field enhancement of delayed fluorescence intensity when such quenching processes are significant regardless of whether optical or redox excitation is used. Bard and his co-workers verified this expectation for optically excited delayed fluorescence of anthracene in the presence of TMPD(+), *p*-benzoquinone(−) and oxygen[110–112].

In principle, one ought to be able to achieve a quantitative correlation between the magnitudes of the effect on this delayed fluorescence and that arising from ECL. However, the required quencher levels for optical experiments are extremely low ($\sim 10^{-7}$ M), and quantitatively significant results are elusive because trace solutions of these reactive substances are not stable. The interpretation of the field effect (or the lack of one) on ECL must therefore rest substantially on the plausibility of the argument given above. In the considerable body of reported data, no obstacles to its acceptance presently exist.

Tachikawa and Bard have performed three studies of special interest, but we shall reserve discussion of one of them (field effects on hetero-excimer emission) for Section 6.4.3. Perhaps the most important of their efforts concerns the rubrene homomolecular ion annihilation in DMF, acetonitrile, benzonitrile and THF[8,116a]. We noted above that this process was the sole field-dependent example among energetically marginal or sufficient systems; hence the interpretation outlined above suggests that a substantial fraction of its emission arises via the T-route. Figure 6.8 shows interestingly that the magnitude of the effect declines with decreasing solvent polarity. Tachikawa and Bard noted a possible trend in $\Delta G^{O}_{ex}$ toward more negative values, perhaps because ionic solvation energies decline. They have therefore suggested that the change in the field effect with decreasing solvent polarity may indicate a mixed ST mechanism in which the S component increases as the system moves toward (or past) the threshold for energy sufficiency.

Tachikawa and Bard also examined field effects on the monomer and excimer bands from TMPD(+)/pyrene(−) and found identical enhancements[116b]. In a supporting study of pyrene's delayed fluorescence (without added quencher) they again observed identical effects on emission from the two bands, but, of course, the field decreased the intensities. Though this body of evidence does not confirm the T-route for ECL, it certainly is consistent with that mechanism.

#### *6.4.2.4 Analysis of luminescence transients*

Work in this category involves an examination of the shapes of single pulses of light emanating from electrochemical step experiments like that described in Section 6.3.4. It is generally based upon a theoretical analysis of diffusion and reaction dynamics which was first introduced by Feldberg[70,71] and later refined by Bezman and Faulkner[73]. The treatment depends chiefly upon the assumptions that (a) the generating potentials yield mass transfer limited currents; (b) changes in potential are established instantly; and (c) the redox process is sufficiently fast that it exerts no effect on the luminescence curve shape. Assumptions (a) and (b) can be satisfied experimentally with moderate ease, and (c) has been shown by Van Duyne and Drake[6,7] to be valid for a forward step duration $t_f$ greater than *ca.* 100 μs.

A unified treatment arises from the following composite (ST) reaction scheme:

$$R^+ + R^- \xrightarrow{\phi_s} {}^1R^* + R \qquad (6.26a)$$

$$R^+ + R^- \xrightarrow{\phi_t} {}^3R^* + R \qquad (6.26b)$$

$$R^+ + R^- \longrightarrow 2R \qquad (6.26c)$$

$${}^3R^* + {}^3R^* \xrightarrow{k_a} \xrightarrow{k_s} {}^1R^* + R \qquad (6.26d)$$

$${}^3R^* + {}^3R^* \xrightarrow{k_a} \xrightarrow{k_t} {}^3R^* + R \qquad (6.26e)$$

$${}^3R^* + {}^3R^* \xrightarrow{k_a} \xrightarrow{k_g} 2R \qquad (6.26f)$$

$${}^3R^* \ (+\ Q) \xrightarrow{1/\tau} R(+\ Q) \qquad (6.26g)$$

$${}^1R^* \xrightarrow{\phi_f} R + h\nu \qquad (6.26h)$$

where the $k$'s, $\phi$'s and $\tau$'s are the pertinent rate constants, branching efficiencies and lifetimes, respectively. The calculation of luminescence intensity is complicated vastly by the non-uniform concentrations of all key substances along the axis perpendicular to the generating electrode, but Feldberg simplified the problem by breaking it into two segments[71].

Since the rate at which (6.26a–c) occurs depends only on the relevant diffusion coefficients and the electrode boundary conditions, the time profile of the reaction rate $N$ (mol s$^{-1}$) is accessible without reference to (6.26d–h). It has been obtained through a finite difference method; hence dimensionless variables have been employed consistently. For example, time is studied as $t_r/t_f$, where the actual time is measured into the second step is $t_r$. The rate curve can usually be linearised for $t_r/t_f > 0.2$ by

$$\log \omega_n = a_n + b_n(t_r/t_f)^{\frac{1}{2}} \qquad (6.27)$$

where the dimensionless redox rate $\omega_n$ is $Nt_f^{1/2}/AD^{1/2}C$, with $A$ as the electrode area and $D$ as the diffusion coefficient of the chemiluminescent substrate, which is present initially at concentration $C$. For stable reactants derived from the same precursor, $a_n = 0.71$ and $b_n = -1.45$, and essentially the same values apply for separate precursors if their initial concentrations are equal. If a reactant is unstable, the redox rate naturally falls off faster; hence $b_n$ becomes more negative. Actually, $\omega_n$ is a multicomponent exponential in $(t_r/t_f)^{1/2}$ with important fast-decaying terms, but (6.27) is a good approximation at the longer times indicated[119].

The second segment of the problem is treated by regarding (6.26a–c) as occurring at the instantaneous rate given[71] by (6.27) in a reaction zone of volume $fAD^{1/2}t_f^{1/2}$. The parameter $f$ can be regarded as a dimensionless zone thickness and probably has the order of $(t_f/\tau)^{1/2}$. Concentrations inside the zone are taken as uniform, and the absolute emission rate $I$ arising from the T-route is calculated via (6.26d–h) using the steady-state approximation for $^3R^*$. The latter simplification has been defended[73]. For the S-route, the contribution to the intensity is simply $\phi_s\phi_f N = \gamma N$, so the overall total dimensionless intensity is

$$\omega_i = \beta[1 - (1 + \alpha\omega_n/\beta)^{\frac{1}{2}}] + 0.5\alpha\omega_n + \gamma\omega_n \tag{6.28}$$

where $\omega_i = It_f^{1/2}/AD^{1/2}C$, $\alpha = \phi_f\phi_{tt}\phi_t/(1-g)$, $\beta = \phi_f\phi_{tt}ft_f/8k_a(1-g)^2\tau^2C$, $\phi_{tt} = k_s/k_a$, and $g = k_t/2k_a$. Note that $\alpha$ is a gauge of T-route efficiency and that $\beta$ reflects the effectiveness of quenching as a means for triplet removal.

When values for $\alpha$, $\beta$ and $\gamma$ are specified, one can straightforwardly calculate the intensity–time curve via (6.27) and (6.28). Ordinarily the plot of $\log \omega_i$ *vs.* $(t_r/t_f)^{1/2}$ (Feldberg plot) is virtually linear[73] for $t_r/t_f > 0.2$; hence

$$\log \omega_i = a_i + b_i(t_r/t_f)^{\frac{1}{2}} \tag{6.29}$$

Data analysis for transient experiments has almost universally involved these linearised curves. If $\alpha$ or $\gamma$ is zero, and one therefore deals with a purely S- or T-route system, experimental parameters from Feldberg plots can, in principle, be correlated with $\gamma$ or with $\alpha$ and $\beta$, as the case may be. However, numerical correlation requires absolute luminescence measurements, which have been reported for transients only by Bezman and Faulkner[73,120,121].

The slopes of Feldberg plots, which may be extracted from relative measurements, have been examined in several studies because they have a diagnostic value[71]. For an S-route system, $b_i = b_n$ always, but if a T component is important, $b_i$ depends markedly on $\beta$ and may be much more negative than $b_n$. When triplet quenching is not competitive with triplet–triplet annihilation, $b_i = b_n$ for both the T and the ST cases. As $\beta$ becomes larger, $|b_i|$ rises. For a T-route system, the rise is monotonic to a strong-quenching asymptote of $|2b_n|$, but the ST case shows a maximum in $|b_i|$ and a return in the strong quenching limit to $|b_i| = |b_n|$.

The first experimental examination of transients were reported simultaneously by Hercules and co-workers[122] and by Visco and Chandross[77]. Both concerned the rubrene homomolecular ion annihilation in benzonitrile. The former authors observed Feldberg slopes that varied systematically from $-1.45$ to $-2.90$ with step durations ($t_f$) decreasing from 9.2 to 0.043 s. In contrast, the latter investigators reported slopes distributed randomly in the interval from $-1.41$ to $-1.76$ for $t_f$ from 0.28 to 14.0 s. Bezman and Faulkner later studied the same system and found slopes that were independent of $t_f$ and rubrene concentration[121]. However, the slopes differed with the sequence of reactant generation: for $t_f = 1$ s, they were $-2.3$ and $-2.6$ with initial anion generation and $-1.9$ to $-2.2$ with the opposite order. The observation of slopes more negative than $-1.45$ has been taken as evidence for an important T component in the luminescence scheme. The differences among the three studies are not fully understood, but they seem partially ascribable to differences in triplet quenching rates. These results seem to support the conclusions drawn from magnetic effects, and they may even support the hypothesis of Tachikawa and Bard[116a] that the system at hand features an ST mechanism. For example, Bezman and Faulkner found that a T-route analysis of absolute transients yielded $\alpha$ values that differed with the generation sequence[121]. A more recent, but preliminary, treatment based on the ST mechanism eliminates this seemingly inexplicable inference[119]. It is also interesting that the Feldberg slopes from experiments with DMF solutions of rubrene seem too high to allow significant S component; thus the emission from this solvent may arise virtually via the T-route, as Tachikawa and Bard have suggested from other evidence.

The DPA homomolecular ion annihilation in DMF appears much less complex. For $t_f = 1$ s, Bezman and Faulkner observed Feldberg slopes ranging from $-1.84$ to $-2.28$, but they could show that the enhanced magnitudes (above 1.45) were wholly attributable to the instability of DPA(+)[120]. Thus they concluded that the system takes the S-route to emission, in accord with conclusions drawn from magnetic effects.

Van Duyne and Drake have also studied this system via a very sophisticated fast-step method which features acquisition of decay curves by the single-photon timing method[6,7]. With step times of the order of 10 μs, the redox process may not be regarded as infinitely fast and the decay curve for an S-route system adopts a shape which depends on its rate constant[70]. These workers have observed the predicted shape variations with changing $t_f$, and they have extracted a rate constant for (6.26a).

Still other studies have been carried out by Bezman and Faulkner for fluoranthene(−)/10-MP(+) in DMF[73] and by Grabner and Brauer for TPTA(+)/perylene(−), perylene(+)/benzil(−), and perylene(+)/perylene(−) in benzonitrile[123,124]. The first two reactions produced Feldberg slopes much steeper than $-1.45$; hence the T-route was implicated. They are both energy deficient and are uncomplicated electrochemically, so the conclusions are not surprising. Similar inferences were drawn for the latter two systems, but the interpretations are clouded by the very complicated electrochemistry surrounding perylene(+)[125].

The future will probably see wider use of transient analysis because it offers both diagnostic and quantitative features. Unfortunately the quantitative consistency of its results is not yet known, and one must still regard the theoretical treatment as incompletely tested. Much more work with well-behaved systems and more direct data analysis is needed.

### 6.4.3 Heteroexcimer emission from reactions in non-polar media

Any excited molecule is a stronger oxidant and a stronger reductant than the same molecule in the ground state; hence one may expect the excited state to undergo processes that are not spontaneous for the unexcited species. In 1963, Leonhardt and Weller confirmed this expectation and showed that such excited state donor–acceptor interactions could lead to emission from a charge-transfer complex[54]. Their work was expanded by others in Weller's group and elsewhere, and a firm picture of the phenomenon has evolved[16,55–57,126–129].

Spectroscopic evidence has arisen from studies of fluorescence quenching by electron donors (D) such as *N*,*N*-diethylanilene (DEA) or acceptors (A) such as 1,2,4,5-tetracyanobenzene. Anthracene quenching by DEA follows a pattern that applies to more than 100 reported systems[16,55,57]. It depends sharply on solvent polarity[54]. With toluene as the solvent, addition of DEA introduces a broad emission band situated some 6000 $cm^{-1}$ to the red of the structured anthracene component. The latter declines in intensity with rising DEA concentration, and an isoemissive point is observed. Repeating the experiment in a more polar solvent such as DME yields similar behaviour, but the longer-wavelength emission has a much reduced intensity. If $\varepsilon > 30$, as for acetonitrile, no longer-wavelength component appears at all; only quenching of anthracene emission is recorded.

Weller and his co-workers have rationalised these observations through the following scheme:

$$\begin{array}{ccccc} & & \phi_f' & & \\ {}^1A^* + D & \rightleftharpoons & (A^-D^+) & \rightarrow & A^- + D^+ \\ & \swarrow & & \searrow & \\ A + D & & & & A + D + h\nu' \end{array} \qquad (6.30)$$

Quenching in non-polar solvents apparently proceeds almost exclusively through the intermediate heteroexcimer $(A^-D^+)$, which may give rise to its own emission with efficiency $\phi_f'$. In polar media the ions are stabilised with respect to the complex; hence ionic dissociation of $(A^-D^+)$ becomes more important and $\phi_f'$ is consequently reduced. Moreover, a direct path to the ions, excluding the heteroexcimer, seems increasingly important with rising polarity[16]. Evidence for a dominant charge-transfer character of this process is extensive. It initially included the observations (a) that the radical ion absorption spectra are seen for acetonitrile solutions following microsecond flashes[54] and (b) the emission maxima for a large number of heteroexcimers are linearly related to $E°(D/D^+) - E°(A/A^-)$ [16,57]. A later study of solvent effects on these emission maxima indicated substantial dipole moments (~10 D) for the complexes[56]. Most recently, elegant laser flash experiments have provided absorption spectra for several heteroexcimers and these consistently resemble the spectra of the related radical ions[126,127].

From the spectroscopic studies, it seemed that the ionic association process

$$D^+ + A^- \rightarrow (A^-D^+) \qquad (6.31)$$

might be spontaneous in non-polar media; hence Weller and Zachariasse began to test the idea in 1964 by undertaking a massive study of radical ion chemiluminescence from ether solutions. They examined the reactions of four radical cation perchlorates with a host of radical anions, which were produced by alkali metal reduction in DME, THF or MTHF. The cation precursors were TMPD, DMPD, TPTA, tri-(*p*-dimethylaminophenyl)amine (TPDA) and tri-*p*-anisylamine (TPAA)[15–17,20].

Evidence for process (6.31) comes most straightforwardly from the TPTA(+) reactions[15,17,20,59]. In the studies at hand, the concentrations of A and D were always so low that formation of the complex by association,

$$ {}^1A^* + D \rightarrow (A^-D^+) \qquad (6.32a)$$

$$^{1}D^{*} + A \rightarrow (A^{-}D^{+}) \tag{6.32b}$$

could not be considered. Moreover, almost all the reactions are energy deficient; hence direct formation of $^{1}A^{*}$ or $^{1}D^{*}$ must generally be ruled out. Weller and Zachariasse cite several reasons for regarding (6.31) as being much more efficient in generating $(A^{-}D^{+})$ than the mixed annihilation of redox-excited triplets

$$^{3}A^{*} + {}^{3}D^{*} \rightarrow (A^{-}D^{+}) \tag{6.33}$$

Among them is the observation that heteroexcimer emission frequently occurs when $^{3}D^{*}$ is not accessible to the redox process. In fact, neither triplet can be populated by TPTA(+)/benzophenone(−) in THF, yet heteroexcimer emission is recorded for that case[20a]. Moreover, Zachariasse has pointed out TPTA's poor ability to undergo (6.33) in view of its exceptionally short triplet lifetime (50 ns) in room-temperature ethanol[20a].

An interesting feature of the TPTA(+), TPDA(+) and TPAA(+) reactions is that they display luminescence efficiencies which exceed by one to two orders those typically recorded for systems producing all emission via the T-route[20,63] (see Section 6.5.1). Apparently the yield from (6.31) and the heteroexcimer emission efficiency are large enough to provide a much more probable path to luminescence than the easily intercepted T-route. If so, an intriguing question concerns the origin of luminescence from $^{1}A^{*}$ or $^{1}D^{*}$, which often equals or dominates the heteroexcimer component. Weller and Zachariasse suggested thermal dissociation of the complex as the source[20,59]

$$(A^{-}D^{+}) \underset{k_a}{\overset{k_d}{\rightleftharpoons}} {}^{1}A^{*} + D \tag{6.34a}$$

or alternatively

$$(A^{-}D^{+}) \underset{k_a}{\overset{k_d}{\rightleftharpoons}} A + {}^{1}D^{*} \tag{6.34b}$$

as energy constraints allow. This scheme, which provides a path to locally excited singlets without intermediate triplets, requires a thermally surmountable energy gap between the states represented as the two sides of (6.34). Since the energy of $(A^{-}D^{+})$ is virtually the same as that characterising the pair of free ions (relative to the pair of unexcited neutral species), this dissociative path becomes viable only for systems within a few tenths of an electron volt of energy sufficiency[20].

Weller and Zachariasse have supported their hypothesis with a study of the temperature effect on the relative emission intensities from $(A^{-}D^{+})$ and $^{1}A^{*}$ in reactions of TPTA(+) with DMA(−)[59]. Figure 6.9 displays the trend. Lower temperatures markedly enhance the relative intensity of heteroexcimer luminescence, as one expects if $^{1}A^{*}$ arises solely through dissociation (6.34a). The quantitative results portrayed in Figure 6.10 yield an activation energy of 12.0 kcal mol$^{-1}$.

An interpretation of this parameter can be derived from the following sequence[59], which we may designate as the 'C route':

$$A^{-} + D^{+} \xrightarrow{\phi_c} (A^{-}D^{+}) \underset{}{\overset{k_d}{\rightleftharpoons}} {}^{1}A^{*} + D \tag{6.35}$$

$(A^{-}D^{+}) \xrightarrow{k'_f} A + D + h\nu'$; $(A^{-}D^{+}) \xrightarrow{1/\tau_0} A + D$; $^{1}A^{*} \rightarrow A + D$; $^{1}A^{*} \xrightarrow{\phi_0} A + D + h\nu$

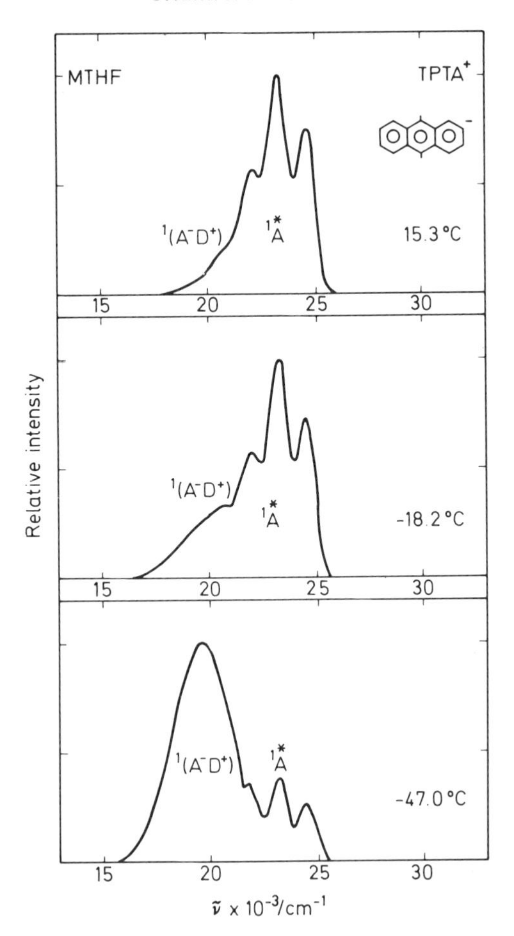


**Figure 6.9** Temperature dependence of the emission spectrum from TPTA(+)/DMA(−) in MTHF (Reproduced from Weller and Zachariasse[59], by courtesy of North Holland Publishing Company)

in which $\phi_c$ is the chemical yield of the heteroexcimer and the other parameters represent customary quantities as indicated. If triplet–triplet annihilation is neglected as an important route to emission, the chemiluminescence efficiencies for the $^1A^*$ and $(A^-D^+)$ components are given as $\chi$ and $\chi'$ by

$$\chi = \phi_c\, k_d\, \phi_o\, /(k_f' + 1/\tau_0 + k_d) \tag{6.36a}$$

$$\chi' = \phi_c\, k_f'/(k_f' + 1/\tau_0 + k_d) \tag{6.36b}$$

Thus the emission ratio of interest is

$$\chi/\chi' = k_d\phi_0/k_f' \tag{6.36c}$$

Over the 210–290 K temperature interval involved, $\phi_0/k_f'$ is probably

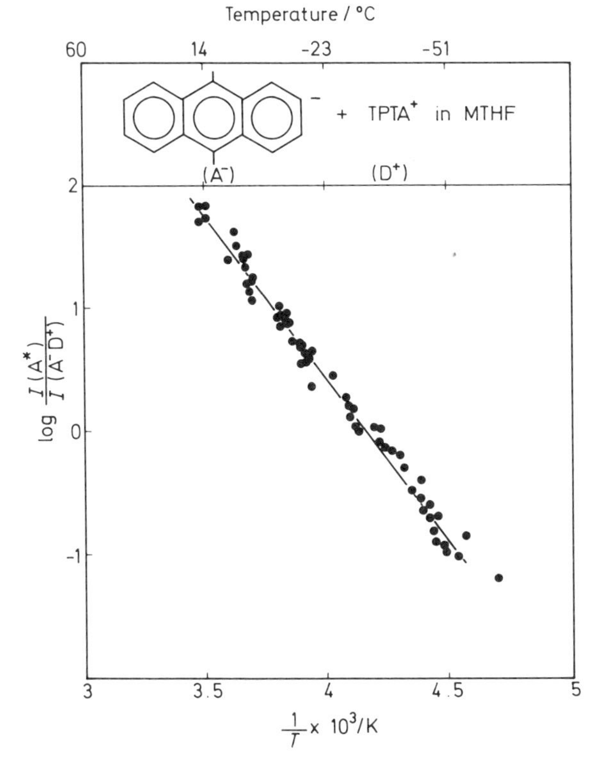


**Figure 6.10** Temperature dependence of the monomer-to-excimer emission ratio for the DMA(−)/TPTA(+) reaction in MTHF (Reproduced from Weller and Zachariasse[59], by courtesy of North Holland Publishing Company)

virtually constant; hence the activation energy associated with $\chi/\chi'$ would be that of $k_d$.

From (6.34) it is evident that

$$k_d = k_a \exp(-\Delta G^\circ_d / RT) \tag{6.37}$$

where $\Delta G_d$ is the free energy of dissociation. If $k_a$ is diffusion limited, as seems likely, the activation energy for $k_d$ is

$$E_d = \Delta H^\circ_d + E_{diff} \tag{6.38}$$

in which the standard enthalpy of dissociation is $\Delta H_d$ and the activation energy for diffusion is $E_{diff}$. For THF and MTHF, the latter quantity is 2.3 kcal mol$^{-1}$. Earlier studies by Knibbe *et al.*[129] provided an estimate of $\Delta H^\circ_d = 4.7$ kcal mol$^{-1}$ for ($TPTA^+DMA^-$) in hexane. For the change in solvation from hexane to the ethers, the Born model gives a correction of 4.8 kcal mol$^{-1}$. Thus Weller and Zachariasse[59] estimate $\Delta H^\circ_d$ for the ether solvents at 9.5 kcal mol$^{-1}$, a value they regard as reliable to $\pm 1$ kcal mol$^{-1}$.

This analysis therefore predicts an activation energy for $\chi/\chi'$ of 11.8 kcal mol$^{-1}$, which is in very good agreement with the observed value. An extension of the treatment has been presented for the anthracene(−)/TPTA(+)

and 9-methylanthracene(—)/TPTA(+) reactions[59]. It produced similarly good agreement; thus scheme (6.35) appears firmly supported as a luminescence mechanism for systems involving TPTA(+). It appears to hold also for the reactions of TPDA(+) and TPAA(+), although the T-route seems to play a larger role when the former oxidant is involved in energy deficient processes.

A markedly contrasting pattern evolved from the work with the much more weakly emitting TMPD(+) oxidations. In almost every case that yielded heteroexcimer emission (about 20 examples), $^3D^*$ and $^3A^*$ were both accessible to the redox process[17,20]. Accordingly Weller and Zachariasse postulated the mixed triplet annihilation (6.33) as the usual source of $(A^-D^+)$ and the respective homomolecular triplet annihilation as routes to $^1A^*$ and $^1D^*$[20,62]. Evidence for this view was found in experiments involving addition of TMPD to a system in which the 4,4′-dimethylbiphenyl(—)/TMPD(+) reaction was carried out[20]. In this instance

$$^3A^* + D \rightarrow {}^3D^* + A \tag{6.39}$$

is spontaneous; hence the added TMPD was expected to quench strongly any emission from $^1A^*$ and $(A^-D^+)$. The former component was indeed completely eliminated by submillimolar TMPD concentrations. Emission from $(A^-D^+)$ showed a sharp reduction (*ca.* 50%) in the same quenching range, but further addition of TMPD had little effect. These results were cited as supporting a T-mechanism with a contribution from (6.31) to the $(A^-D^+)$ population. Other evidence for the T-route to $^1A^*$ was outlined in Section 6.4.2 above.

The reasons for the low level of heteroexcimer participation in luminescence from the reactions of TMPD(+) are apparently related to the small energy gap between $(A^-TMPD^+)$ and the no-bond ground state. Zachariasse has examined this problem in terms of overlap integrals and the current theories of radiationless processes[20]. His analysis suggests that the consequent substantial interaction between these two states destabilises the heteroexcimer with respect to both the ground state and the solvent-separated ion pair.

In recent months, Bard *et al.* have appended to the story of the DMA(—)/TPTA(+) reaction some interesting studies of supporting electrolyte effects. Electrogeneration of luminescence from THF solutions with supporting TBAP yielded intensities which rose monotonically with decreasing electrolyte concentration[13,115,130]. An extrapolation to zero electrolyte gave an intensity which could be translated into a luminescence efficiency comparable with that obtained by Weller and Zachariasse via bulk reaction. However, 200 mM TBAP reduced the intensity by an order of magnitude. These observations paralleled the variation in the magnetic field effect with TBAP concentration: at 6.15 kG, enhancements of 12.4%, 9.2% and 6.3% were recorded for solutions with 200 mM, 100 mM and 10 mM TBAP, respectively. Bard *et al.* have proposed on the basis of these results that ionic association processes interfere with formation or emission of $(A^-D^+)$. In their view, the T-route probably contributes most luminescence from solutions with high TBAP concentrations, but the C-route may become dominant in more dilute media.

However, even in polar solvents the TPTA(+)/DMA(−) and TPTA(+)/ 9-MA(−) reactions exhibit long-wavelength emissions which may result from the heteroexcimers. Work by Tachikawa and Bard[116b] has shown further that the magnetic effects on the long-wavelength and parent hydrocarbon bands are not the same (Table 6.3). These observations seem inconsistent with formation of $(A^-D^+)$ by association of TPTA with $^1A^*$ arising via the T-route, and they may indicate a significant extension of the C-route into polar media. Very recently, Bard has bolstered this idea by presenting examples of exciplex emission from reactions, in polar solvents, which are incapable of locally exciting a product molecule to any state[18].

## 6.5 PROCESS EFFICIENCIES

### 6.5.1 Definitions and experimental aspects

Quantitative understanding of the factors which promote or retard redox chemiluminescence lies at the present frontier of the field. Recent years have seen several efforts at evaluating efficiencies, but much of the work has necessarily emphasised experimental design, rather than collection of data. A full review here seems premature, and we shall attempt below only an overview of existing results and a brief discussion of their salient implications. More detailed examinations of basic considerations and experimental methods are available[13,19,131].

Unquestionably the most widely sought parameter is the emission efficiency, $\phi_{ecl}$, which is the ratio of total photonic output to the number of redox events. The latter quantity is straightforwardly obtainable from the amounts of material used in bulk reaction or from the faradaic currents involved in electrogeneration; hence the crux of the measurement problem is determining the absolute emission.

Only two especially bright systems have lent themselves to such a direct method as actinometry with ferrioxalate. Weller and Zachariasse, in their study of the DMA(−)/TPTA(+) bulk reaction in THF, were first to implement the technique successfully[63]. Later, Bard *et al.* found it useful for evaluating $\phi_{ecl}$ for the thianthrene(+)/DPA(−) reaction, which they drove by electrolysis at the rotating ring-disc[13,131].

Because most processes yield somewhat lower light levels than these, every other report has involved some sort of calibrated detection system. This kind of procedure is rendered painfully difficult by the anisotropic nature of the source, the low intensities involved, the dependence of the detector sensitivity on wavelength, and a frequent need for fitting complicated apparatus inside a detection chamber. Several approaches have recently been detailed[19].

Weller and Zachariasse calibrated their apparatus with the DMA(−)/ TPTA(+) luminescence that they had previously studied by direct actinometry[63]. Their $\pm 20\%$ measurement precision was determined largely by the reproducibility of the flow rate in the vessel used for bulk reaction[20b]. Corrections were applied for the wavelength dependence of the detector sensitivity.

Keszthelyi and Bard tried to eliminate the need for such corrections by using a PIN photodiode whose sensitivity was determined by its response to a laser beam of known intensity[13,131]. It possesses a flat ($\pm 10\%$) power sensitivity over the 400–900 nm range, and its 1.00 $cm^2$ active area is sufficiently small that it can be used to intercept a known solid angle at any reasonable distance from a source. Since Keszthelyi and Bard had ascertained that the light generated at the rotating electrode assembly was uniformly distributed in a hemisphere bounded by the electrode surface, this latter feature handily solved the geometry problem for measurements via that technique.

For their transient experiments, Bezman and Faulkner calibrated a system which achieved constant geometry by means of an integrating sphere[19,120]. A Rhodamine B quantum counter interposed between the sphere and a photomultiplier rendered the detector's sensitivity independent of wavelength. The entire unit was standardised with a small argon lamp which had an actinometrically calibrated output. Pighin employed a similar approach in his study of the rubrene anion–cation annihilation, though he used solar cells as the light sensitive elements[79].

Table 6.4 displays most of the reported results. Space limitations prohibit complete coverage of the extensive work by Weller and Zachariasse, so a few representative elements of their tables have been selected[20,63]. Because cross-checking the measurement systems with well-characterised emissions (such as photoluminescence) is ordinarily not feasible, the accuracy of the tabulated data must be judged by the reasonableness of the procedures involved and by interlaboratory comparisons of results from common reactions. Unfortunately, the scatter in such comparisons cannot always be taken as a measure of experimental precision because alternative generating conditions may yield inherently different luminescence probabilities. Such an effect would likely to be especially noticeable with T-route systems, for which the reaction zone size and its triplet and radical concentrations would be important aspects. Probably the most apt comparisons are made for TPTA(+)/DMA(−) in THF without supporting electrolyte and for DPA(+)/DPA(−) in DMF under conditions for which the effect of ion instability is negligible. In neither case are easily quenched triplet intermediates thought to play a serious role. For the former reaction, the extrapolation by Bard *et al.* to infinite dilution yields a $\phi_{ecl}$ value of 7%[115], which compares favourably with the more direct determination of 7.5% reported by Weller and Zachariasse[63]. Bezman and Faulkner, in transient studies of the latter case, maintain that their asymptotic limit for high DPA concentrations manifests no effects of ion instability[120]. The same argument applies to the fast rotation limit in the ring–disc method employed by Bard *et al.*[13,50,131]. Table 6.4 shows that the former authors reported 1.4% and the latter 1.5% and 3% for these limits. The comparison is fairer when one divides the last two results by the factor of 1.4 which was used by Bard *et al.*[13,131] to correct for losses by imperfect reflectivity at the electrode. The three figures of 1.4%, 1.0% and 2% are close enough together to provide confidence in their approximate veracity, yet they are far enough apart to inhibit overinterpretation of the numerical data.

It probably is also true that raw results are biased toward negative errors. The matter of electrode reflectivity is one reason, but the size of a proper correction factor is still debatable. The value used by Bard *et al.* is probably near the upper limit for visible light reflected from a polished Pt electrode.

**Table 6.4 Emission efficiencies for redox chemiluminescence**

| Reactants[a] Oxidant | Reductant | Solvent[b] | $\phi_{ecl} \times 10^2$ [c] | Method[d] | Ref. |
|---|---|---|---|---|---|
| DPA(+), [2.0] | DPA(−), [2.0] | DMF | 1.5 | 1 | 13, 50 |
| [0.27–2.42] | [0.27–2.42] | DMF | 0.25–0.75(1.4)[e] | 2 | 120 |
| [1.0] | [1.0] | DMF | 3 | 3 | 13 |
| [2.20] | [2.20] | $AN/C_6H_6$ | 10 | 3 | 13 |
| [2.2–7.8] | [2.2–7.8] | ABT | 4–8 | 3 | 131 |
| {Thianthrene(+), [11.11]; DPA(+), [7.77]} [g] | DPA(−), [7.77] | ABT | 4,20[f] | 4,3 | 131 |
| $Ru(bipy)_3^{3+}$ | $e^-_{aq}$ | $H_2O$[l] | 1.5 | 8 | 29 |
| $Ru(bipy)_3^{3+}$, [1.0] | $Ru(bipy)_3^{+}$, [1.0] | DMF[m] | 5–6 | 3 | 28 |
| TPTA(+), [1.00] | DMA(−), [1.00] | THF[h] | 0.07[h], 7[i] | 3 | 13, 115 |
| TPTA(+) | DMA(−) | THF | 7.5 | 5 | 59 |
| TPTA(+) | Anthracene(−) | THF | 5 | 6 | 20a |
| TPTA(+) | Naphthalene(−) | DME | 0.7 | 6 | 20a |
| TPTA(+) | *trans*-Stilbene(−) | DME | 0.8 | 6 | 20a |
| TPTA(+) | Pyrene(−) | THF | 3 | 6 | 20a |
| TPTA(+) | Fluoranthene(−) | DME | 0.2 | 6 | 20a |
| TPTA(+) | Benzophenone(−) | MTHF | 0.01 | 6 | 20a |
| TPDA(+) | Naphthalene(−) | THF | 0.5 | 6 | 20a |
| TPDA(+) | DMA(−) | DME | 0.4 | 6 | 20a |

[a] TPTA = tri-(*p*-tolyl)amine; TPDA = tri-(*p*-dimethylaminophenyl)amine; TPAA = tri-(*p*-anisyl)amine; bipy = 2,2′-bipyridine. Other abbreviations are defined in notes (*a*) to Tables 6.1–6.3. Concentrations [mM] of reactant precursors are bracketed
[b] See note (*b*) to Table 6.3. $AN/C_6H_6$ and $DMF/C_6H_6$ are 1 : 1 mixtures. ABT = 0.5 : 0.33 : 0.17 acetonitrile–benzene–toluene; THF = tetrahydrofuran; DME = 1,2-dimethoxy-ethane; MTHF = 2-methyltetrahydrofuran. Unless otherwise noted, DME, THF and MTHF are without supporting electrolyte, but other media have 0.1 M TBAP
[c] Emitted photons per reaction event
[d] 1 = RRDE with calibrated photomultiplier; 2 = single pulse, transient analysis; 3 = RRDE with calibrated photodiode; 4 = RRDE with actinometry; 5 = bulk reaction, actinometry; 6 = bulk reaction, calibrated photomultiplier; 7 = pulse trains from multistep generation with detection by calibrated solid state device, reported efficiency is photons emitted per faradaic electron; 8 = pulse radiolytic generation of aquated electrons
[e] Inverse of $\phi_{ecl}$ is linear with $[DPA]^{-1}$. Extrapolation to large [DPA] gives the parenthesised limiting value
[f] The actinometric figure at left is the average $\phi_{ecl}$ for 16 min electrolysis. The right-hand figure applied to the maximum instantaneous emission rate
[g] Thianthrene and DPA oxidations overlap almost completely, so the reaction is not well defined
[h] [TBAP] = 0.1 M; $\phi_{ecl}$ depends strongly on [TBAP]
[i] Extrapolation to [TBAP] = 0
[j] Essentially independent of substrate concentrations
[k] Depends strongly on [TMPD]
[l] pH = 4.5
[m] Tetra-n-butylammonium tetrafluoroborate used as supporting electrolyte

**Table 6.4**—*continued*

| *Reactants*[a] *Oxidant* | *Reductant* | *Solvent*[b] | $\phi_{ecl} \times 10^2$ [c] | *Method*[d] | *Ref.* |
|---|---|---|---|---|---|
| TPDA(+) | Anthracene(−) | DME | 0.07 | 6 | 20a |
| TPAA(+) | Naphthalene(−) | THF | ⩾0.2 | 6 | 20a |
| TPAA(+) | DMA(−) | THF | ⩾0.3 | 6 | 20a |
| TPAA(+) | Anthracene(−) | THF | 0.3 | 6 | 20a |
| Rubrene(+), [1.0] | Rubrene(−), [1.0] | DMF | 0.1 | 1 | 13, 50 |
| [0.4–1.0] | [0.4–1.0] | DMF | 0.26[j] | 2 | 121 |
| [1.0] | [1.0] | DMF | 0.51 | 7 | 79 |
| [1.0] | [1.0] | DMF/$C_6H_6$ | 0.97 | 7 | 79 |
| [0.6–1.25] | [0.6–1.25] | BZN | 0.62[j] | 2 | 121 |
| [2] | [2] | BZN | 8.7 | 7 | 42 |
| [2] | [2] | BZN | 1.9 | 7 | 131 |
| 10-MP(+), [0.16–1.45] | Fluoranthene, [0.16–1.45] | DMF | 0.04[j] | 2 | 74 |
| TMPD(+), [1.0] | DPA(−), [1.0] | DMF | 0.1 | 1 | 13, 50 |
| TMPD(+), [1.0] | Pyrene(−), [5.0] | DMF | 0.01[k] | 1 | 13, 50 |

[a] TPTA = tri-(*p*-tolyl)amine; TPDA = tri-(*p*-dimethylaminophenyl)amine; TPAA = tri-(*p*-anisyl)amine; bipy = 2,2′-bipyridine. Other abbreviations are defined in notes (*a*) to Tables 6.1–6.3. Concentrations [mM] of reactant precursors are bracketed

[b] See note (*b*) to Table 6.3. AN/$C_6H_6$ and DMF/$C_6H_6$ are 1 : 1 mixtures. ABT = 0.5 : 0.33 : 0.17 acetonitrile–benzene–toluene; THF = tetrahydrofuran; DME = 1,2-dimethoxyethane; MTHF = 2-methyltetrahydrofuran. Unless otherwise noted, DME, THF and MTHF are without supporting electrolyte, but other media have 0.1 M TBAP

[c] Emitted photons per reaction event

[d] 1 = RRDE with calibrated photomultiplier; 2 = single pulse, transient analysis; 3 = RRDE with calibrated photodiode; 4 = RRDE with actinometry; 5 = bulk reaction, actinometry; 6 = bulk reaction, calibrated photomultiplier; 7 = pulse trains from multistep generation with detection by calibrated solid state device, reported efficiency is photons emitted per faradaic electron; 8 = pulse radiolytic generation of aquated electrons

[e] Inverse of $\phi_{ecl}$ is linear with $[DPA]^{-1}$. Extrapolation to large [DPA] gives the parenthesised limiting value

[f] The actinometric figure at left is the average $\phi_{ecl}$ for 16 min electrolysis. The right-hand figure applied to the maximum instantaneous emission rate

[g] Thianthrene and DPA oxidations overlap almost completely, so the reaction is not well defined

[h] [TBAP] = 0.1 M; $\phi_{ecl}$ depends strongly on [TBAP]

[i] Extrapolation to [TBAP] = 0

[j] Essentially independent of substrate concentrations

[k] Depends strongly on [TMPD]

[l] pH = 4.5

[m] Tetra-n-butylammonium tetrafluoroborate used as supporting electrolyte

Work reviewed by McIntyre[132] suggests that a factor near 1.2 is more appropriate. Solution reabsorption also produces low results, but the importance of this interference varies. If, as with DPA solutions, the absorber has a near-unit fluorescence efficiency the effect is negligible, but in other situations it may assume great importance.

From a fundamental standpoint, the excitation yields of the redox reaction are much more significant parameters than $\phi_{ecl}$ because they pertain solely to the elementary process of interest. For an S-route system, the singlet yield $\phi_s$ is given by $\phi_{ecl}/\phi_f$; hence it may be had straightforwardly. Triplet and heteroexcimer yields are unfortunately much less accessible. The latter can be obtained from some of the data of Weller and Zachariasse if an assumption about the heteroexcimer fluorescence efficiency is made[59]. In addition, triplet yields, $\phi_t$, have been estimated by Bezman and Faulkner from $\alpha$-values extracted from light transients emanating from T-route systems (see Section 6.4.2). These estimates are rough because two other parameters in $\alpha$, viz. $\phi_{tt}$ and $g$, must be approximated from spectroscopic data.

More reliable measurements of $\phi_t$ seem available from a technique developed by Freed and Faulkner, which relies on complete interception of triplet intermediates by *trans*-stilbene[82,83]:

$$^3R^* + trans\text{-Stilbene} \rightarrow R + {}^3\text{Stilbene}^* \quad (6.40a)$$

$$^3\text{Stilbene}^* \begin{cases} \xrightarrow{1-\alpha_s} cis\text{-Stilbene} \\ \xrightarrow{\alpha_s} trans\text{-Stilbene} \end{cases} \quad (6.40b)$$

The amount of *cis*-stilbene resulting from this sensitised isomerisation gauges the total number of intercepted triplets if the branching efficiency, $1-\alpha_s$, for decay of stilbene triplets to the *cis* isomer is known. The $\phi_t$ value can then be obtained as the ratio of the number of intercepted triplets to the number of redox events, which is controlled by coulometric reactant generation. In published work, $1-\alpha_s$ was taken as the literature value (0.59) for hexane solvent[81]. No strong dependence on solvent has been noted, but inefficiencies in (6.40a) sometimes create less than a unit yield of stilbene triplets[133]. In these cases the overall quantum efficiency for *cis* isomer production is less than $1-\alpha_s$. For this and other reasons, Freed and Faulkner have regarded their figures as lower bounds, but they have ruled out serious problems from incomplete interception. The main limitation of this method is that only a few systems satisfy the electrochemical and spectroscopic requirements (*cf.* Section 6.4.2.1); hence data have been recorded only for three oxidations of fluoranthene(−).

The excitation yields which have been determined by each of these methods are collected in Table 6.5.

### 6.5.2 Theoretical considerations

We may provide a context for our interpretations of the available yield data by first considering the predictions of theory. Common to several treatments

**Table 6.5 Excitation yields from chemiluminescence measurements**

| Reactants[a] | | | | | |
|---|---|---|---|---|---|
| Oxidant | Reductant | Solvent[b] | Quantity reported | Yield (× $10^2$) | Ref. |
| DPA(+) | DPA(−) | DMF | $\phi_s$[c] | 1.5–3 | 13, 120 |
| DPA(+), [2.20] | DPA(−), [2.20] | AN/$C_6H_6$ | $\phi_s$[c] | 11 | 13 |
| DPA(+), [7.77] | DPA(−), [7.77] | ABT | $\phi_s$[c] | 4–9 | 131 |
| $Ru(bipy)_3^{3+}$ | $e^-_{aq}$ | $H_2O$ | $\phi_t$[d] | 75 ± 20 | 29 |
| TPTA(+) | DMA(−) | THF | $\phi_c$[e] | 10 ± 1 | 59 |
| TPTA(+), [15.3] | Fluoranthene(−), [25.0] | DMF | $\phi_t$[f] | 9 ± 1 | 84 |
| 10-MP(+), [5.6–20.9] | Fluoranthene(−), [18.8–49.4] | DMF | $\phi_t$[f] | 2 ± 1 | 82, 84 |
| 10-MP(+), [0.16–1.45] | Fluoranthene(−), [0.16–1.45] | DMF | $\phi_t$[g] | 20–100[h] | 74 |
| 10-PP(+), [9.4–21.3] | Fluoranthene(−), [18.7–41.5] | DMF | $\phi_t$[f] | 80 ± 2 | 83 |
| Rubrene(+), [0.38–0.99] | Rubrene(−), [0.38–0.99] | DMF | $\phi_t$[g] | 50[i] | 121 |

[a] See note (*a*) to Table 6.4. Reactant precursor concentrations [mM] are bracketed
[b] See note (*b*) to Table 6.4
[c] $\phi_{ecl}/\phi_f$, where $\phi_f$ is taken as 0.9. See Ref. 52
[d] $\phi_{ecl}/\phi_p$, where $\phi_p$ is the phosphorescence efficiency of the directly-populated charge-transfer state
[e] From $\chi$ and $\chi'$ values via the estimate that $\phi_0' = 0.15 \pm 0.10$. See Section 6.2.3.3
[f] From interception experiments. $\phi_t$ seems independent of substrate concentrations
[g] Estimated from $\alpha$ values determined by T-route transient analysis. Rough value only
[h] $\alpha$, hence $\phi_t$, seems to increase with decreasing concentrations of substrates
[i] Independent of substrate concentrations

of homogeneous electron transfer is the hypothesis that the Franck–Condon principle influences any redox event[4,7,14,75,134]. Each approach must therefore evaluate the probability that the pair of reactants occupies a region of configuration space which is also allowed for the products; hence the zone in which the potential surfaces for reactants and products intersect is of primary interest. Of course, the configuration space must have sufficient dimensionality to describe the potential energy as a function of all the internuclear coordinates defining the reacting system (including solvation spheres), but one can represent the essence of these surfaces in a two-dimensional plot with one reaction coordinate as shown in Figure 6.11. Because

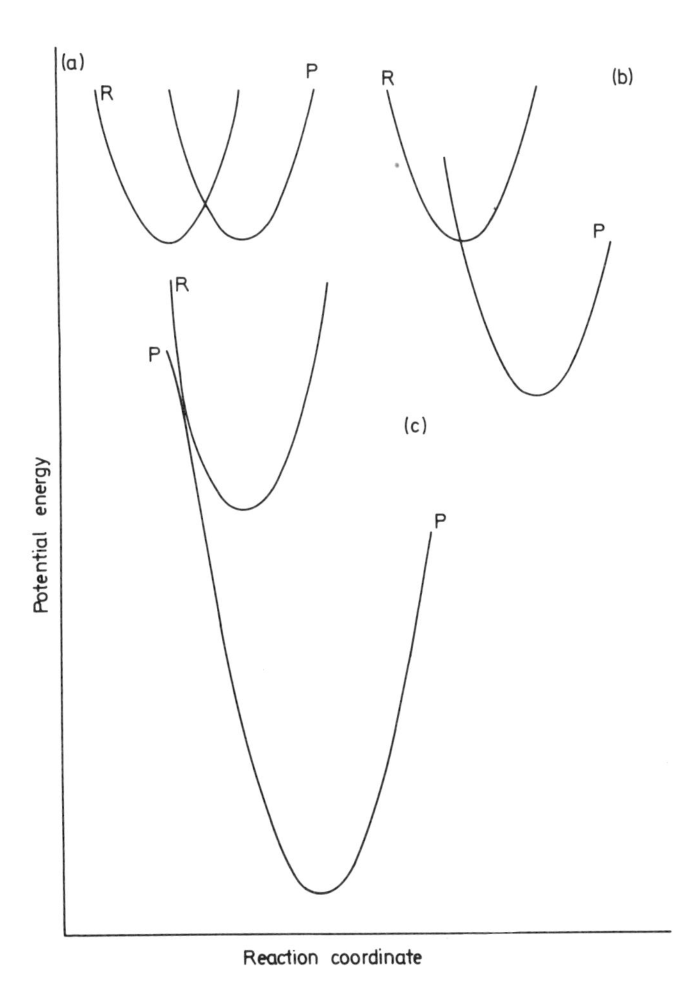


**Figure 6.11** Schematic diagrams of potential surfaces for reactants (R) and products (P). (a) For $\Delta G^\circ = 0$. (b) For $\Delta G^\circ \approx \lambda$. (c) For $\Delta G^\circ < -\lambda$. The intersection region for (b) will also feature a surface splitting which is not depicted here (Adapted from Marcus[4c], by courtesy of the American Institute of Physics)

the reactants must usually be distorted from their equilibrium configuration to achieve a format that is disposed toward reaction (Figure 6.11a), the probability that a collision will find them ready for electron transfer is ordinarily less than unity.

Marcus, who has developed perhaps the most workable electron transfer theory[4], has chosen to evaluate this configurational probability as the classical likelihood that nuclear motion will achieve the required reactant distortion. Thus he writes for the transfer rate constant

$$k_{et} = (8\pi kT/m^*)^{\frac{1}{2}} R^2 \rho\kappa \exp(-\Delta G^*/kT) \tag{6.41}$$

where $R$ is the mean distance between the centres of the reactant masses at the time of the transfer and $m^*$ is their reduced mass. The parameter $\kappa$, which describes the adiabaticity of the reaction, depends strongly on the degree of electronic interaction between reactants separated by $R$. For the systems at hand, substantial splitting in the surface crossing region is expected; hence $\kappa$ is probably near unity. A unit value is also taken for $\rho$, a subtle parameter which describes the relative precision with which certain co-ordinates are defined. The activation free energy $\Delta G^*$ characterises the work required to bring widely separated reactants to the activated configuration.

With the assumption that the reactants and products follow harmonic potential surfaces, Marcus demonstrated that this energy barrier could be related to the standard free energy for reaction and a single reorganisation parameter $\lambda$ by

$$\Delta G^* = \frac{w^r + w^p}{2} + \frac{\lambda}{4} + \frac{\Delta G^\circ}{2} + \frac{(\Delta G^\circ + w^p - w^r)^2}{4\lambda} \tag{6.42}$$

The work terms $w^r$ and $-w^p$, which express the energy required to create the reactant pair and separate the product pair, are usually of lesser importance. The value of $\lambda$ is governed by the differences in the equilibrium product and reactant configurations and by the resistance of each system to distortion from equilibrium. Electron exchange reactions of the type

$$A^- + A \rightarrow A + A^- \tag{6.43a}$$

$$B^+ + B \rightarrow B + B^+ \tag{6.43b}$$

have $\Delta G^\circ = 0$; hence their rate constants provide easy access to $\lambda_{aa}$ [for (6.43a)] and $\lambda_{bb}$ [for (6.43b)]. Since Marcus has shown that $\lambda_{ab}$ for

$$A^- + B^+ \rightarrow A + B \tag{6.44}$$

is simply $(\lambda_{aa} + \lambda_{bb})/2$, the recombination rate constant can be estimated without additional information. Several experimental tests of the theory have found it rather successful at predicting and correlating kinetic behaviour for thermoneutral or slightly exothermic ($\Delta G^\circ > -10$ kcal mol$^{-1}$) redox processes[135]. Existing data suggest that $\lambda$ values for aromatic systems are typically 0.4–0.8 eV.

Of course, the most facile transfer will feature $\Delta G^* = 0$, which is the situation shown in Figure 6.11b. The surface intersection occurs in the equilibrium region for the reactants; hence achievement of the activated complex requires no distortion. Note that this optimum process corresponds to an exothermicity $\Delta G° \approx -\lambda$. If the reaction is less exothermic, a barrier develops for reactant distortion to the required compromise configuration (Figure 6.11b), and a lower rate constant is predicted. Figure 6.11c displays the interesting situation for a reaction which is even more exothermic than $-\lambda$ (the 'abnormal' free energy region). The activated complex then resides in a part of configuration space which is remote from either equilibrium configuration, and a substantial barrier $\Delta G^*$ applies. This manifestation of the Franck–Condon principle would portend markedly slower rates for very energetic redox processes. Moreover, the effect is reinforced by the nature of the avoided crossing in the abnormal region. A non-adiabatic reaction is required there, so $\kappa$ becomes less than unity.

These are important consequences for the chemiluminescence work, because the corollary is that a reaction will populate its highest accessible product state with virtually a unit reaction efficiency. The only serious exceptions would involve excitation pathways with unusually large $\lambda$ values or several closely situated accessible product states. However, subsequent treatments of excitation by Marcus[4d], by Hoytink[14,75] and by Fischer and Van Duyne[7] have considered non-classical effects which tend to reduce the rigidity of this prediction.

Hoytink was first to introduce factors accounting for a postulated conservation of spin in the doublet–doublet annihilations which produce radical ion chemiluminescence[14]. On this basis, triplet product states are three times as likely as singlets; hence a weighting factor of 3/4 or 1/4 was inserted into the Marcus rate expression (6.41) to achieve a new rate equation for triplet or singlet product states, respectively. If this approach is valid, spin effects exert only minor influence over the distribution of product states because the rate expressions are still dominated by the exponential factors. Other authors have given spin conservation a major role by predicting total yields which are rigidly fixed at 25% for all singlet channels and 75% for all triplet channels. In this view, the product distribution within a manifold is influenced strongly by Franck–Condon factors, but spin completely rules the division between manifolds.

Fischer and Van Duyne have very recently advanced a new theory of electron transfer which takes explicit account of nuclear tunnelling probabilities and internal rearrangement effects within the actual reacting species[7]. They have concluded that the rates of very exothermic ($\Delta G° < -\lambda$) processes depend exponentially on essentially a linear function of $\Delta G°$, rather than on a quadratic one such as (6.42). If a sufficient number of mechanical modes are available to take up excess energy, processes populating the lower product states would then be competitive with that populating the highest accessible state. For a reaction as energetic as DPA(+)/DPA(−), direct generation of the ground state is deemed unlikely within the present treatment; however, these authors have outlined several modifications which might ultimately accommodate this possibility.

### 6.5.3 Evaluation of existing results

The firmest conclusions one can presently draw pertain to the points *en route* to chemiluminescence at which energy is wasted. The values of $\phi_{ecl}$, which reach upward to *ca.* 15% (see Table 6.4), are often quite large by chemiluminescence standards, but they appear stratified to some degree according to mechanism.

Yields above 1% arise exclusively from the S and C paths and depend only on the probability with which the emitter is formed and on its quantum efficiency for radiative deactivation. There are no obvious barriers to the existence of perfectly efficient S- and C-route systems, unless spin conservation applies rigorously. In that event, the upper limit to $\phi_{ecl}$ from doublet–doublet annihilation would be 25%.

In contrast, chemiluminescence efficiencies from the T-route cases extend downward from 1%. They likewise depend on the excitation yield and the emitter's fluorescence efficiency, but they are additionally influenced by the efficiency with which triplet intermediates undergo annihilation (as opposed to being quenched) and the probability with which the triplet–triplet annihilation generates emitting species [see reactions (6.26d–f). Table 6.5 shows that redox processes can produce triplet states with very substantial efficiencies; hence neither of the first two factors appears to limit $\phi_{ecl}$ in any general sense. One might instead expect triplet quenching to reduce the triplet concentrations to such low levels that their mutual annihilation is unimportant. However, Bezman and Faulkner, on the basis of their transient studies of rubrene(+)/rubrene(−) and 10-MP(+)/fluoranthene(−), con-concluded that about half the triplets participate in annihilation[74,121]. Apparently the triplet concentration in the reaction zone produced by a step experiment can be impressively high. They consequently suggested that the efficiency of light generation by this technique would not be generally limited by triplet quenching. Rather, the unavoidable point of energy waste in the T-route seems to be the poor efficiency with which triplet–triplet annihilation produces excited singlets. Studies mostly by Parker and Koizumi and their co-workers indicate figures for aromatic hydrocarbons which average less than 10%[52,103]. It therefore seems unlikely that an emission efficiency greater than a few percent will ever be recorded for a system which achieves luminescence via this process.

Correlating the yield data with theory is a much more speculative exercise, because one must examine the excitation efficiencies, which are more sparse and less reliable. Moreover, little of the theoretical work takes hetero-excimers into account, and it is not clear how the existence of these states would affect the treatments of full electron transfer. Possibly one is justified in disregarding this contingency if discussion is restricted to results from polar media. On this basis, one is tempted to conclude (a) that the highest accessible states are not always strongly favoured and (b) that triplet and singlet paths are not followed in a 3 : 1 ratio. However, the chemiluminescence evidence justifies neither position in an obligingly unambiguous way.

In the case of DPA(+)/DPA(−), for example, the yield of the emitter in any solvent is no more than a few percent (Table 6.4). Thus it is possible that most events, as Fischer and Van Duyne suggest[7a], populate the lower-energy states, particularly $^3$DPA*. The issue is clouded somewhat by the likelihood that DPA's second triplet, which apparently lies very near $^1$DPA* [104], is also accessible. However, Van Duyne, Fischer and Drake have recently examined this point, and they have reaffirmed their prediction that direct population of the lowest triplet is the most frequently followed path[7b].

Even so, the low triplet yields from 10-MP(+)/fluoranthene(−), (2–3 %) and TPTA(+)/fluoranthene(−) (~9 %), as determined by interception in DMF (Table 6.5), might seem to back both (a) and (b) persuasively. Note, however, that the 10-PP(+)/fluoranthene(−) reaction in DMF generates triplets with high efficiency (80 %). Changing *N*-methyl to *N*-phenyl on phenothiazine therefore increases the yield by a substantial factor, even though it alters the spectroscopic and electrochemical properties hardly at all. No explanation for this puzzling result has been supported experimentally, but it has been suggested that the 10-MP system may feature the formation of a dimer cation which undergoes a dark reaction with fluoranthene(−)[74]. If such association is inhibited for phenothiazines by phenyl substitution, only the latter figure may truly represent an elementary process. It is sufficiently close to unity or 75 % to weaken one's faith in either (a) or (b).

By no means are these brief comments meant as a complete survey of present quantitative knowledge concerning redox excitation Instead, they are intended to underscore the embryonic state of experimentally derived views concerning either the range of factors influencing excitation efficiencies or the mechanisms by which they operate. Considerable clever, quantitative effort defining the effects of environment, reaction energy and reactant structure will certainly have to precede the emergence of firm conclusions in this fundamental area.

## 6.6 APPENDIX: REFERENCE LIST OF ABBREVIATED COMPOUNDS

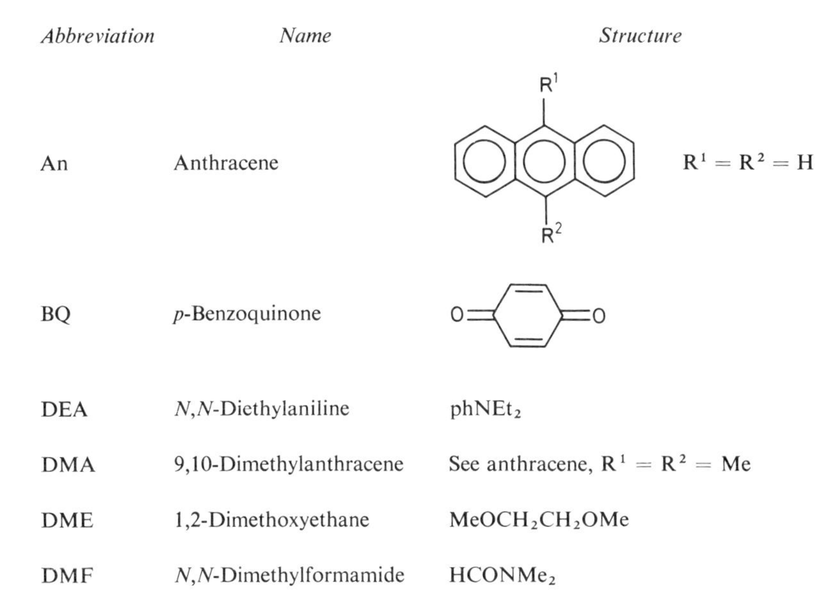

| *Abbreviation* | *Name* | *Structure* |
|---|---|---|
| An | Anthracene | $R^1 = R^2 = H$ |
| BQ | *p*-Benzoquinone | |
| DEA | *N*,*N*-Diethylaniline | $phNEt_2$ |
| DMA | 9,10-Dimethylanthracene | See anthracene, $R^1 = R^2 = Me$ |
| DME | 1,2-Dimethoxyethane | $MeOCH_2CH_2OMe$ |
| DMF | *N*,*N*-Dimethylformamide | $HCONMe_2$ |

| *Abbreviation* | *Name* | *Structure* |
| --- | --- | --- |
| DMPD | *N,N*-Dimethyl-*p*-phenylenediamine |  $R^1_2N$–C$_6$H$_4$–$NR^2_2$ $R^1 = H$, $R^2 = Me$ |
| DNA | 9,10-Di-α-naphthylanthracene | See anthracene, $R^1 = R^2 =$ 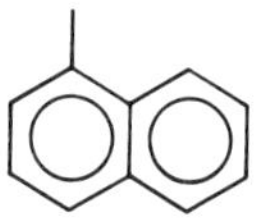 1-naphthyl |
| DPA | 9,10-Diphenylanthracene | See anthracene, $R^1 = R^2 = Ph$ |
| Fluoranthene | Fluoranthene | 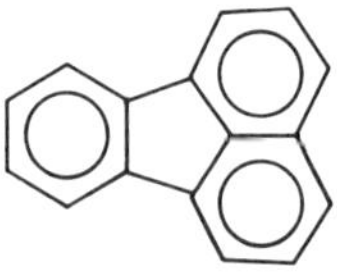 |
| 9-MA | 9-Methylanthracene | See anthracene, $R^1 = H$, $R^2 = Me$ |
| 10-MP | 10-Methylphenothiazine | 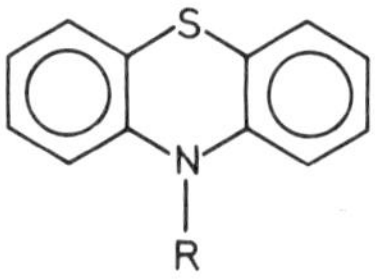 S, N, R; R = Me |
| Perylene | Perylene | 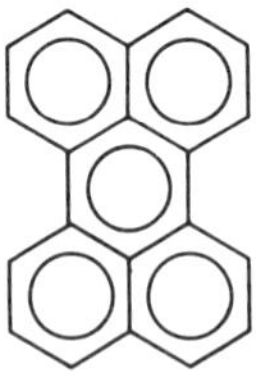 |
| 10-PP | 10-Phenylphenothiazine | See 10-MP, R = Ph |
| PPD | 2,5-Diphenyl-1,3,4-oxadiazole | 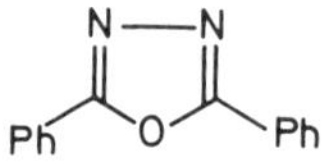 N—N, Ph, O, Ph |
| Py | Pyrene | 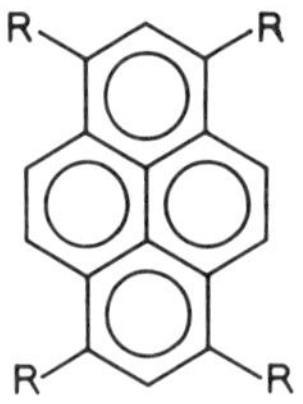 R, R, R, R; R = H |
| Rubrene | Rubrene | See tetracene, R = Ph |
| $Ru(bipy)_3{}^{3+}$ | Tris(bipyridyl)-ruthenium(II) | bipy 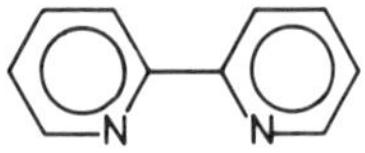 = N, N |
| *trans*-Stilbene | *trans*-Stilbene | 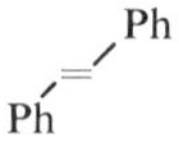 Ph, Ph |

| *Abbreviation* | *Name* | *Structure* |
|---|---|---|
| TBAP | Tetra-n-butylammonium perchlorate | $Bu_4N^+ ClO_4^-$ |
| Tetracene | Tetracene | R = H |
| THF | Tetrahydrofuran | |
| Thianthrene | Thianthrene | |
| TMPD | *N,N,N′,N′*-Tetramethyl-*p*-phenylenediamine | See DMPD, $R^1 = R^2 = Me$ |
| TPAA | Tri-(*p*-anisyl)amine | See TPTA, R = OMe |
| TPDA | Tri-(*p*-dimethylamino-phenyl)amine | See TPTA, R = $Me_2N$ |
| TPP | 1,3,6,8-Tetraphenylpyrene | See pyrene, R = Ph |
| TPTA | Tri-(*p*-tolyl)amine | $[R-C_6H_4-]_3N$ R = Me |

## Acknowledgements

The author wishes to express his appreciation for the financial support of the National Science Foundation and for the timely comments offered by Drs. A. J. Bard, J. T. Maloy, R. A. Marcus, H. Tachikawa, R. P. Van Duyne, A. Weller, T. Wilson and K. Zachariasse.

## References

1. (a) McCapra, F. (1970). *Prog. Org. Chem.*, **8,** 231; (b) Vassil'ev, R. F. (1970). *Russ. Chem. Rev.*, **39,** 529; (c) White, E. H. and Roswell, D. F. (1970). *Accounts Chem. Res.*, **3,** 54; (d) Rauhut, M. M. (1969). *Accounts Chem. Res.*, **2,** 80; (e) Gundermann, K. D. (1968). *Chemilumineszenz Organische Verbindungen* (Berlin: Springer-Verlag); (f) McCapra, F. (1966). *Quart. Rev.*, 485
2. (a) Kuwana, T. (1966). *Electroanalytical Chemistry*, Vol. 1, Chap. 3 (A. J. Bard, editor) (New York: Dekker); (b) Bard, A. J., Santhanam, K. S. V., Cruser, S. A. and Faulkner, L. R. (1967). *Fluorescence*, Chap. 14 (G. G. Guilbault, editor) (New York: Dekker); (c) Zweig, A. (1968). *Advances in Photochemistry*, Vol. 6, 425 (W. A. Noyes, Jr., G. S. Hammond and J. N. Pitts, editors) (New York: Interscience); (d) Chandross, E. A. (1969). *Trans. N.Y. Acad. Sci., Ser. 2*, **31,** 571; (e) Hercules, D. M. (1969). *Accounts Chem. Res.*, **2,** 301; (f) Hercules, D. M. (1970). *The Current Status of Liquid Scintillation Counting*, Chap. 32 (E. D. Bransome, Jr., editor) (New York: Grune

and Stratton); (g) Hercules, D. M. (1971). *Physical Methods of Organic Chemistry*, Part II, Chap. 13 (A. Weissberger and B. Rossiter, editors) (New York: Academic)
3. Turro, N. J., Lechtken, P., Shore, N. E., Schuster, G., Steinmetzer, H. C. and Yekta, A. (1973). *Accounts Chem. Res.*, **7,** 97
4. (a) Marcus, R. A. (1964). *Ann. Rev. Phys. Chem.*, **15,** 155 and references cited therein; (b) Marcus, R. A. (1965). *J. Chem. Phys.*, **43,** 679; (c) Marcus, R. A. (1965). *J. Chem. Phys.*, **43,** 2654; **52,** 2803; (d) Marcus, R. A. (1972). *Proceedings of the International Summer School on Quantum Mechanical Aspects of Electrochemistry* (P. Kirkov, editor) (Skopje, Yugoslavia: Centre for Radioisotope Application in Science and Industry)
5. Arai, S., Kira, A. and Imamura, M. (1971). *J. Chem. Phys.*, **54,** 5073
6. Van Duyne, R. P. and Drake, K. F. (1973). Presented at the 143rd meeting of the Electrochemical Society, Chicago
7. (a) Van Duyne, R. P. and Fischer, S. F. (1974). *Chem. Phys.*, **5,** 183; (b) Van Duyne, R. P., Fischer, S. F. and Drake, K. F. (1974). Presented at the 145th Meeting of the Electrochemical Society, San Francisco
8. Tachikawa, H. (1973). *Ph.D. Dissertation*, University of Texas, Austin
9. Hercules, D. M. (1964). *Science*, **145,** 808
10. Chandross, E. A. and Sonntag, F. I. (1964). *J. Amer. Chem. Soc.*, **86,** 3179
11. Visco, R. E. and Chandross, E. A. (1964). *J. Amer. Chem. Soc.*, **86,** 5350
12. Santhanam, K. S. V. and Bard, A. J. (1965). *J. Amer. Chem. Soc.*, **87,** 139
13. Bard, A. J., Keszthelyi, C. P., Tachikawa, H. and Tokel, N. E. (1973). *Chemiluminescence and Bioluminescence*, 193 (D. M. Hercules, J. Lee and M. Cormier, editors) (New York: Plenum)
14. Hoytink, G. J. (1973). Ref. 13, p. 147 and references cited therein
15. Weller, A. and Zachariasse, K. A. (1973). Ref. 13, p. 169, 181
16. Weller, A. (1968). *Pure Appl. Chem.*, **16,** 115 and references cited therein
17. Zachariasse, K. A. (1974). Presented at the International Exciplex Conference, London, Ontario, Canada; proceedings in preparation
18. Bard, A. J. (1974). Ref. 17
19. Faulkner, L. R. and Bard, A. J. (1975). *Electroanalytical Chemistry*, Vol. 10 (A. J. Bard, editor) (New York: Dekker) in Press
20. (a) Zachariasse, K. A. (1972). *Thesis*, Free University, Amsterdam; (b) Zachariasse, K. A. (1974). Personal communication
21. Zweig, A., Maurer, A. H. and Roberts, B. G. (1967). *J. Org. Chem.*, **32,** 1322
22. Zweig, A., Metzler, G., Maurer, A. and Roberts, B. G. (1967). *J. Amer. Chem. Soc.*, **89,** 4091
23. Zweig, A., Hoffmann, A. K., Maricle, D. L. and Maurer, A. H. (1968). *J. Amer. Chem. Soc.*, **90,** 261
24. Janzen, E. G., Harrison, W. B. and Dubose, C. M., Jr. (1972). *J. Organometal. Chem.*, **40,** 281
25. Hercules, D. M. and Lytle, F. E. (1966). *J. Amer. Chem. Soc.*, **88,** 4745
26. Lytle, F. E. and Hercules, D. M. (1971). *Photochem. Photobiol.*, **13,** 123
27. Tokel, N. E. and Bard, A. J. (1972). *J. Amer. Chem. Soc.*, **94,** 2862
28. Tokel-Takvoryan, N. E., Hemingway, R. E. and Bard, A. J. (1973). *J. Amer. Chem. Soc.*, **95,** 6582
29. Martin, J. E., Hart, E. J., Adamson, A. W., Gafney, H. and Halpern, J. (1972). *J. Amer. Chem. Soc.*, **94,** 9238
30. Mayeda, E. A. and Bard, A. J. (1973). *J. Amer. Chem. Soc.*, **95,** 6223
31. Chandross, E. A. and Sonntag, F. I. (1966). *J. Amer. Chem. Soc.*, **88,** 1089
32. Livingston, R. and Leonhardt, H. R. (1968). *J. Phys. Chem.*, **72,** 2254
33. Hercules, D. M., Lansbury, R. C. and Roe, D. K. (1966). *J. Amer. Chem. Soc.*, **88,** 4578
34. Zweig, A., Hoffmann, A. K., Maricle, D. L. and Maurer, A. H. (1967). *Chem. Commun.*, 107
35. Maricle, D. L. and Maurer, A. (1967). *J. Amer. Chem. Soc.*, **89,** 188
36. Maricle, D. L., Zweig, A., Maurer, A. H. and Brinen, J. S. (1968). *Electrochim. Acta*, **13,** 1209
37. Chandross, E. A. and Visco, R. E. (1968). *J. Phys. Chem.*, **72,** 378
38. Zweig, A. and Maricle, D. L. (1968). *J. Phys. Chem.*, **72,** 378
39. Malbin, M. D. and Mark, H. B., Jr. (1969). *J. Phys. Chem.*, **73,** 2786

40. Malbin, M. D. and Mark, H. B., Jr. (1969). *J. Phys. Chem.*, **73,** 2992
41. (a) Siegel, T. M. and Mark, H. B., Jr. (1971). *J. Amer. Chem. Soc.*, **93,** 6281; (b) Siegel, T. M. and Mark, H. B., Jr. (1972). *J. Amer. Chem. Soc.*, **94,** 9020
42. Schwartz, P. M., Blakeley, R. A. and Robinson, B. B. (1972). *J. Phys. Chem.*, **76,** 1868
43. Rapaport, E., Cass, M. W. and White, E. H. (1972). *J. Amer. Chem. Soc.*, **94,** 3153
44. Rapaport, E., Cass, M. W. and White, E. H. (1972). *J. Amer. Chem. Soc.*, **94,** 3160
45. Brundrett, R. B., Roswell, D. F. and White, E. H. (1972). *J. Amer. Chem. Soc.*, **94,** 7536
46. Keszthelyi, C. P., Tachikawa, H. and Bard, A. J. (1972). *J. Amer. Chem. Soc.*, **94,** 1522
47. Faulkner, L. R. and Bard, A. J. (1968). *J. Amer. Chem. Soc.*, **90,** 6284; **91,** 2411
48. Werner, T. C., Chang, J. and Hercules, D. M. (1970). *J. Amer. Chem. Soc.*, **92,** 763
49. Maloy, J. T., Prater, K. B. and Bard, A. J. (1968). *J. Phys. Chem.*, **72,** 4348
50. Maloy, J. T. and Bard, A. J. (1971). *J. Amer. Chem. Soc.*, **93,** 5968
51. Förster, Th. and Kasper, K. (1955). *Z. Elektrochem.*, **59,** 976
52. Parker, C. A. (1968). *Photoluminescence of Solutions* (Amsterdam: Elsevier) and references contained therein
53. Birks, J. B. (1970). *Photophysics of Aromatic Molecules* (New York: Wiley-Interscience) and references contained therein
54. Leonhardt, H. and Weller, A. (1963). *Ber. Bunsenges. Phys. Chem.*, **67,** 791
55. Rehm, D. and Weller, A. (1969). *Ber. Bunsenges. Phys. Chem.*, **73,** 834
56. Beens, H., Knibbe, H. and Weller, A. (1967). *J. Chem. Phys.*, **47,** 1183
57. Weller, A. (1974). Presented at the International Exciplex Conference, London, Ontario, Canada; proceedings in preparation
58. Weller, A. and Zachariasse, K. A. (1969). *Molecular Luminescence*, 895 (E. C. Lim, editor) (New York: Benjamin)
59. Weller, A. and Zachariasse, K. A. (1971). *Chem. Phys. Lett.*, **10,** 590
60. Michaelis, L. and Granick, S. (1932). *J. Amer. Chem. Soc.*, **65,** 1747
61. Walter, R. I. (1955). *J. Amer. Chem. Soc.*, **77,** 5999
62. Weller, A. and Zachariasse, K. A. (1967). *J. Chem. Phys.*, **46,** 4984
63. Weller, A. and Zachariasse, K. A. (1971). *Chem. Phys. Lett.*, **10,** 424
64. Nicholson, R. S. and Shain, I. (1964). *Anal. Chem.*, **36,** 706
65. Nicholson, R. S. and Shain, I. (1965). *Anal. Chem.*, **37,** 178
66. Saveant, J. M. and Vianello, E. (1967). *Electrochim. Acta*, **12,** 629
67. Maloy, J. T., Prater, K. B. and Bard, A. J. (1971). *J. Amer. Chem. Soc.*, **93,** 5959
68. Levich, V. G. (1972). *Physicochemical Hydrodynamics* (Englewood Cliffs, N.J.: Prentice Hall)
69. Albery, W. J. and Hitchman, M. L. (1971). *Ring-Disc Electrodes* (Oxford: Clarendon Press) and references contained therein
70. Feldberg, S. W. (1966). *J. Amer. Chem. Soc.*, **88,** 390
71. Feldberg, S. W. (1966). *J. Phys. Chem.*, **70,** 3928
72. Feldberg, S. W. (1969). *Electroanalytical Chemistry*, Vol. 3, Chap. 4 (A. J. Bard, editor) (New York: Dekker)
73. Bezman, R. and Faulkner, L. R. (1972). *J. Amer. Chem. Soc.*, **94,** 3699
74. Bezman, R. and Faulkner, L. R. (1972). *J. Amer. Chem. Soc.*, **94,** 6331; **95,** 3083
75. Hoytink, G. J. (1968). *Discuss. Faraday Soc.*, **45,** 14 and references cited therein
76. Faulkner, L. R., Tachikawa, H. and Bard, A. J. (1972). *J. Amer. Chem. Soc.*, **94,** 691
77. Visco, R. E. and Chandross, E. A. (1968). *Electrochim. Acta*, **13,** 1187
78. Van Duyne, R. P. and Reilley, C. N. (1972). *Anal. Chem.*, **44,** 142
79. Pighin, A. (1973). *Can. J. Chem.*, **51,** 3467
80. Freed, D. J. and Faulkner, L. R. (1971). *J. Amer. Chem. Soc.*, **93,** 2097
81. Hammond, G. S., *et al.* (1964). *J. Amer. Chem. Soc.*, **86,** 3197
82. Freed, D. J. and Faulkner, L. R. (1971). *J. Amer. Chem. Soc.*, **93,** 3565
83. Freed, D. J. and Faulkner, L. R. (1972). *J. Amer. Chem. Soc.*, **94,** 4790
84. Freed, D. J. (1972). *Ph.D. Thesis*, Harvard University
85. Stevens, B. and Walker, M. S. (1964). *Proc. Roy. Soc.*, **A281,** 420
86. Chandross, E. A., Longworth, J. W. and Visco, R. E. (1965). *J. Amer. Chem. Soc.*, **87,** 3260
87. Tanaka, C., Tanaka, J., Hutton, E. and Stevens, B. (1963). *Nature*, **198,** 1192
88. Parker, C. A. and Hatchard, C. G. (1963). *Trans. Faraday Soc.*, **59,** 284
89. Birks, J. B. (1963). *J. Phys. Chem.*, **67,** 2199; **68,** 439

90. Parker, C. A. and Short, G. D. (1967). *Trans. Faraday Soc.*, **63,** 2618
91. Parker, C. A. and Joyce, T. A. (1967). *Chem. Commun.*, 744
92. Weller, A. and Zachariasse, K. A. (1971). *Chem. Phys. Lett.*, **10,** 197
93. Johnson, R. C., Merrifield, R. E., Avakian, P. and Flippen, R. B. (1967). *Phys. Rev. Lett.*, **19,** 285
94. Merrifield, R. E. (1968). *Accounts Chem. Res.*, **1,** 131
95. Avakian, P. and Merrifield, R. E. (1968). *Mol. Cryst.*, **5,** 37
96. Merrifield, R. E. (1968). *J. Chem. Phys.*, **48,** 4318
97. Johnson, R. C. and Merrifield, R. E. (1970). *Phys. Rev. B*, **1,** 896
98. Groff, R. P., Merrifield, R. E. and Avakian, P. (1970). *Chem. Phys. Lett.*, **5,** 168
99. Suna, A. (1970). *Phys. Rev. B*, **1,** 1716
100. Kearns, D. R. and Stone, A. J. (1971). *J. Chem. Phys.*, **55,** 3383
101. Geacintov, N. E. and Swenberg, C. E. (1972). *J. Chem. Phys.*, **57,** 378
102. (a) Swenberg, C. E. and Geacintov, N. E. (1973). *Organic Molecular Photophysics*, Vol. 1 (J. B. Birks, editor) (New York: Wiley); (b) Geacintov, N. E. and Swenberg, C. E. (1974). *ibid.*, Vol. 2
103. (a) Kikuchi, K., Kokubun, H. and Koizumi, M. (1968). *Bull. Chem. Soc. Jap.*, **41,** 1545; (b) Kikuchi, K., Kokubun, H. and Koizumi, M. (1971). *ibid.*, **44,** 1527
104. Kellogg, R. E. (1966). *J. Chem. Phys.*, **44,** 411
105. Geacintov, N. E., Pope, M. and Vogel, F. (1969). *Phys. Rev. Lett.*, **22,** 593
106. Merrifield, R. E., Avakian, P. and Groff, R. P. (1969). *Chem. Phys. Lett.*, **3,** 155
107. Faulkner, L. R. and Bard, A. J. (1969). *J. Amer. Chem. Soc.*, **91,** 6495
108. Avakian, P., Groff, R. P., Kellogg, R. E., Merrifield, R. E. and Suna, A. (1971). *Organic Scintillators and Liquid Scintillation Counting*, 499 (New York: Academic Press)
109. Ern, V. and Merrifield, R. E. (1968). *Phys. Rev. Lett.*, **21,** 609
110. Faulkner, L. R. and Bard, A. J. (1969). *J. Amer. Chem. Soc.*, **91,** 6497
111. Tachikawa, H. and Bard, A. J. (1973). *J. Amer. Chem. Soc.*, **95,** 1672
112. Tachikawa, H. and Bard, A. J. (1974). *Chem. Phys. Lett.*, **26,** 10
113. Faulkner, L. R. and Bard, A. J. (1969). *J. Amer. Chem. Soc.*, **91,** 209
114. Tachikawa, H. and Bard, A. J. (1973). *Chem. Phys. Lett.*, **19,** 287
115. Keszthelyi, C. P., Tokel-Takvoryan, N. E., Tachikawa, H. and Bard, A. J. (1973). *Chem. Phys. Lett.*, **23,** 219
116. Tachikawa, H. and Bard, A. J. (1974). *Chem. Phys. Lett.*, **26,** 246
117. Santhanam, K. S. V. (1971). *Can. J. Chem.*, **49,** 3577
118. Periasamy, N., Shah, S. J. and Santhanam, K. S. V. (1973). *J. Chem. Phys.*, **58,** 821
119. Tachikawa, H. and Faulkner, L. R. (1974). Presented at the 145th Meeting of the Electrochemical Society, San Francisco
120. Bezman, R. and Faulkner, L. R. (1972). *J. Amer. Chem. Soc.*, **94,** 6317; **95,** 3083
121. Bezman, R. and Faulkner, L. R. (1972). *J. Amer. Chem. Soc.*, **94,** 6324; **95,** 3083
122. Chang, J., Hercules, D. M. and Roe, D. K. (1968). *Electrochim. Acta*, **13,** 1197
123. Grabner, E. W. and Brauer, E. (1972). *Ber. Bunsenges. Phys. Chem.*, **76,** 106
124. Grabner, E. W. and Brauer, E. (1972). *Ber. Bunsenges. Phys. Chem.*, **76,** 111
125. Werner, T. C., Chang, J. and Hercules, D. M. (1970). *J. Amer. Chem. Soc.*, **92,** 5560
126. Potashnik, R., Goldschmidt, C. R., Ottolenghi, M. and Weller, A. (1971). *J. Chem. Phys.*, **55,** 5344
127. Schomberg, H., Staerk, H. and Weller, A. (1973). *Chem. Phys. Lett.*, **21,** 433
128. Goldschmidt, C. R., Potashnik, R. and Ottolenghi, M. (1971). *J. Phys. Chem.*, **75,** 1025
129. Knibbe, H., Rehm, D. and Weller, A. (1969). *Ber. Bunsenges. Phys. Chem.*, **73,** 839
130. Keszthelyi, C. P. and Bard, A. J. (1974). *Chem. Phys. Lett.*, **24,** 300
131. Keszthelyi, C. P., Tokel-Takvoryan, N. E. and Bard, A. J. (1975). *Anal. Chem.*, **47,** 249
132. McIntyre, J. D. E. (1972). *Advances in Electrochemistry and Electrochemical Engineering*, Vol. 9 (R. H. Muller, editor) (New York: Interscience)
133. Caldwell, R. A. and Gajewski, R. P. (1971). *J. Amer. Chem. Soc.*, **93,** 532
134. (a) Levich, V. G. and Dogonadze, R. R. (1960). *Dokl. Akad. Nauk SSSR*, **133,** 158; (b) Levich, V. G. (1966). *Advances in Electrochemistry and Electrochemical Engineering*, Vol. 4 (P. Delahay and C. W. Tobias, editors) (New York: Interscience) and references cited therein.

135. (a) Dorfman, L. M. (1970). *Accounts Chem. Res.*, **3,** 224; (b) Winograd, N. and Kuwana, T. (1971). *J. Amer. Chem. Soc.*, **93,** 4343; (c) Kowert, B. A., Marcoux, L. and Bard, A. J. (1972). *ibid.*, **94,** 5538

# 7
# Chemiluminescence in the Liquid Phase: Thermal Cleavage of Dioxetanes

**T. WILSON**
Harvard University

---

**Abbreviated compounds**

| | |
|---|---|
| A | acetone |
| D | a dioxetane |
| DBA | 9,10-dibromoanthracene |
| *c*-DCE | *cis*-dicyanoethylene |
| *t*-DCE | *trans*-dicyanoethylene |
| DED | *cis*-diethoxy-1,2-dioxetane (9; Table 7.1) |
| DPA | 9,10-diphenylanthracene |
| $EuT_3Ph$ | europium tris(thenoyltrifluoroacetonate)-1,10-phenanthroline |

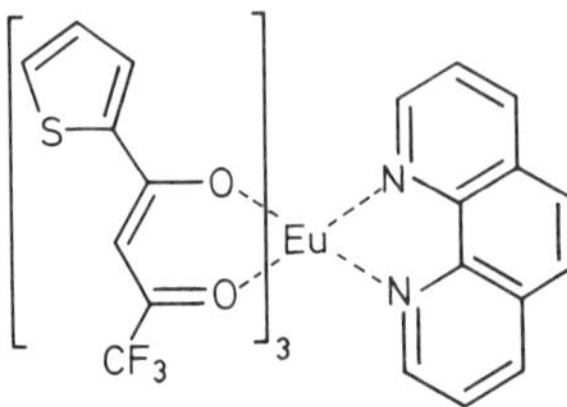


| | |
|---|---|
| F | ethyl formate |
| P | a carbonyl fragment |
| PTD | penta-1,3-diene (piperylene) |
| TMD | tetramethyl-1,2-dioxetane (5; Table 7.1) |

## 7.1 INTRODUCTION

### 7.1.1 Dioxetanes as a model for solution chemiluminescence

Until quite recently, review articles of chemiluminescence in solution read somewhat like catalogues of beautiful, but unrelated and complex reactions, usually prefaced by a few mechanistic comments illustrated with examples from the gas phase. Today, some order is discernible in the chaos. In particular, two extremely simple models of chemiluminescence in the liquid phase are emerging as fairly well understood prototypes of perhaps whole classes of reactions. These are the transfer of one electron from powerful reductant to oxidant, and the unimolecular cleavage of four-membered ring peroxides, called 1,2-dioxetanes or α-peroxylactones [equation (7.1)]. The

$$R^1R^2C(O{-}O)CR^3R^4 \;\;(\text{or } R^1R^2C(O{-}O)C{=}O) \xrightarrow[k_1]{\Delta} R^1R^2C{=}O^* + R^3R^4C{=}O \;\;(\text{or } CO_2) \tag{7.1}$$

first case, which involves no change in chemical bonding at all, is the subject of a companion paper by Faulkner (Chapter 6). In the second case, only four bonding electrons are simultaneously or nearly simultaneously rearranged.

Dioxetanes are intriguing molecules. Many of them are stable for days or months at room temperature. Yet, when heated in solution, say around 50–70 °C, they smoothly decompose into two carbonyl compounds, one of which has a good chance of being formed in an excited electronic state. This decomposition is sufficiently uncomplicated by side reactions that elementary mechanisms can be analysed in depth. Since quite a few examples have now been investigated, fairly general conclusions begin to emerge. An important one is that relatively low quantum yields of chemiluminescence, i.e. number of photons emitted per molecule of dioxetane decomposed, may in fact camouflage very high primary yields of electronic excitation, but to triplet rather than singlet states and hence with a low probability of radiative deactivation. With a convergence that may be significant, such high yields of triplets, inefficiently expressed in low yields of singlet emission, have been observed both from electron transfer reactions and from dioxetane cleavage. The opportunity is therefore at hand to focus on the kind of processes capable of efficiently generating electronically excited molecules, and the rules dictating their spin multiplicities.

Two requirements for efficient electronic excitation are obvious: high exoergicity and availability of an excited electronic state in one of the products. In addition, a Franck–Condon type of reasoning calls for little structural change during the reaction, and for the carbonyl product to be formed with a geometry close to that of its excited electronic state, unlike that of the ground state. Are these the only requirements? Or are orbital symmetry constraints helping to channel the energy freed in the reaction into electronic excitation? This is the type of question dioxetanes may yet answer. Here chemiluminescence proves to be a magnificent probe of molecular processes, since the frequency of the luminescence emitted (or the triplet transition involved) is a direct indication of the minimum exoergicity of the chemical process involved, whereas the luminescence intensity is a measure of the reaction rate. Thanks to the detectors now available, both types of information can be acquired in an instant and with a degree of sensitivity which leaves conventional — and demanding — analytical methods orders of magnitude behind. Following the emission of $10^{-15}$ mol $s^{-1}$ is really routine and, no doubt, such a sensitive and selective tool should benefit chemical dynamics in general, quite outside of chemiluminescent reactions.

This article will outline the results acquired so far in the area of dioxetane chemiluminescence, and the experimental methods employed to obtain them; the emphasis will be on the reliability of these methods. For the five years since the first dioxetane was isolated, the literature consists mostly of short notes, excellent at suggesting the excitement of the field, but hardly adequate for conveying a critical and balanced picture of the situation. The mechanism of excited state generation will then be discussed on the basis of the available data, even though we realise that conclusions cannot be but ephemeral in such a field of lively research.

Besides the intrinsic interest of dioxetanes, our limited focus on them is justified by the trend, which becomes more and more apparent in today's literature, to try to explain the mechanism of more complex chemi- and bio-luminescences on the basis of either the dioxetane or the electron-transfer model. Obviously, not all chemiluminescence in solution resorts to one of these two schemes for excitation. It is, however, quite likely that these and other nearly 'dark' chemical pathways of excitation will someday lead to the basic understanding of all bright chemiluminescences. Besides, though most dioxetanes are not efficient generators of light, they are nevertheless proven sources of electronically excited, long-lasting triplet products, with vast unappreciated possible roles in photochemistry and photobiology. In return, should not one wonder if other systems remain to be discovered which are able to generate triplet products?

The general field of chemiluminescence (of organic compounds in solution) has been the subject of several recent reviews[1–6], which is an additional justification for the deeper but narrower scope of the present article. Although our approach and that of Faulkner in Chapter 6 are basically similar, we have not tried to unify them. The ways of the electrochemist and the organic photochemist are different. These differences have been preserved, but we hope that workers in these fields will have a look at the other side of the coin.

### 7.1.2 History and highlights

Whereas inductive reasoning leads quite directly to the electron transfer model of chemiluminescence, the dioxetane model was clearly offered as an attempt to account for the ubiquity of oxygen involvement among the familiar chemiluminescent processes in solution. Bioluminescence, with quantum yields sometimes reaching unity[7], is the foremost example[8,9]. As no mechanism of matching efficiency seemed available to account for large quanta by pooling the energies of two or more elementary reaction steps, the relative weakness of the peroxidic link ($\sim$35 kcal $mol^{-1}$) was recognised early as of key importance — perhaps as the main explanation of the privileged role of oxygen in chemiluminescence. Credit goes to McCapra for proposing in 1968 that relatively short lived dioxetanes and α-peroxylactones [equation (7.1)] could be common intermediates in many chemiluminescent systems[10].

Although not yet isolated in any form at that time, the hypothetical dioxetane structure seemed uniquely qualified for the role of energy-rich precursor, for the following reasons. (a) Its thermal cleavage into two carbonyl fragments ought to be very exoergic, considering the release of ring strain and the formation of two strong carbonyl bonds at the expense of only one C—C bond and a weak O—O bond. (b) In the dioxetane structure, the C—O group is out of the $R^1$—C—$R^2$ plane; in the excited electronic states of aldehydes and ketones, the carbonyl group is also bent, while planar in the ground state[11,12]. This should favour formation of excited carbonyls, if energetically feasible, since less energy would then need be dissipated in vibrational excitation[2]. (c) Two new carbonyls are often found among the products of chemi- and bio-luminescent reactions: carbon dioxide evolution, for example, accompanies many bioluminescences[1–8]. (d) Unproven, yet at least historically important, the intriguing possibility exists that the cleavage of the four-membered ring may be concerted and occur along a symmetry-allowed path leading to an excited state, whereas the path to the ground state is forbidden[10]. This would offer a possible rationale for efficient generation of excited products.

These were, briefly, the major reasons why dioxetanes appeared appealing as candidates for the role of energy-rich precursor in several well known intense luminescent systems. High chemiluminescence intensities at room temperature imply, however, short lifetimes for the hypothetical dioxetanes; hence a difficult species to isolate.

The breakthrough came with the synthesis of trimethyl-1,2-dioxetane by Kopecky and Mumford[13]. Unbelievably to some, this dioxetane turned out to be quite a stable compound, which upon heating in solution to *ca.* 50 °C decomposed smoothly into acetone and acetaldehyde, with emission of a faint bluish light; or with emission of a more intense chemiluminescence in the presence of good fluorescers like DPA, which acquired their excitation energy via energy transfer from the excited carbonyls. Moreover, evidence was presented which indicated the availability of both singlet and triplet excited carbonyl donors[14].

Thus here is a uniquely simple chemiluminescent system: the unimolecular decomposition of a relatively stable molecule into two stable carbonyl products, one of them in an electronically excited state. More than two dozen such dioxetanes have now been synthesised, by either of two methods briefly outlined in the next section. We will review here the results of studies of these dioxetanes, mainly from the standpoint of decomposition rates, yields and spin-multiplicity of the excited products. The major findings are the fairly high yields of excited products (10% or more) and the general preference for triplet excitation. The different experimental approaches used to arrive at this unexpected result will first be critically discussed. Rcaders mainly interested in a survey of the field may wish to skip Section 7.2 and turn directly to the results (Section 7.4) and their discussion.

### 7.1.3 Synthetic routes

Depending on the nature of the substituents of the double bond, there are essentially two methods for preparing dioxetanes from the corresponding olefin. (a) In the first route, of more general applicability, the bromohydroperoxide is first prepared by treating the olefin with 1,3-dibromo-5,5-dimethylhydanthoin in the presence of excess hydrogen peroxide in ether, at $-40\,°C$ [equation (7.2)][15]. The bromohydroperoxide is then treated with NaOH in water–methanol (for trimethyldioxetane) (7.3)[14,16] or with silver acetate in benzene or methylene chloride (for tetramethyldioxetane and others) (7.4)[17,18].

ether, $-40\,°C$; $H_2O_2$ (7.2)

NaOH, MeOH (7.3)

Yellow oil, m.p. 4–7 °C

or $MeCO_2Ag$; $CH_2Cl_2$ or $C_6H_6$ (7.4)

Yellow crystals m.p. (sealed tube) 73–76 °C

(b) Dioxetanes can also be formed by direct 1,2-addition of singlet oxygen to certain electron-rich olefins, provided that the possibility of the 'ene' reaction (1,3-cycloaddition of $^1O_2$) is unimportant or precluded by the absence of α-hydrogens or for steric reasons[19–21], e.g.

EtO OEt — $h\nu$, sens., $O_2$; $CCl_3F$, $-78\,°C$ → EtO OEt (7.5)

This reaction has been shown to be stereospecific[20]. A high concentration of dissolved oxygen is necessary during the complete course of the reaction, in order to minimise the possibility of reaction between excited sensitiser and dioxetane (see Section 7.5.1).

(c) The report of a new and versatile method of dioxetane synthesis via the ozonolysis of the olefin in certain aldehydes and ketones as solvent (such as in pinacolone)[22] was later independently disproven by several authors[23,24].

## 7.2 EXPERIMENTAL PROCEDURES FOR YIELDS AND KINETIC PARAMETERS

### 7.2.1 General approach to the determination of yields of excitation

The foremost question is: what is the yield of excited carbonyl P* from a dioxetane D, and in what spin states is P* generated? In other words, what are the yields $^1\phi$ and $^3\phi$ of singlet excited $^1P^*$ and of triplet $^3P^*$? Considering the vast literature dealing with the photochemistry of carbonyl compounds[25–29], the interpretation of data and hypothetical mechanisms of dioxetane decomposition should in principle rest on a firm and wide basis.

If the chemiluminescence spectrum matches the fluorescence spectrum of $^1P^*$, then the observed yield of $^1P^*$, $^1\phi_{obs}$, can be determined from the measured intensity of this emission as a function of time. Carbonyl compounds, however, undergo fast and efficient intersystem crossing. Therefore, even if all the excited carbonyls were initially formed in the singlet state (i.e. $^1\phi = 1$), the observed yield of $^1P^*$ would only be a fraction, $k_F/(k_{isc} + k_F)$, of $^1\phi$. $k_F/(k_{isc} + k_F)$ is the quantum efficiency of fluorescence of $^1P$, $\phi^P{}_F$.

$$\begin{array}{l} D \xrightarrow{k_1} P + {}^1P^* \xrightarrow{k_F} h\nu_F \\ \quad \downarrow k_{isc} \\ \quad\; P + {}^3P^* \longrightarrow h\nu_P \end{array} \qquad (7.6)$$

$$^1\phi_{obs} = {}^1\phi \times \phi^P{}_F \text{ with } \phi^P{}_F \ll 1$$

The quantum yield $^3\phi$ of directly formed $^3P^*$ is, similarly, not $^3\phi_{obs}$ (assuming it is easily measurable), but $^3\phi_{obs}$ corrected for the amount of $^3P^*$ formed via intersystem crossing from $^1P^*$. With some carbonyl derivatives, such as acetone, the yield of intersystem crossing is very large, and therefore $^3\phi_{obs}$ is a good approximation for the *total* yield of excited products generated.

The experimental methods which have been used to estimate these yields belong to four distinct categories: (a) Measurement of the direct emission from the excited carbonyls. (b) Transfer of energy from P* to suitable acceptors, followed either by measurement of the luminescence of a chemically stable acceptor or the yield of product from a reactive acceptor undergoing, for example, a sensitised isomerisation. (c) Reaction of P* with an interceptor, not preceded by transfer of energy, e.g. hydrogen abstraction from a solvent. (d) Intramolecular reaction of the excited carbonyl.

These methods will now be discussed in detail, with relevant examples. The focus will be on the validity of the basic, although often implicit, assumptions made in each case. The techniques most commonly used for absolute light intensity measurement will be briefly outlined first.

### 7.2.2 Calibration of light intensities

Low levels of chemiluminescence, as is usually the case in dioxetane studies, cannot be directly measured by thermopiles[25] or by chemical actinometry[30]. To obviate the tedious calibration of photomultiplier tubes against standard lamps (from the National Bureau of Standards, for example)[31] and the often difficult assessment of a system's geometry, two different secondary liquid standards have been developed, the luminol standard of Lee and Seliger[31,32] and the radioactive standard of Hastings and Weber[33,34]. Because of their extreme convenience, most of the recent quantum yield determinations in solution chemiluminescence have been based on either one or the other of these secondary standards. The luminol standard takes advantage of the high reproducibility of the total chemiluminescence emitted by a given amount of luminol reacting with hydrogen peroxide and haemin in water ($\lambda_{max}$ = 430 nm) or with potassium t-butoxide in aerated dimethyl sulphoxide ($\lambda_{max}$ = 484 nm), at well specified concentrations. The radioactive standard, usually made of carefully degassed solutions of [1-$^{14}$C]hexadecane and a scintillator (PPO and POPOP, $\lambda_{max}$ = 416 nm) has been shown to maintain a constant photon flux over years.

The validity of the luminol standard rests, of course, on the reliability and reproducibility of the reported quantum yield of the luminol reaction, as well as on the implied insensitivity of this yield to small procedural changes. The reaction itself is very fast (half of the total light within one minute) and a fast recorder is therefore necessary. If large volumes of solutions are called for to duplicate the conditions of a chemiluminescence experiment, sufficiently quick mixing of the reagents may be a problem. Whenever a particular reaction cell (vial, tube, cuvette) can be replaced by an identical but sealed cell containing the degassed Hastings–Weber standard, the convenience of repeated standardisations is invaluable. The original determination of the photon flux of a particular standard may not be easy to check at the present time, but if, in the future, more accurate determinations were made, they could always be used to revise earlier published values, whereas one can never check or know how accurate a luminol determination may have been.

For all the convenience of these two secondary standards, it is unfortunate that they do not, apparently, give mutually consistent results. Quantum yields differing by a factor of about 2.5 have been reported for chemiluminescent reactions measured against the two standards, the luminol method yielding the lower values[35,36]. An independent reassessment of these standards supports the higher value[35], but the question is certainly not settled. With chemiluminescent reactions of low efficiency, such as the autoxidation of aldehydes or hydrocarbons where the quantum yields are of the order of $10^{-8}$–$10^{-10}$ $h\nu$ per molecule[37,38], a factor of 2 is not very significant. But the situation is entirely different with dioxetanes, where yields of primary excited products between 10 and 100% have been claimed and where a factor of 2 could be mechanistically meaningful. Thus it would be very desirable to eliminate the discrepancy between these two most commonly used standards, or to develop a new one of high reliability. In the meantime, this potential source of systematic error should be kept in mind and no quantum yield be reported without specific reference to the standard employed.

### 7.2.3 Direct emission from carbonyl products

Aerated solutions of trimethyldioxetane and of tetramethyldioxetane emit faint but measurable blue luminescences when heated. Kopecky and Mumford[14] did not offer a definitive assignment for the luminescence from trimethyldioxetane, which peaks at 430–440 nm. They suggested as emitter either singlet acetone or singlet acetaldehyde, or an excimer of both. The emission from tetramethyldioxetane (hereafter TMD) deserves a thorough discussion. Turro *et al.* identified its observed emission maximum at 395–405 nm with acetone fluorescence[39]:

$$\text{O—O} \xrightarrow[k_1]{\Delta} \text{O}^* + \text{O} \qquad (7.7)$$

If oxygen is removed by nitrogen purging, the luminescence intensity reportedly increases ~100 times, and the peak of emission shifts from 395–405 to 430–440 nm (Figure 7.1)[40]. The new emission is attributed to acetone phos-

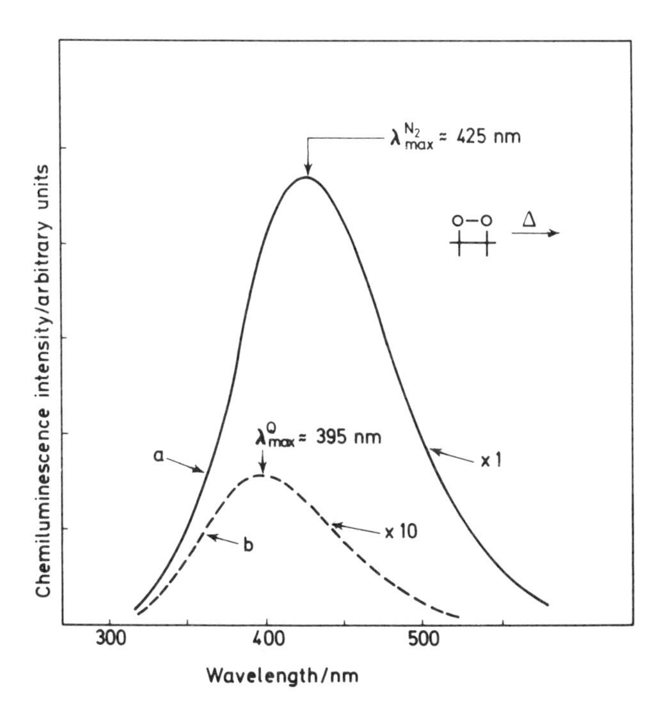


**Figure 7.1** Chemiluminescence spectrum from the thermolysis of tetramethyldioxetane ($10^{-2}$ M) at 40 °C in acetonitrile or Freon-113. Curve (a) (chemiexcited phosphorescence) was obtained by purging the solutions with nitrogen; curve (b) (chemiexcited fluorescence) was obtained for aerated solutions containing $10^{-1}$ M acrylonitrile (a quencher of triplet acetone) (From Turro *et al.*[40], by courtesy of the American Chemical Society)

phorescence. The spectral shift between fluorescence and phosphorescence is of the right magnitude. This assignment is supported by the following observations: the long wavelength emission is quenched by oxygen (reversibly), by acrylonitrile and penta-1,3-diene (all known quenchers of triplet acetone), whereas these quenchers have negligible effect on the short wavelength luminescence.

The quantitative interpretation of these data must take into account the dual origin of triplet acetone ($^3A^*$) in this system (Figure 7.2). In acetone,

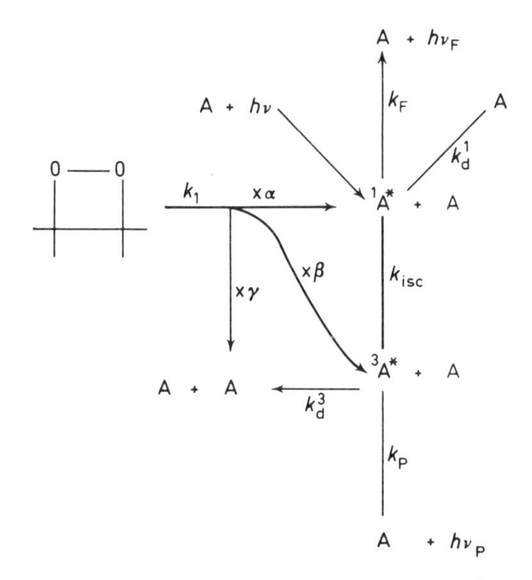


**Figure 7.2** Pathways of excitation and deactivation of singlet and triplet acetone, generated by the thermolysis of tetramethyldioxetane or by light absorption

$k_{isc} \gg k_F + k^1{}_d$. The fraction $\alpha/(\alpha + \beta)$ of excited acetone generated in the excited singlet state is given by

$$\frac{\alpha}{\alpha + \beta} = \frac{I_F}{I_p} \times \frac{k_p\, k_{isc}}{k_F(k_d{}^3 + k_p)} \tag{7.8}$$

where $I_F$ is the intensity of luminescence in the presence of $O_2$ and/or other triplet quenchers (i.e. its fluorescence) and $I_p$ the intensity in the absence of oxygen and quenchers (i.e. phosphorescence).

Degassing acetone solutions was found by Turro *et al.*[40] to produce the same red shift in the *photo*luminescence, with a concomitant increase in intensity, although much smaller (a factor of ~2 only) than in the chemiluminescence case. Assuming with Turro that the new long wavelength emission from degassed fluid solutions at room temperature is indeed acetone phosphorescence (the lifetime is 2–20 s, in agreement with this hypothesis), we can measure the ratio $I^*{}_p/I^*{}_F$ of these two photoluminescences. From the scheme of Figure 7.2:

$$\frac{I^*{}_p}{I^*{}_F} = \frac{k_p k_{isc}}{k_F(k_p + k^3{}_d)} \tag{7.9}$$

If oxygen is removed by the same method in both experiments, i.e. chemiluminescence of TMD and photoluminescence of acetone, then

$$\frac{\alpha}{\alpha + \beta} = \frac{I_F}{I_p} \times \frac{I^*{}_p}{I^*{}_F} \tag{7.10}$$

From Turro's data, $I_F/I_p \approx 0.01$–$0.03$ (Ref. 40 : figure 1 and text) and $I^*{}_p/I^*{}_F$

$\approx$ 1.5 (figure 2[40], after subtracting curve b from curve a). Hence $\alpha/(\alpha+\beta) \approx$ 0.01–0.04.

This is a very important result, since it shows that tetramethyldioxetane generates more excited acetone *directly* in the triplet state than in the singlet state. The fraction of the $^3A^*$ due to intersystem crossing from $^1A^*$ is actually quite negligible here.

If the absolute value of the chemiluminescence intensity in the presence of oxygen, $I_F$, is measured, the quantum yield $\phi^*_F$ of this chemiluminescence is given by

$$\phi^*_F = \frac{I_F}{k_1[D]} = \frac{k_F[^1A^*]}{k_1[D]} = \alpha\phi_F \tag{7.11}$$

with $\phi_F$ = fluorescence quantum yield of acetone. Turro and Lechtken[39] report $\alpha = 0.0016$ (from ferrioxalate actinometry). The absolute quantum yield of singlet and triplet acetone can thus be obtained. Nevertheless, one should keep in mind that this analysis is based on the assumption that $^1A^*$ generated chemically and photophysically behave indistinguishably, i.e. that their respective values of $k_F$, $k_{isc}$ and $k^1{}_d$ are identical. There is no experimental basis at present to dispute this hypothesis. However, it is conceivable that excited singlet acetone is formed from the dioxetane cleavage with an excess of vibrational energy, and that this 'hot' $^1A^*$ could cross-over to the triplet manifold faster. Chemiluminescence and photoluminescence should in any case always be performed at the same temperature and in the same solvent, to ensure the fairest possible comparison.

### 7.2.4 Energy transfer from excited carbonyls to luminescent acceptors

9,10-Diphenylanthracene (DPA) and 9,10-dibromoanthracene (DBA) make a choice pair of fluorescers, when the energies of the singlet and triplet carbonyl donors are $\geqslant$72 kcal mol$^{-1}$ [$E_s$(DPA) = 70.1 kcal mol$^{-1}$ and $E_s$(DBA) = 70.2 kcal mol$^{-1}$][29]. DPA is a more efficient fluorescer than DBA[29], whereas DBA is a better acceptor of triplet energy than DPA, because of the bromine atoms' effect on spin–orbit coupling[41,42].

Let us consider the four energy transfer reactions which could yield singlet excited anthracenes, and hence their fluorescence:

$$^1P^* + DPA \xrightarrow{k^{DPA}{}_{SS}} {}^1DPA^* + P \tag{7.12}$$

$$^1P^* + DBA \xrightarrow{k^{DBA}{}_{SS}} {}^1DBA^* + P \tag{7.13}$$

$$^3P^* + DPA \xrightarrow{k^{DPA}{}_{TS}} {}^1DPA^* + P \tag{7.14}$$

$$^3P^* + DBA \xrightarrow{k^{DBA}{}_{TS}} {}^1DBA^* + P \tag{7.15}$$

At equal concentrations of DPA and DBA, transfer from singlet donor $^1P^*$ should result in higher fluorescence yields from DPA, because $k^{DPA}{}_{SS}$ as well as $k^{DBA}{}_{SS}$ corresponds to spin-allowed transfer, but $\phi e^{DPA}{}_F \approx 10 \times \phi^{DBA}{}_F$ (at 20 °C)[29]. On the contrary, transfer from $^3P^*$ may be considerably more efficient to DBA than to DPA, i.e. $k^{DBA}{}_{TS} \gg k^{DPA}{}_{TS}$, hence results in higher singlet emission from DBA, in spite of the lower fluorescence yield of DBA. Therefore, if the addition of DBA brings about a more intense sensitised fluorescence than the addition of DPA at equal concentration, it means that more triplet than singlet excited donors must be present[43]. This qualitative conclusion appears inescapable. Quantitatively, however, to derive the absolute yield of triplet excited carbonyls from the measured yield of sensitised DBA fluorescence requires knowledge of the efficiency of the energy transfer step involved

$$\phi_{TS} = \frac{k_{TS}}{k_{DA}}$$

where $k_{DA} = k_{TS} + k_{TT} + k'_{TS}$, the sum of the rate constants for the following processes:

$$^3P^* + {}^1DBA \begin{cases} \xrightarrow{k_{TS}} {}^1DBA^* + {}^1P \\ \xrightarrow{k_{TT}} {}^3DBA^* + {}^1P \\ \xrightarrow{k'_{TS}} {}^1DBA + {}^1P \end{cases} \qquad (7.16)$$

Vassil'ev and his co-workers[41], who first used DBA as an amplifier of weak chemiluminescences, reported $\phi_{TS} = 2.7 \times 10^{-2}$ for triplet acetophenone as donor, assuming $k_{DA} = 1.4 \times 10^{10}$ l mol$^{-1}$ s$^{-1}$. For the same donor, but with DPA as acceptor, these authors reported triplet–singlet energy transfer efficiencies two orders of magnitude smaller. Thus they concluded that the presence of the bromine atoms indeed relaxes the spin restriction rules limiting this energy transfer rate.

While qualitatively supporting this conclusion, Berenfel'd *et al.*[44] reported that measurements of the quantum yield of DBA fluorescence sensitised by acetophenone (under steady illumination at 313 nm in benzene) showed a much higher efficiency of triplet–singlet transfer, $\phi^{DBA}{}_{TS} = 0.3$. From the rate of decay of the sensitised fluorescence of DBA, under $10^{-7}$ s pulsed photoexcitation, they derived the rate constant $k_{DA}$ for the sum of all quenching rates of triplet acetophenone by DBA, $k_{DA} = (5 \pm 0.5) \times 10^9$ l mol$^{-1}$ s$^{-1}$; hence $k_{TS} = 1.5 \times 10^9$ l mol$^{-1}$ s$^{-1}$. This is a remarkably high rate value, an order of magnitude higher than reported by Belyakov and Vassil'ev. This discrepancy is not discussed by Berenfel'd, nor by Vassil'ev.

An independent determination of $\phi_{TS}$ in the acetophenone–DBA system, using essentially the static method of Berenfel'd *et al.*, gave $\phi_{TS} = 0.20$ as an upper limit in degassed benzene[45]. Therefore there is no doubt that triplet–singlet transfer from acetophenone to DBA is indeed a very efficient process, if not quite as efficient as reported by Berenfel'd.

The reasons for the discrepancy with the value of Belyakov and Vassil'ev are not immediately obvious. It may represent the cumulation of several successive small uncertainties. Their value for $\phi_{TS}$ was derived from very elegant studies of the chemiluminescence accompanying the autoxidation of hydrocarbons, where the excitation step is the disproportionation of peroxyl radicals leading, among other products, to an excited carbonyl compound. The emitter is either this carbonyl itself, predominantly in the triplet state, or singlet DBA excited by energy transfer. Vassil'ev and Belyakov's value is based on the quantum yield of emission of the excited carbonyl, which they derived from a Stern–Volmer analysis of an (assumed triplet–triplet) energy transfer, in the same autoxidation system, to a luminescent chelate of europium as acceptor. We feel that the quantitative aspects of this particular energy transfer step are somewhat open to question (see below, this Section). The interpretation of these data may well be the origin of the discrepancy with Berenfeld on $\phi^{DBA}{}_{TS}$.

If one accepts $\phi_{TS} = 0.20 \pm 0.05$ for triplet acetophenone–singlet DBA transfer, one must now assume that the efficiency of this type of transfer is not greatly dependent on the nature of the carbonyl donor. The rate of TS transfer to DBA from cyclohexanone, for example, is reported to be 0.63 of that from acetophenone[41].

The quantum yield of triplet carbonyl generated by a dioxetane is then

$$^{3}\phi = \frac{\phi^{DBA}{}_{obs}}{\phi^{DBA}{}_{F} \times \phi_{TS}} \tag{7.17}$$

where $\phi^{DBA}{}_{F}$ is the fluorescence yield of DBA in the solvent and at the temperature of the experiment, and $\phi^{DBA}{}_{obs}$ is the observed quantum yield in the presence of DBA, i.e. the total number of photons emitted per dioxetane molecule decomposed. Here $\phi^{DBA}{}_{obs} = I^{max}{}_{DBA}/k_1[D]$ where $I^{max}{}_{DBA}$ is the initial intensity of chemiluminescence at infinite concentration of DBA, $k_1$ is the first order rate of dioxetane decomposition, and [D] is the dioxetane concentration. $I^{max}{}_{DBA}$ is the $Y$ axis intercept of a plot of $1/I^{0}{}_{DBA}$ *vs.* 1/[DBA], based on the following relation at steady-state:

$$\frac{1}{I^{0}{}_{DBA}} = \frac{k_d}{\phi^{DBA}{}_{F}\, k_1\, k_{TS}[D]} \frac{1}{[DBA]} + \frac{k_{TS} + k_{TT} + k'_{TS}}{\phi^{DBA}{}_{F}\, k_1\, k_{TS}[D]} \tag{7.18}$$

($I^{0}{}_{DBA}$ is the initial chemiluminescence intensity in the presence of DBA and $k_d$ is the rate of decay of $^3P^*$ in aerated solution in the absence of DBA).

Figure 7.3 shows a typical plot for tetramethyldioxetane[46]. Its linearity is consistent with the assumption that DBA intercepts one excited state only. The slope/intercept ratio =

$$\frac{k_d}{k_{TS} + k_{TT} + k'_{TS}} \quad (\approx 2.6 \times 10^{-3}, \text{ in Figure 7.3}) \tag{7.19}$$

yields $k_d \approx 10^7\ s^{-1}$, if $(k_{TS} + k_{TT} + k'_{TS}) \approx 5 \times 10^9\ l\ mol^{-1}\ s^{-1}$. This value of $k_d$, too small to correspond to singlet acetone as donor[29], is compatible with a triplet lifetime limited by dissolved oxygen quenching. DBA thus provides a sensitive and quantitative method, within the assumptions outlined above, for monitoring triplet carbonyls and estimating their yields.

The same approach can be used with DPA for counting singlet excited carbonyls. The energy transfer is spin-allowed and, if exothermic, should occur with unit efficiency at infinite concentration of DPA. Besides, the fluorescence quantum yield of DPA is near unity over a useful temperature range. Therefore $^{1}\phi = \phi^{DPA}{}_{obs}$, where $\phi^{DPA}{}_{obs}$ is the quantum yield of chemiluminescence calculated from the intensity at the intercept of a plot of $1/I^{0}{}_{DPA}$ *vs.* 1/[DPA]. Such plots may show curvature at low DPA concentrations[47], where SS transfer no longer dominates and some TS transfer therefore shows. This will be determined by the ratio $\alpha/\beta$ (see Figure 7.2, for

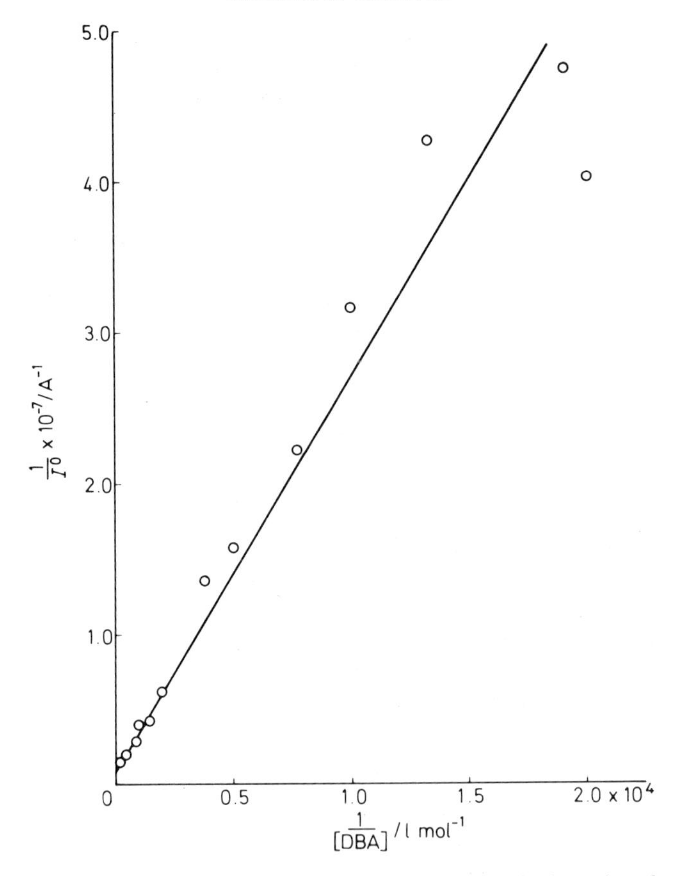


**Figure 7.3** Effect of DBA concentration on the initial intensity of chemiluminescence from tetramethyldioxetane ($5 \times 10^{-3}$ M) in benzene at 57 °C. (Chemiluminescence intensity calibrated with reference to the Hastings–Weber standard[33]: 1 A $\approx 6 \times 10^{17}$ $h\nu$ $s^{-1}$ $ml^{-1}$)

the case of TMD). If $\alpha k_1$ is the rate constant for dioxetane cleavage into excited singlet carbonyls, and $\beta k_1$ into triplet carbonyls, the chemiluminescence intensity in the presence of DPA is given by:

$$I_{DPA} = k_1[D]\left[\frac{\alpha k_{SS}[DPA]}{k_{SS}[DPA] + k^1{}_d} + \frac{\beta k_{TS}[DPA]}{(k_{TS} + k_{TT})\,[DPA] + k^3{}_d}\right] \tag{7.20}$$

According to Turro, the SS energy transfer from singlet acetone to DPA proceeds faster than the diffusion-controlled rate, by a long-range, Förster-type mechanism[48]. In two recent communications describing energy transfer experiments in a polystyrene matrix, Turro and Steinmetzer[49] suggest that the 'forbidden' transfer from triplet acetone to singlet DBA may also occur via a long-range, through space, mechanism. This interesting suggestion underlines the need for further very careful work in this whole area, where obviously little can be taken for granted at this point.

As mentioned briefly earlier, a chelate of europium, europium tris(thenoyl-trifluoroacetonate)-1,10-phenanthroline (hereafter $EuT_3Ph$, see p. 266) has been used fairly extensively for monitoring triplet intermediates[41,50]. The assumption here is that the lanthanide ion $Eu^{3+}$ is excited via intramolecular energy transfer from the triplet excited ligand[51], excited in turn via intermolecular energy transfer from the triplet carbonyl[52], a process which Vassil'ev and Belyakov report to be diffusion controlled[41]. $Eu^{3+}$ then deactivates with emission of the $^5D_0 \rightarrow {}^7F_2$ transition, concentrated mostly in a remarkably sharp band at $\sim$ 612 nm.

However attractive $EuT_3Ph$ may appear in view of its ability to accept triplet energy and convert it fairly efficiently into visible light, we feel that results obtained with this sensitiser require very cautious interpretation, for the following reasons. (a) The fluorescence of $EuT_3Ph$ has a large negative temperature coefficient: at 50 °C, for example, the quantum yield of fluorescence in benzene[53,54] is only $\sim$0.5 of that at 20 °C (plots of log $I_F$ *vs.* $1/T$ are not linear). This yield is also quite solvent dependent and reliable values, in benzene for example, do not seem available in the literature[55,56]. (b) It is not really established whether the emission (fluorescence) quantum yield is the same in systems where $EuT_3Ph$ is excited by intermolecular energy transfer as when it is excited by light. Indeed, in fluorescence measurements the exciting light is absorbed by the ligand (the excitation spectrum matches the absorption spectrum of the ligand) which then undergoes efficient intersystem crossing from its lowest excited singlet ultimately to its lowest triplet. From this triplet state the energy is then intramolecularly transferred to $Eu^{3+}$, with a well defined probability. In chemiluminescent situations, where the *singlet* states of the ligand are altogether by-passed, more efficient intramolecular transfer could conceivably occur if, for instance, a higher triplet state of the chelate was populated via intermolecular transfer from the carbonyl donor[57]. This point should be investigated. (c) The stability of the chelate is questionable above room temperature. The presence of free ions, which could themselves intercept the triplet donors, should be taken into account[58].

In conclusion, we feel that the use of $EuT_3Ph$ as an activator should at present be restricted to qualitative testing for triplet excited products. In that capacity, it is indeed a very elegant and sensitive tool, which has the added advantage of a lower excitation energy requirement (ligand $E_T \approx$ 58 kcal $mol^{-1}$)[41] than DBA ($E_S = 70$ kcal $mol^{-1}$).

### 7.2.5 Energy transfer from excited carbonyls to reactive acceptors

White and his co-workers[16,59,60] were the first to attempt counting the excited carbonyl products from trimethyldioxetane by intercepting them with compounds whose chemical behaviour after photoexcitation is established, such as *trans*-stilbene.

In principle, at *trans*-stilbene concentration high enough to ensure complete interception (i.e. energy transfer yield $\phi_{ET} = 1$), the quantum yield of carbonyl $^3P^*$ from the dioxetane is simply

$$^{3}\phi = \frac{\phi_{obs}}{\phi_{photo}} \tag{7.21}$$

$\phi_{obs}$ and $\phi_{photo}$ are the ratios of *cis*-stilbene molecules produced to dioxetane molecules decomposed and to photoexcited triplet stilbene, respectively. However, a complication arises due to the significant self-quenching of triplet stilbenes by ground state *trans*-stilbene[61]. This may require extrapolation of data acquired at low stilbene concentration. Besides, this relation is expected to be valid only at low conversion yields, where excitation of *cis*-stilbene is negligible, if the excited carbonyls acting as donors have sufficient energy to excite efficiently both *cis*- and *trans*-stilbene triplets, as in the case of trimethyldioxetane (*trans*-stilbene $E_T = 50$, *cis*-stilbene $E_T = 57$ kcal $mol^{-1}$)[27].

Whenever possible, the photochemical and the dioxetane experiments should be performed in the same solvent and at the same temperature. The partitioning of triplet stilbene into ground state *cis* and *trans* isomers may be quite different at 100 °C, where White *et al.* performed the dioxetane experiments, and at 60 °C or below, where the photosensitisation experiments of Hammond *et al.* were conducted[61]. The fact that the composition of the 'stationary *cis–trans* mixture' reached in the dioxetane experiments[16] favoured the *trans* isomer instead of the *cis*, as in the photostationary state, reflects perhaps some of these difficulties[62]. Nevertheless, the main cause of error here is specific to the dioxetanes themselves. These were shown by Wilson and Schaap[43] and by Turro and Lechtken[39,63], to be excellent quenchers of triplet states, via a most interesting induced decomposition process of the type (see Section 7.5.1):

$$^{3}T^{*} + D \rightarrow {}^{3}P^{*} + P + T \tag{7.22}$$

At concentrations of dioxetane higher than that of stilbene, this process is likely to be a major pathway in the quenching of triplet carbonyl and triplet stilbene.

For all these reasons, we believe that the literature values of $^{3}\phi$ obtained so far from stilbene isomerisation experiments should only be considered as lower bounds. The importance of the stilbene and other 'photo'-chemical systems studied by White and co-workers (dimerisation of acenaphthylene and rearrangement of 4,4-diphenylcyclohexa-2,5-dienone, for example)[16] is in the striking demonstration they provide of the formation of triplet products from dioxetanes.

Turro and Lechtken have studied the isomerisation of *trans*-dicyanoethylene (*t*-DCE) into *c*-DCE by triplet acetone from tetramethyldioxetane[39]. Photo-generated triplet acetone has previously been shown[64] to interact at a diffusion controlled rate with *t*-DCE, probably via exothermic energy transfer (acetone $E_T = 78$, NCCH=CHCN $E_T = ?$, but $CH_2{=}CHCN$ $E_T \approx 60$ kcal $mol^{-1}$) to generate triplet DCE, which then decays either to *cis* or to *trans* ground state DCE. However, this system is complicated by a direct reaction of *singlet* excited acetone with *t*-DCE. Turro and Lechtken have elegantly turned this to their advantage, for counting simultaneously both singlet and triplet acetone from tetramethyldioxetane. This will be discussed in the next section.

Isomerisable quenchers like hexa-2,4-diene or penta-1,3-diene (PTD) have also been used to determine the lifetime of the excited carbonyls responsible for the excitation of a fluorescer[47]. For example, in the system TMD + DBA, three parallel reactions now decide the fate of triplet acetone, the presumed energy donor:

$$^3A^* + DBA \xrightarrow{k_{TS}} {}^1DBA^* + A$$
$$\xrightarrow{k_{TT}} {}^3DBA^* + A \qquad (7.16)$$
$$\xrightarrow{k'_{TS}} DBA + A$$

$$^3A^* + PTD \xrightarrow{k_q} {}^3PTD^* + A \qquad (7.23)$$

$$^3A^* \xrightarrow{k_d} A \qquad (7.24)$$

($k_d$ represents the sum of all pseudo-first-order decay of $^3A^*$ in aerated solution). The slope of the Stern–Volmer $1/I^0$ *vs.* PTD concentration is

$$\frac{k_q}{k_d + (k_{TT} + k_{TS} + k'_{TS})[DBA]}$$

or, if DBA is small, simply $k_q/k_d$. For example, one series of runs with TMD (in aerated benzene at 57 °C, at [DBA] = $5 \times 10^{-5}$ mol $l^{-1}$) gave[46] $k_d = 2.7 \times 10^7$ $s^{-1}$, assuming $k_q = 5 \times 10^9$ l $mol^{-1}$ $s^{-1}$. This is in satisfactory agreement with the $k_d$ value obtained earlier [Section 7.2.4, equation (7.19)] from the dependence of the chemiluminescence intensity on the concentration of DBA, and certainly strengthens the case for a triplet donor.

In contrast, when the enhancer is DPA, quenchers like PTD have very little effect[47]. Whereas $5 \times 10^{-3}$ M [PTD] is enough to reduce the intensity with DBA by 50%, 0.2 M [PTD] is required to achieve the same effect with DPA[46]. This is consistent with a singlet carbonyl donor in this case. Interestingly, the plots of $1/I^0$ *vs.* [PTD], with DPA as fluorescer, are curved; the slope decreases as the concentration of PTD increases, as one may expect if there is a contribution from TS transfer to the predominantly SS transfer excitation mode of DPA.

### 7.2.6 Reaction of the excited carbonyl with interceptors, without the intermediacy of an energy transfer step

Such direct reactions are appealing, because they promise to by-pass the often tricky problem of energy transfer efficiency. As mentioned above (Section 7.2.5), Turro and Lechtken[39] made clever use of the photochemical reactions of acetone with *trans*-dicyanoethylene. Singlet acetone cycloadds to *t*-DCE to give exclusively the corresponding oxetane, whereas triplet acetone yields only *c*-DCE[64].

$$A + h\nu \longrightarrow {}^1A^* + t\text{-DCE} \longrightarrow \text{oxetane (CN, CN)} \quad (7.25)$$

$$\downarrow k_{isc}$$

$$ {}^3A^* + t\text{-DCE} \longrightarrow c\text{-DCE} \quad (7.26)$$

At low concentration (0.05–0.1 M), *t*-DCE reacts mostly with $^3A^*$ and the yield of the *cis* isomer increases with *t*-DCE concentration. However, when interception of $^1A^*$ becomes competitive with intersystem crossing, the yield of *c*-DCE decreases and that of the oxetane increases. The maximum yields of oxetane and of *c*-DCE were found to be nearly equal: $\phi^{photo}{}_{ox} \approx 0.08$ and $\phi^{photo}{}_{c\text{-DCE}} \approx 0.10$ per photon absorbed[64].

In contrast, with tetramethyldioxetane as the source of excited acetone, only traces of the oxetane were produced: $\phi^{obs}{}_{ox} \approx 2 \times 10^{-4}$ and $\phi^{obs}{}_{c\text{-DCE}} \approx 4.8 \times 10^{-2}$, per dioxetane mole decomposed[39]. At the high concentration of *t*-DCE used in these experiments (1.3 M), most $^1A^*$ should be intercepted before they cross over to $^3A^*$; hence practically no isomerisation should occur. It is therefore clear that TMD generates $^3A^*$ directly, without the intermediacy of $^1A^*$, a very important result indeed which entirely confirms the direct luminescence results[40] (Section 7.2.3). Thus

$$^1\phi = \frac{\phi^{obs}{}_{ox}}{\phi^{photo}{}_{ox}} \approx 0.002 \quad \text{and} \quad ^3\phi = \frac{\phi^{obs}{}_{c\text{-DCE}}}{\phi^{photo}{}_{c\text{-DCE}}} \approx 0.48$$

or, in the notation of Figure 7.2

$$\alpha \approx 0.002 \quad \text{and} \quad \beta \approx 0.5$$

For all its elegance, this method has nevertheless two important drawbacks. First, the low yields of the photochemical reaction are a strain on the analytical procedures (no information is available on this point, but the error limits are estimated at $\pm 50\%$)[39]. Secondly, the causes of the basic inefficiency of the photochemical cycloaddition to make the oxetane are not understood. It has been blamed on two possibilities[64]: (a) formation of a complex (exciplex) between $^1A^*$ and *t*-DCE which would partition between either oxetane or acetone and *t*-DCE, or (b) competition between oxetane formation and quenching of $^1A^*$. Turro *et al.* favoured the first of these interpretations, without offering decisive arguments. Either of these alternatives could, it seems, be quite sensitive to solvent and temperature, and it is regrettable therefore that the dioxetane and the photochemical experiments were carried out in different conditions (70 °C instead of room temperature, benzene instead of acetonitrile). Moreover, as high concentration of the reagents (~M) were necessary to boost the products' yields, quenching of $^1A^*$ and $^3A^*$ by the dioxetane itself is certain to take place to some extent, as Turro and Lechtken acknowledged.

A potentially simpler and therefore attractive reaction, but one which makes no claim to distinguishing between singlet and triplet, hence certainly not between direct or indirect generation of triplet products, is the 'photo' reduction of excited carbonyls by hydrogen donors (Type II reaction)[26,27,62,65].

$$\mathrm{C{=}O^*} + \mathrm{RH} \xrightarrow{k_{abst}} \dot{\mathrm{C}}\text{–O} + \mathrm{R\cdot} \quad \text{(ketyl radical)} \tag{7.27}$$

The literature data[66] on the rate constants for hydrogen abstraction show unfortunately that most hydrogen donors have little chance of successfully competing with the quenching of excited carbonyls by the dioxetane itself (the rate of this quenching process is *ca.* $10^9$ l mol$^{-1}$ s$^{-1}$, see Section 7.5.1). Hexane[67], for example, with $k_{abst} \approx 10^6$ l mol$^{-1}$ s$^{-1}$, would be fighting a losing battle, at dioxetane concentration compatible with analytical accuracy. On the other hand, hydroxylic solvents are plagued by a catalytic impurity problem (see Section 7.5.3)[68].

Cyclohexa-1,4-diene does not suffer from this complication and, besides, is an excellent donor of its doubly allylic hydrogens. Hence, very meaningful yields of products were obtained from the decomposition of TMD in cyclohexa-1,4-diene as solvent[68,69] (it was checked that the rate of TMD decomposition in this solvent is the same as in benzene; hence there is no impurity problem):

$$\text{TMD} \xrightarrow[55-65\,^{\circ}\mathrm{C}]{\text{cyclohexa-1,4-diene}} \underset{90\%}{\text{acetone}} + \underset{7.5\%}{\text{propan-2-ol}} + \underset{7\%}{\text{benzene}} + \underset{2.5\%}{\text{2-(cyclohexadienyl)propan-2-ol}} + \text{bi(cyclohexadienyl)} \tag{7.28}$$

The yields of propan-2-ol and benzene obtained correspond to primary yields of at least ~20% excited acetone per dioxetane decomposed. Unfortunately, no quantum yields are available at present for the photoreduction of acetone by cyclohexa-1,4-diene. However, the reaction products and their relative yields have been shown to be identical with those above from the thermal reaction of TMD, a strong argument indeed in favour of excited acetone as a common precursor.

The 20% yield of excited acetone per dioxetane decomposed should be considered as a lower bound only, for the following reasons. First, as the rate constant $k_{abst}$ with cyclohexa-1,4-diene is not known, there is no guarantee that reduction of excited acetone completely dominates its decay by other quenching purposes. A Stern–Volmer analysis of the quenching of triplet acetone from TMD (monitored by chemiluminescence with DBA) by cyclohexa-1,4-diene or by *cis*-pentadiene [PTD, Section 7.2.5, equation (7.23)] indicates that their rates of *quenching* are very similar, probably diffusion controlled in both cases[46]. However, this does not necessarily mean that hydrogen abstraction proceeds at the same high rate. Arguments for reversible reduction of triplet acetone to ketyl radical have recently been discussed[70,71]. In any case, the results obtained with this system confirm that high yields of excited carbonyls are indeed generated.

### 7.2.7 Intramolecular reaction of the excited carbonyl

Darling and Foote[72] have recently followed a clever variant of the photoreduction approach and their results are intriguing. Excited hexan-2-one from 3,4-dimethyl-3,4-di-n-butyl-1,2-dioxetane can undergo *intra*molecular $\gamma$-hydrogen abstraction:

(7.29)

1.1 % 3 %

*Photo*excited hexan-2-one yields acetone and cyclobutanol from both the singlet and the triplet states of the ketone, but the ratio of these two products depends on the multiplicity of the excited ketone[73,74]. On the basis of these photochemical results, obtained at the same temperature and in the same solvent as in the dioxetane experiments, Darling and Foote calculated the primary excitation yields of hexan-2-one: $^1\phi = 0.05$ and $^3\phi = 0.035$. Their study of the quenching by hexa-1,3-diene of cyclobutanol formation confirmed these low yields of excited products, as well as the relatively higher yield of singlet than triplet products. These results contrast with those obtained so far with other dioxetanes (see Table 7.1, Section 7.4.1). If confirmed by other, independent methods, they may throw light on the factors which influence the production of excited carbonyls.

This experiment illustrates strikingly the obstacles encountered in the determination of excitation yields. The first problem here is the low photochemical yields, which not only make for analytical difficulties (yields of less than 1 % cannot be very accurate), but also force us to question the validity of the approach. What are the reasons for this inefficiency of product formation? Is it due to reversibility of the initial hydrogen abstraction? Can one be sure that photoexcited and chemically excited hexanone behave identically? The rate of intersystem crossing of hexan-2-one, for example, is apparently temperature dependent[72]; should not this be regarded as a warning that the ketone generated from the dioxetane, perhaps with an excess of vibrational energy, may react differently from a thermally equilibrated excited ketone? Then there is the possibility that the dioxetane decomposition occurs via an intermediate biradical, the intramolecular reactivity of which may be mistaken in part for that of the excited ketone.

As another interesting example of intramolecular reaction, type I cleavage of dibenzyl ketone, was found by Richardson *et al.*[75] to accompany the decomposition of 3,3-dibenzyl-1,2-dioxetane; the yields of bibenzyl reported by these authors are, however, disappointingly low:

Ph(O—O)Ph —60 °C→ PhCH$_2$COCH$_2$Ph + PhCH$_2$CH$_2$Ph + CH$_2$O + CO (7.30)

88 % 2.2 %

According to Engel[76], the photodecarbonylation of dibenzyl ketone is actually a very efficient process, with $\phi_{\text{photo}} = 0.7$ at 313 nm. It proceeds from the ketone's triplet state which, for this reason, has a remarkably short lifetime ($10^{-10}$ s, shorter than its singlet). The small amount of bibenzyl formed from the dioxetane may not reflect a correspondingly low yield of triplet products since the triplet state of formaldehyde is well below that of the ketone (72.5 compared with ~79 kcal mol$^{-1}$)[77]; the possibility that this asymmetric dioxetane yields preferentially excited formaldehyde should be tested for. A study of the symmetric tetrabenzyldioxetane, if feasible, would avoid this difficulty.

### 7.2.8 Determination of reaction rates and activation parameters

The determination of $k_1$ (the rate of decomposition of the dioxetane) can be accomplished by monitoring either the concentration of the dioxetane (for example by n.m.r., i.r. or iodometry) or that of the carbonyl fragments. If the excited carbonyls can be chemically intercepted, $k_1$ can of course be obtained by following the concentration of that reaction's products. Otherwise $k_1$ can be obtained from the rate of decay of the direct fluorescence (or phosphorescence) of the carbonyls, or of the indirect fluorescence of an activator (different activators should give the same rate). These luminescence techniques are by far the most sensitive.

The important point here, which could not be overemphasised, is that all these methods should give the same value of $k_1$, and the reaction should be first order. Any disagreement between rates obtained by different methods, or any deviation from a first-order course, signals complications, such as induced decomposition or catalytic side reactions (see Section 7.5). (Faster rates and deviation from first order could also result from a free radical attack on the dioxetane, although this has not yet been observed)[63].

Similarly, Arrhenius plots of $k_1$ should give values of the activation energy $E_a$ of reaction (7.1) which are independent of the method used to determine the rates. Satisfactory agreements between different methods has been reported in the case of *cis*-diethoxy- and tetramethyl-dioxetanes[17,43,78].

Taking fuller advantage of the chemiluminescence and the high sensitivity of photometric devices compared with standard analytical methods, one can use the fact that the intensity of emission is proportional to $k_1$. In the simplest case, where the fluorescence of P* is measured

$$\mathrm{D} \xrightarrow{\alpha k_1} {}^1\mathrm{P}^* + \mathrm{P}$$

$${}^1\mathrm{P}^* \xrightarrow{k_F} \mathrm{P} + h\nu$$

$${}^1\mathrm{P}^* \xrightarrow{k_d} \mathrm{P}$$

the intensity of chemiluminescence is

$$I_{chl} = \frac{\alpha\, k_1\, k_F[\mathrm{D}]}{k_F + k_d} = \alpha\, k_1\, \phi^P{}_F[\mathrm{D}]$$

where $\alpha$ is the fraction of total P generated in the singlet state (Figure 7.2) and $\phi^P{}_F$ is the emission quantum yield of $^1$P*. If one assumes that the activation energy of the fluorescence of P (which can be measured independently, if not known) and of $\alpha$ are both negligible compared with $E_a$, then the activation energy $E_{chl}$ of the chemiluminescence intensity is

$$E_a = E_{chl} \tag{7.31}$$

In the presence of DBA, the activation energy now depends, besides on $E_a$, on the sum of the temperature coefficients of (a) the efficiency of triplet–singlet energy transfer between carbonyl and DBA, $E_{ET}$, which should be $\sim 0$ for exothermic energy transfer; (b) the fraction $\beta$ of total carbonyls generated in the triplet state, $E_\beta$, which is assumed negligible compared with $E_a$; and (c) the fluorescence of DBA ($E^{DBA}{}_F = -4.5$ kcal mol$^{-1}$)[53]

$$E_{chl} = E_a + E^{DBA}{}_F \tag{7.32}$$

Wilson and Schaap[43] considered a solution of dioxetane decomposing at temperature $T_1$, then cooled to $T_2$ in a time interval sufficiently short ($\sim$30 s) that the change of dioxetane concentration is negligible; the activation energy of chemiluminescence is then

$$E_{chl} = \frac{R \ln I_1/I_2}{1/T_2 - 1/T_1} \tag{7.33}$$

where $I_1$ and $I_2$ are the chemiluminescence intensities immediately before and after the temperature change. With their experimental set-up, this quick and accurate experiment can be repeated at any pair of temperatures. The agreement between the values of $E_a$ obtained in this manner [equation (7.32)] and the values of $E_a$ obtained from the slopes of intensity decay is excellent. This establishes that the dioxetane decomposes by only one observable path, $k_1$.

On the other hand, when there is a systematic disagreement between the $E_a$ obtained by these two methods, with $E_a < E_{chl}$, it shows that there are now two or more competitive pathways of dioxetane decomposition. For example, tetramethyldioxetane was found to decompose faster in methanol than in benzene. Arrhenius plots of the decay rates of chemiluminescence (or of TMD concentration) give an overall activation energy *ca.* 10 kcal mol$^{-1}$ lower in methanol than in benzene[78,79]. Yet the method of quick temperature change yields the *same* value of $E_{chl}$ in these two solvents[68]. This means that the light emitting reaction is the same, but that a 'dark' reaction (shown to be catalysed by metallic impurities, see Section 7.5.3) causes the faster decomposition of TMD in methanol. A very similar situation had also been observed in the case of *cis*-diethoxydioxetane in the presence of amines[80] (see Figure 7.4 and Section 7.5.2). No example of the opposite case, $E_{chl} < E_a + E^F{}_A$, corresponding to a chemiluminescent catalysis by an activator A, has so far been reported.

Using basically the same idea, Steinmetzer *et al.*[81] have recently demonstrated that the ratio of singlet and triplet acetone ($\alpha/\beta$, in the notations of Figure 7.2) produced by TMD is temperature independent, and that the temperature coefficients of $\alpha$ and $\beta$ must both be equal to zero. Working at temperatures sufficiently low that the overall rate of decomposition of TMD, and hence of luminescence decay, is negligibly small, these authors measured both the intensity of acetone fluorescence, $I_F$, and of phosphorescence, $I_P$, as a function of $1/T$. Arrhenius plots of these data give, in our notation, the activation energy of these two emissions, $E^F{}_{chl}$ and $E^P{}_{chl}$:

$$E^F{}_{chl} = E_a + E_\alpha + E^F{}_A = 27.7 \text{ kcal mol}^{-1}$$
$$E^P{}_{chl} = E_a + E_\beta + E^P{}_A = 25.2 \text{ kcal mol}^{-1}$$

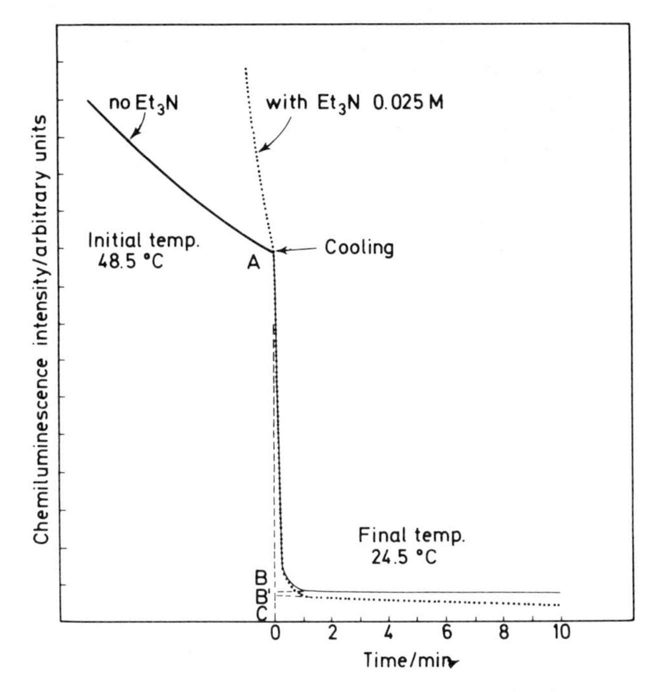


**Figure 7.4** Effect of a temperature drop on the chemiluminescence of a solution of *cis*-diethoxydioxetane, with or without $Et_3N$ (solvent, benzene; fluorescer, DBA; initial conc. of D, 0.015 M) (From Lee and Wilson[80], by courtesy of Plenum)

where $E_\alpha$ and $E_\beta$ are the activation energies for generation of singlet and triplet acetone, and $E^F{}_A$ and $E^P{}_A$ are the activation energies of acetone fluorescence and phosphorescence. Independent measurements gave $E_a = 27.9$, $E^F{}_A \approx 0.0 \pm 0.2$ and $E^P{}_A \approx -2.6$ kcal mol$^{-1}$. Thus, $E_\alpha = E_\beta = 0$ within the limits of experimental error ($\pm 0.3$ kcal mol$^{-1}$, according to the authors)[81]. This result is important, for it establishes that the transition state for disappearance of TMD is above the singlet as well as above the triplet state of acetone.

## 7.3 ENERGETICS

### 7.3.1 Thermochemistry

In spite of the obvious importance of such data for the understanding of dioxetane mechanisms, only one set of calorimetric data is available so far. Lechtken and Hohne[82] measured the decomposition enthalpy of $TMD_{solid} \rightarrow acetone_{liquid}$, $\Delta H_{obs} = -71 \pm 3$ kcal mol$^{-1}$, consistent with a value of $\Delta H_{obs} = -68 \pm 10$ kcal mol$^{-1}$ which they obtained from the heat of combustion of TMD. No explanation is offered for the markedly lower enthalpy which these authors obtained for the same dioxetane dissolved in dibutyl phthalate, $\Delta H_{obs} = -61 \pm 3$ kcal mol$^{-1}$. The significance of this difference should certainly be assessed. In any case, these experimental data support the thermochemical calculations of O'Neal and Richardson[83,84], based on Benson's group additivity method[85]: for TMD, $\Delta H^0_{calc} = -68.8$ kcal mol$^{-1}$. The major uncertainty in this type of calculation is the strain energy of the dioxetane ring, which is assumed to be the same as in cyclobutane (~26 kcal mol$^{-1}$).

Having measured the activation energy for decomposition, one can now compare the total energy available for excitation of the carbonyl fragments

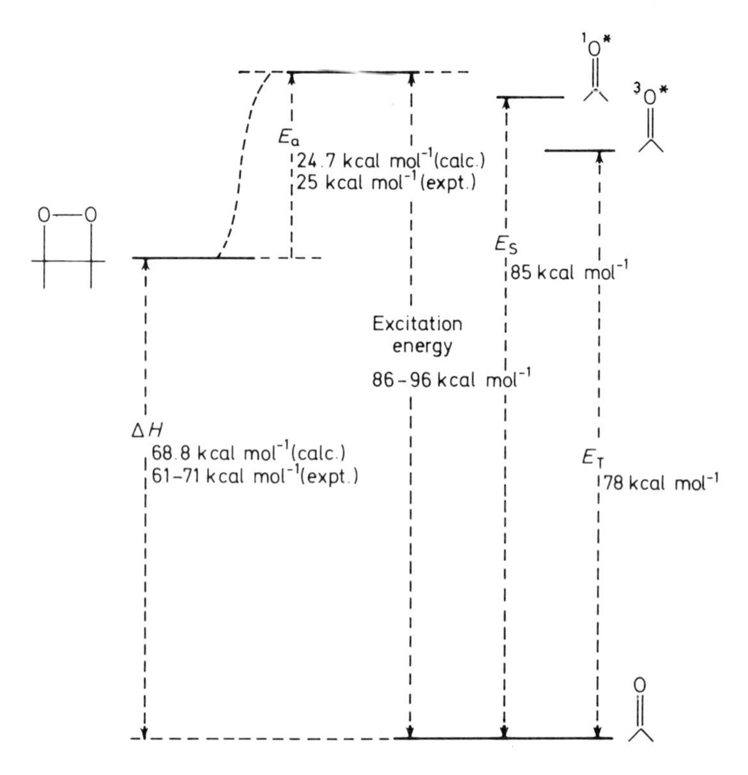


**Figure 7.5** Energetics of the tetramethyldioxetane system

with the levels of their first singlet and triplet states. Figure 7.5 illustrates the case of tetramethyldioxetane. There is enough energy to excite one acetone molecule to either its first singlet or first triplet state, but clearly not to excite both acetones even to their triplet state, a restriction to bear in mind while interpreting the yields of triplet acetone. This conclusion probably applies to all dioxetanes.

According to Adam[86], the calculated value of $\Delta H^0_r$ for dimethylperoxylactone [(15), Table 7.1] is ~80 kcal mol$^{-1}$, i.e. higher than in the case of TMD, but the activation energy is lower (~19 kcal mol$^{-1}$). Therefore the total energy available for electronic excitation is about the same (~99 kcal mol$^{-1}$): enough for excitation of acetone, but not of $CO_2$.

### 7.3.2 Spectroscopy

All the u.v. spectra of dioxetanes reported so far show a long tail of absorption extending into the visible, probably due to a $\pi^* \rightarrow \sigma^*$ transition[25] (Figure 7.6). They generally show a maximum between 270–290 nm, in the region of the $n \rightarrow \pi^*$ band of the carbonyl product. No such maximum appears in the absorption spectrum of *cis*-diethoxyethylene.

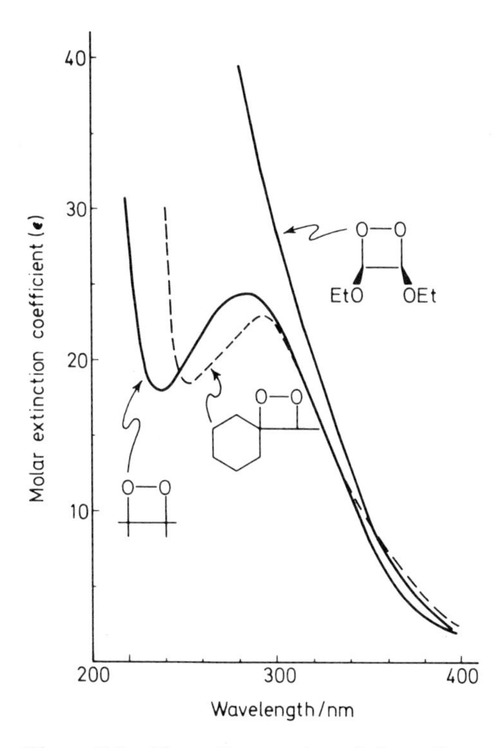


**Figure 7.6** Absorption spectra of three dioxetanes (Data from Kopecky *et al.*[24], Turro and Lechtken[79], Wilson and Schaap[43], by courtesy of the American Chemical Society)

The i.r. spectrum of several dioxetanes has been reported, but there has been no attempt at interpreting the data. The rapid decomposition of TMD in a metal wave guide has foiled a first try at microwave spectroscopy[87]. Thus there is no structural information available now on dioxetanes.

## 7.4 RESULTS

### 7.4.1 Tabulation of data and main conclusions

Table 7.1 summarises the data on isolated (or well-characterised) dioxetanes for which kinetic and/or yield information is now available. Other isolated

**Table 7.1**

| *No.* | *Dioxetane* | *Solvent* | *Rate* $(k_1)$ $/s^{-1}$, 50 °C | $E_a$/kcal $mol^{-1}$ | log *A* | $\lambda_{max}$/nm (*emission*) | $^1\phi$ | $^3\phi$ | *Ref.* |
|---|---|---|---|---|---|---|---|---|---|
| (1) | | $CCl_4$ | $4.0 \times 10^{-4a}$ | $23.0^a$ | $12.2^a$ | | | | 88 |
| (2) | | $CCl_4$ | $4.0 \times 10^{-4a}$ | $22.9^a$ | $12.1^a$ | | | | 88 |
| | | $CCl_4$ | | | | 431 | | | 16 |
| (3) | | $CCl_4$ : benzene (5 : 1) | $9 \times 10^{-4}$ | $23^{d,e}$ | $12.9^{d,e}$ | | | $0.05^g$ | 69, 89 |
| | | benzene | | | | | | $0.034^j$ | 90 |
| (4) | | gas; benzene | | | | 425–440 | | | 14, 16 |
| | | $CCl_4$ | $3.0 \times 10^{-4}$, 55 °C | $23.6^k$ | $12.4^k$ | | | | 91 |
| | | benzene | $1 \times 10^{-4}$ | | | | | $0.04^h$ | 60, 16 |
| | | acetone | $1.3 \times 10^{-4}$ | | | | | $0.14^{g'}$ | 16, 48 |
| | | benzene | | | | | $\sim 0.002^g$ | $0.5^g$ | 92 |
| | | $CCl_4$ | $1.6 \times 10^{-4b,c,d}$ | $23.5 \pm 0.5^{b,d}$ | $12.1^{b,d}$ | | | | 17, 93 |
| | | toluene | | | | | $7 \times 10^{-5g}$ | $0.12 \pm 0.05^g$ | 17, 93 |
| (5) | (TMD) | Aerated $CH_3CN$ | | | | 395 | | | 40 |
| | | Degas. $CH_3CN$ | | | | 425 | | | 40 |
| | | Benzene | | | | | $0.0028^i$ | $0.48^h$ | 39 |
| | | Benzene | | | | | $\sim 0.0025^g$ | $0.50^g$ | 92 |
| | | Benzene or cyclohexane | $6 \times 10^{-5}$, 55 °C[c] | $25.6^{c,k}$ | $13.0^{c,k}$ | | | | 78 |
| | | $CH_3CN$ | $4 \times 10^{-5}$, 55 °C[c] | $30.6^{c,k}$ | $15.6^{c,k}$ | | | | 78 |

| Compound | Solvent | | | | | | Ref. |
|---|---|---|---|---|---|---|---|
| | $CH_3CN$ | | $27^{a,e}$ | | | | 81 |
| | $CCl_4$ | $2.6 \times 10^{-5\,b,d}$ | $25.8^{b,d,k}$ | $12.9^{b,d,k}$ | | | 17, 93 |
| | Toluene | | | | $2.5 \times 10^{-4\,g}$ | $0.22 \pm 0.09^{g}$ | 17, 93 |
| | Benzene | $4 \times 10^{-5}$, 57 °C[d] | $26.5^{e}$ | 13.1 | | | 68 |
| | $CH_3OH$; $C_2H_5OH$ | $4 \times 10^{-5}$, 57 °C[d] | $25 \pm 2^{e}$ | | | | 68 |
| | Cyclohexa-1,4-diene | | | | | $\geqslant 0.2^{i}$ | 68 |
| | Benzene | $2.6 \times 10^{-5\,d}$ | $27.6^{e}$ | 14.1 | $0.002^{g}$ | $0.3^{g}$ | 46 |
| (6) | 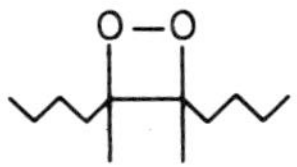 Decalin | | $25.5 \pm 0.03^{k}$ | $12.5^{k}$ | $0.05^{j}$ | $0.035^{j}$ | 72 |
| (7) |  Benzene | $4.0 \times 10^{-4}$, 52 °C[d] | $25.5 \pm 0.7^{d}$ | $12.9^{d}$ | | | 17, 93 |
| | Toluene | | | | $2 \times 10^{-4\,g}$ | $0.17 \pm 0.07^{g}$ | 17, 93 |
| (8) | 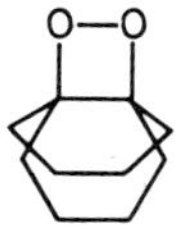 Toluene | $1.9 \times 10^{-4}$, 30 °C[d] | $22.7 \pm 0.7^{d}$ | $12.7^{d}$ | $<10^{-5\,g}$ | $8 \pm 3 \times 10^{-3\,g}$ | 17, 93 |
| 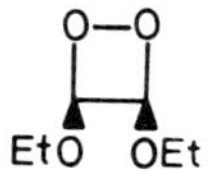 (9) (DED) | Benzene | $2 \times 10^{-3\,a,b,d}$ | $23.6^{a,b,d,e}$ | $13.1^{a,b,d}$ | $<10^{-4\,g}$ | $\sim 0.2^{g}$ | 43, 46 |

**Table 7.1**—*continued*

| *No.* | *Dioxetane* | *Solvent* | *Rate* ($k_1$) /$s^{-1}$, 50 °C | $E_a$/kcal $mol^{-1}$ | log $A$ | $\lambda_{max}$/nm (*emission*) | $^1\phi$ | $^3\phi$ | *Ref.* |
|---|---|---|---|---|---|---|---|---|---|
| (10) | 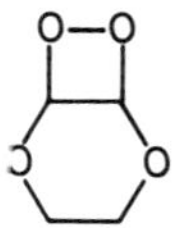 | Acetone | $\geqslant 8 \times 10^{-4}$, 60 °C[a] | | | | | | 94 |
| | | Benzene | $2.2 \times 10^{-3}$[d] | $24 \pm 1$[d,e] | 12.6 | | $\sim 10^{-4}$[g] | $0.3 \pm 0.1$[g] | 46 |
| (11) | 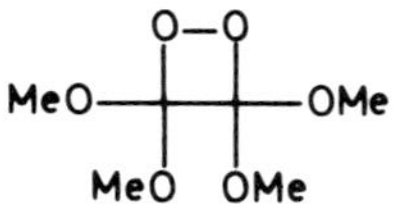 | Benzene | $\sim 1.1 \times 10^{-4}$, 56 °C | | | | | | 21 |
| | | Xylene | $1.4 \times 10^{-5}$, 80 °C[d] | $28.6 \pm 1$[d] | 12.8 | | 0.01[g] | $0.1 \pm 0.1$[g] | 95 |
| (12) | 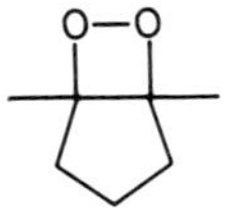 | Benzene | | | | | 0.003[g] | 0.2[g] | 92 |
| (13) | 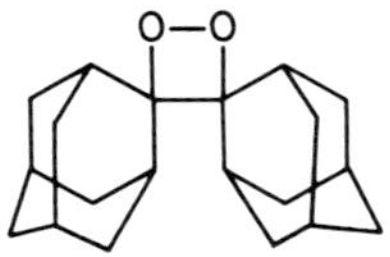 | Xylene | | $35 \pm 2$[a] | $14 \pm 1$ | | 0.02[i] | 0.15[g,h] | 96, 97 |
| | | $CH_3CN$ | | $39 \pm 2$[e] | | | | | 97 |
| (14) | 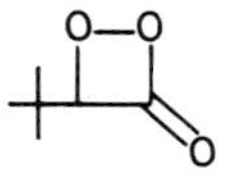 | $CCl_4$ | $\sim 2\text{–}3 \times 10^{-3}$, 22 °C | | | | | | 98 |
| | | $CCl_4$ | | | | | | | |
| | | Benzene | $1.3 \times 10^{-3}$, $23 \pm 2$ °C | 19.4[a,k] | 11.3[a,k] | | | | 99, 100 |

| | | | | | | | | |
|---|---|---|---|---|---|---|---|---|
| (15) | | $CCl_4$ | | | | ~0.01 | 0.2 | 98 |
| | | $CCl_4$ | | | | | | |
| | | Benzene | $\sim 10^{-3}$, 23 ± 2 °C | | | | | 101 |
| (16) | | $CCl_4$ | $\sim 2 \times 10^{-3}$, 22 °C[a] | 14.3[a,k] | 8.0[a,k] | | | 99 |

[a] Decay rate by n.m.r. or i.r.
[b] Decay rate by iodometric titration
[c] Rate of decay of chemiluminescence intensity, no fluorescer added
[d] Rate of decay of chemiluminescence intensity, with added fluorescer (usually DPA or DBA).
[e] Activation energy from chemiluminescence intensity (by temperature drop method or by step analysis method, see Section 7.2.8)
[f] Yields from measurements of direct emission of excited carbonyls (Section 7.2.3)
[g] Yields from energy transfer to chemically stable, luminescent acceptor (usually DPA or DBA) (Section 7.2.4)
[g'] Same as above, but energy transfer to $EuT_3Ph$ (Section 7.2.4)
[h] Yields from energy transfer to reactive acceptor (Section 7.2.5)
[i] Yields from direct reaction of excited carbonyl with scavenger (no intermediate energy transfer step, Section 7.2.6)
[j] Yields from intramolecular reaction of the excited carbonyl (Section 7.2.7)
[k] Activation parameters calculated from the equations $E_a = \Delta H\ddagger + RT$ and $\log A = (\Delta S\ddagger/4.576) + 10.753 + \log T$.

**Table 7.2**

| *No.* | *Dioxetane* | *Solvent* | *Rate* ($k_1$) /$s^{-1}$ | $\lambda_{max}$/nm (*emission*) | *Additional information* | *Ref.* |
|---|---|---|---|---|---|---|
| (17) |  O—O | Cyclohexane | | 418 | | 16 |
| (18) |  O—O | Cyclohexane | | 424 | | 16 |
| (19) | 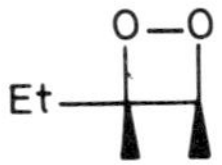 O—O, Et | Cyclohexane | | 431 | | 16 |
| (20) | 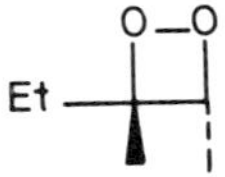 O—O, Et | Cyclohexane | | 425 | | 16 |
| (21) | 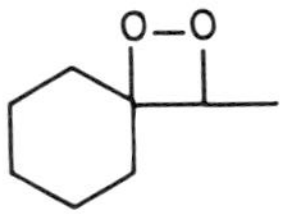 O—O | Benzene | ~8 × $10^{-4}$, 50 °C | | Bluish-white luminescence strongly enhanced by DBA | 24 |
| (22) | 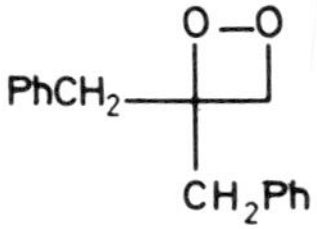 O—O, $PhCH_2$, $CH_2Ph$ | Benzene | | | | 75 |

| | | | | | | |
|---|---|---|---|---|---|---|
| (23) | O—O | $CCl_4$ | $\sim 8 \times 10^{-4}$, 44 °C | 426 | | 102 |
| (24) | O—O, O, O | Acetone | $\geqslant 5 \times 10^{-4}$, 60 °C | | | 94 |
| (25) | O—O, EtO, OPh, *cis* and *trans* | | Similar to (9) | | | 103, 104 |
| (26) | O—O | | | | M.P. 129–130 °C decomposes ~200 °C | 105 |
| (27) | O—O, MeO, OMe, Ph, Ph, *cis* and *trans* | | | | At room temp., complete decomp. takes minutes in $CH_2Cl_2$ or $CHCl_3$, hours in MeOH, EtOH, ether; *cis*-dioxetane is more stable than *trans* | 106 |
| (28) | O—O, Ph, Ph | $CHCl_3$ | $3.1 \times 10^{-5}$, 35 °C | | Chemiluminescence in $CHCl_3$, with DPA or DBA; m.p. 57–61 °C (dec.) | 107 |

**Table 7.2**—*continued*

| *No.* | *Dioxetane* | *Solvent* | *Rate* ($k_1$) /$s^{-1}$ | $\lambda_{max}$/nm (*emission*) | *Additional information* | *Ref.* |
|---|---|---|---|---|---|---|
| (29) | 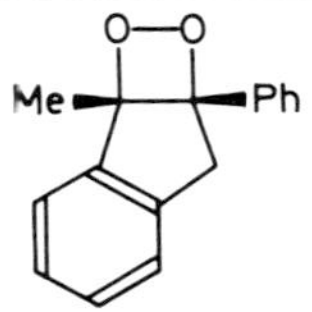 Me, Ph | | | | M.P. 69–70 °C (dec.) | 107 |
| (30) | 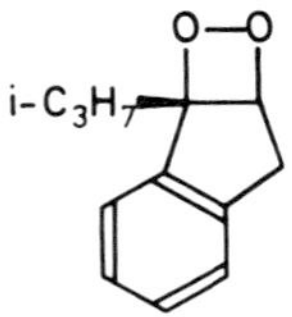 i-$C_3H_7$ | | | | M.P. 49.5–51 °C (dec.) | 107 |
| (31) | 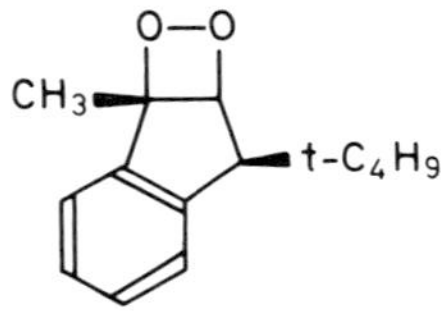 $CH_3$, t-$C_4H_9$ | | | | M.P. 56–58 °C (dec.) | 107 |

| | | | |
|---|---|---|---|
| (32) | 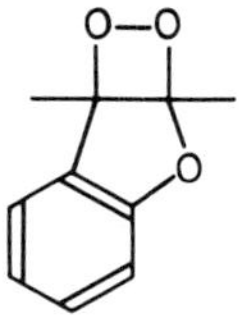 | | 108 |
| (33) | 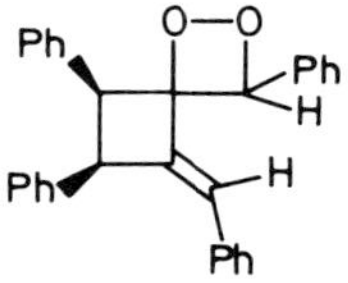 | Reportedly stable <200 °C | 109 |
| (34) | 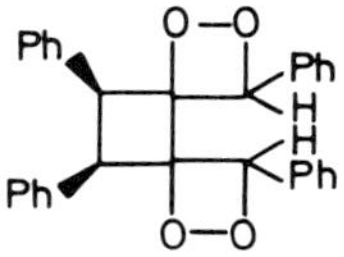 | Reportedly stable <200 °C | 109 |

dioxetanes are listed in Table 7.2. These tables suggest four generalisations. (1) The kinetic parameters of most of these isolable, simple dioxetanes are remarkably similar ($E_a \approx$ 23–27 kcal mol$^{-1}$, log $A \approx$ 12–13). (2) No dioxetane has yet been reported which does not emit light, either directly or in the presence of a fluorescer. (3) With but one exception (6) reported so far, the excited carbonyls formed are found predominantly in the triplet rather than in the singlet state. A ratio of triplets to singlets of 100 or more appears to be fairly general. (4) The primary yield of excited carbonyls is high: 10% or more per dioxetane molecule. These salient points now deserve a closer examination.

### 7.4.2 Stability

Looking at dioxetanes (1)–(5), we see that increasing substitution of the ring hydrogens by methyl groups stabilises the dioxetane, with the biggest gain between the trimethyl- (4) and the tetramethyl-dioxetane (5) (Figure 7.7). Replacing one methyl by one phenyl group brings very little change: see (1), (2) and (3). The cyclic dioxetanes (7) and (10) do not behave very differently from their non-cyclic counterparts (5) and (9). On the other hand, the bicyclic dioxetane (8) appears quite a bit less stable than the monocyclic equivalent (7) or than TMD (5). The strain of the decomposition product of (8), the ten-membered ring diketone, may be important. Note that dioxetanes (5), (7) and (8) were studied by the same authors and the same methods; hence the observed differences in rate are probably significant.

Substitution by two bulky, rigid groups brings about a much more drastic change in stability. Any decomposition mechanism will have to account for the high activation energy (~37 kcal mol$^{-1}$) of the adamantylideneadamantanedioxetane (13; m.p. 163–164 °C!), at least 10 kcal mol$^{-1}$ higher than the average $E_a$. The bis-norbornanedioxetane (26) is again extremely stable (m.p. 129–130 °C) and should be investigated.

The α-peroxylactones stand at the other end of the stability scale. The activation energy of the 1-adamantyl derivative (16) is reported to be only ~14 kcal mol$^{-1}$; all three α-peroxylactones (14), (15) and (16) synthesised thus far by Adam and his co-workers have lifetimes of the order of only 10 min at room temperature, compared with days or years for the other dioxetanes of Table 7.1.

### 7.4.3 Yields and multiplicities of excited products

There is no doubt that the cleavage of many of the simple dioxetanes of Table 7.1 generates high yields of excited carbonyl fragments. The simplest and best studied example is TMD (5), for which a quantum yield of 0.2–0.5 has now been determined by four different methods, and in three laboratories. This point is very important, as every method has its hidden or overt pitfall, and standardisation of light intensities, for example, is a notorious source of errors. Table 7.1 itself does not reveal, for instance, that the yields of triplet acetone from TMD obtained by Turro's group by chemical titra-

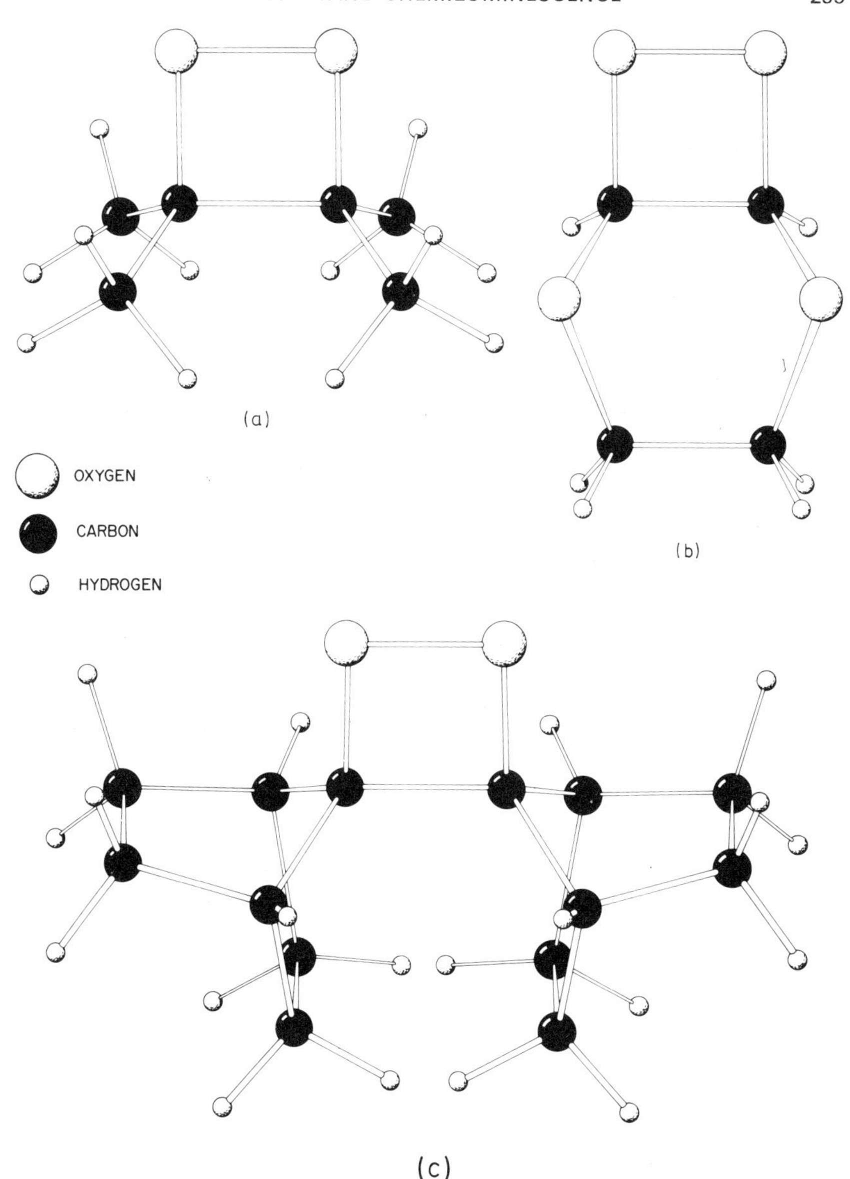


**Figure 7.7** Models of dioxetanes: (a) tetramethyldioxetane; (b) *p*-dioxenedioxetane; (c) bis-norbornanedioxetane

tion ($^3\phi = 0.48$)[39] and by the DBA method ($^3\phi = 0.50$)[92] are actually not independent results, since the DBA method was calibrated against the previous chemical results. Thus, it is rewarding to see the convergence between the yields determined by luminescence methods or, totally independently, from the photoreduction of excited acetone by cyclohexa-1,4-

diene. The latter method gives a lower bound of ~0.2, which can probably be trusted for the reasons discussed earlier (Section 7.2.6).

Trimethyldioxetane (4) has also been studied both by chemical and by luminescence methods. The agreement here is much less satisfactory. However, White *et al.*'s result of 0.04[16,60], based on stilbene isomerisation, should be considered only as a lower limit (Sections 7.2.5 and 7.5.1); we have already expressed earlier (Section 7.2.4) our misgivings concerning the use of $EuT_3Ph$ for quantitative yield determination. Trimethyl- and tetramethyl-dioxetanes were found to give identical yields of excited products in the hands of Turro *et al.*[92] (by the DBA/DPA technique), whereas using essentially the same method, Kopecky and Filby report yields differing by a factor of 2[17,93].

Judging by the results of Table 7.1, these two dioxetanes (4) and (5) give the highest yields of excited products. None of the other dioxetanes have thus far been studied by more than one method. Nevertheless, the smaller yields of ~0.2 obtained for dioxetanes (7), (9), (12), (13) and (15) are probably reliable. The very significant lower yield of 0.03 obtained for the bicyclic compound (8) is particularly notable.

With but one exception (6), all dioxetanes have been found to generate at least two orders of magnitude more triplet than singlet excited products. As discussed at length earlier, this conclusion rests on several lines of evidence: (1) direct luminescence (in the case of tetramethyldioxetane, where both acetone fluorescence and phosphorescence can be observed, see Section 7.2.1); (2) higher chemiluminescence from DBA than DPA, in spite of the lower fluorescence yield of DBA compared with DPA (Section 7.2.4); (3) photochemical reactions characteristic of triplet carbonyls (Sections 7.2.4–7.2.7).

The only exception to this preferred triplet excitation was reported by Darling and Foote[72] for dioxetane (6), from their study of the Type II photoproducts of hexan-2-one. Not only did these authors find a significantly lower total yield of excited products (0.08–0.10), but according to their results, these are predominantly in the singlet state ($^3\phi/^1\phi \approx 0.4$–$0.7$). (See, however, Section 7.2.7 for a discussion of the method.) The results on this dioxetane stand in such striking contrast to the general pattern that it is important that they be confirmed by other methods.

### 7.4.4 Effect of solvent

Apart from impurity effects, there seems to be little solvent effect on the kinetic parameters of tetramethyldioxetane (5). The same obtains probably for the yields of excited products. In a note, Turro and Lechtken[78] reported a substantial decrease of the activation enthalpy, accompanied by a large negative activation entropy, in methanol. This, however, has been shown to be a spurious effect[68] (Section 7.5.3). The opposite behaviour in acetonitrile (higher $\Delta H^* = 30$ kcal mol$^{-1}$ and large positive $\Delta S^*$) reported by the same authors awaits confirmation; a value of $E_a = 27$ kcal mol$^{-1}$ in acetonitrile was later reported by Turro *et al.*[81] without discussion of their previous result.

## 7.5 REACTIVITY OF DIOXETANES: COMPETITIVE PATHWAYS OF DECOMPOSITION

We have mentioned on several occasions the interpretative hazards created by possible side reactions of dioxetanes which, if not properly diagnosed, may lead to erroneously high rates and low yields. Each of the three 'catalysed' modes of decomposition outlined below was discovered fortuitously; yet each throws new light on the stability of the dioxetanes and on the mechanism of generation of excited states.

### 7.5.1 Induced decomposition

Expecting to increase their chemiluminescence yields from DED (9), Wilson and Schaap[43] discovered that degassing had the opposite effect. The removal of oxygen brought about no significant change in initial intensity, but a marked increase in rate and therefore a decrease in quantum yield. Instead of the expected beneficial effect of increasing the efficiency of TS energy transfer to DBA by lengthening the life of the triplet formate donors, it is the harmful consequences of increasing the lifetime of triplet DBA which dominates the picture. Unquenched by oxygen, triplet DBA attacks the dioxetane and 'sensitises' its decomposition [equation (7.22)].

$$\text{DED (EtO, OEt dioxetane)} \xrightarrow[K_1]{\Delta} {}^3\text{O}^*{=}\text{CH–OEt} + \text{O}{=}\text{CH–OEt} \qquad (7.34)$$

$$(\text{DED} \longrightarrow {}^3\text{F}^* + \text{F})$$

$$^3\text{F}^* + \text{DBA} \longrightarrow {}^1\text{DBA}^* + \text{F} \qquad (7.16a)$$

$$\phantom{^3\text{F}^* + \text{DBA}} \longrightarrow {}^3\text{DBA}^* + \text{F} \qquad (7.16b)$$

$$^3\text{DBA}^* + {}^3\text{O}_2 \longrightarrow \text{DBA} + {}^1\text{O}_2 \qquad (7.35)$$

$$^3\text{DBA}^* + \text{DED} \longrightarrow \text{F}^{?*} + \text{F} + \text{DBA} \qquad (7.36)$$

At the concentration of dioxetane in these experiments (~0.1 M), the rate was increased by a factor of 2.5–3 and remained of apparent first order. If no fluorescer (DBA, DPA or biacetyl) was present, degassing brought no increase in rate (measured by n.m.r. or iodometry, since the formate does not fluoresce). The best interpretation of these results calls for a chain decomposition carried on by triplet DBA, the chains being short on account of the low concentration of the dioxetane; this, and the presumed short lifetime of triplet formate, is the likely reason for the failure to observe induced decomposition in the absence of fluorescer. Arguments in support of this interpretation will be discussed in Section 7.6.

Turro and Lĕchtken[63] observed a similar but even more striking effect in degassed solutions of tetramethyldioxetane (~1 M), which they interpreted essentially as follows:

$$\text{TMD} \xrightarrow{\Delta} {}^3\text{A}^* + \text{A} \quad (7.37)$$

$${}^3\text{A}^* + \text{TMD} \xrightarrow{k_{id}} x\,{}^3\text{A}^* + (3 - x)\,\text{A} \quad (7.38)$$

The best fit of their kinetic data corresponds to $x = 1$. With 1 M TMD in degassed benzene solution at room temperature, the half-life of the reaction was of the order of 1–2 min, compared with $10^3$–$10^4$ min in aerated solution. The rate constant for induced decomposition $k_{id}$ is $\sim 10^9$ l $\text{mol}^{-1}$ $\text{s}^{-1}$. Because of this chain character of the reaction at high dioxetane concentration, the decomposition then follows second-order kinetics[68].

### 7.5.2 Amine catalysis

Different amines, such as $Et_3N$, $Et_2N$ and DABCO (1,4-diazabicyclo[2.2.2]-octane), were found to catalyse the decomposition of *cis*-diethoxydioxetane (9), via a competitive 'dark' pathway of lower activation energy ($\sim$12 instead of $\sim$24 kcal $\text{mol}^{-1}$), superposed on the normal chemiluminescent decomposition[80]. This duality of mechanisms was best demonstrated in quick temperature-drop experiments (Section 7.2.8 and Figure 7.4). The product of the reaction is still ethyl formate, as in the uncatalysed decomposition, and the amine remains unchanged. Yet *no* excited carbonyls are generated. Typically, at 43.5 °C the observed rate with 0.015 M $Et_3N$ in equimolar concentration with DED is $\sim$35 times larger than the uncatalysed rate.

The electron-rich olefin *cis*-diethoxyethylene, which paradoxically is the starting material for the synthesis of this dioxetane via addition of $^1O_2$ (Section 7.1.3), was itself found to be a catalyst of decomposition, like the amines. In fact, it also has a fairly low ionisation potential; this points to a possible charge transfer interaction between dioxetane and catalyst, weakening the O—O bond, as a likely mechanism. There is indeed a rough correlation between the efficiency of a catalyst and its ionisation potential[80].

Other dioxetanes [such as (10) and (11), Table 7.1] have since been found to be also susceptible to amine catalysis, whereas TMD is apparently not[46,95].

### 7.5.3 Metal catalysis

The very powerful catalytic effect of trace amounts of impurities, probably of transition metal compounds, commonly present in hydroxylic solvents, has already been mentioned. Very much like that of an amine, their overall effect is to induce a parallel path of decomposition of the dioxetanes, a path which does not generate excited states. This results in a lowering of the apparent activation energy, a considerable though irreproducible increase in rate and a corresponding decrease in chemiluminescence yield. Traces of cupric chloride ($\sim 10^{-5}$ M) increase the rate of TMD decomposition in methanol by a factor of $\sim$100, whereas in the presence of the chelating agent ethylenediaminetetra-acetic acid (EDTA), rates, yields and activation energy in methanol or ethanol become the same as in benzene[68]. Further studies showed that many transition metals ($Cu^{2+}$, $Ni^{2+}$, $Co^{2+}$, $Zn^{2+}$, $Mn^{2+}$, $Cd^{2+}$) are active as catalysts of TMD decomposition to acetone, $Cu^{2+}$ having the greatest activity[110].

The cupric chloride catalysed decomposition of TMD is first order with respect to both TMD and metal ion, and a counter ion effect is observed. A linear free energy correlation was found between the logarithms of the second-order rate constants for the catalysed decomposition and the Lewis acidities of the metal ions (as measured by the negative logarithms of the dissociation constants of the metal malonates). On this basis, Bartlett *et al.*[110] postulated a coordination mechanism involving the metal ion as a Lewis acid. They reasoned that complexation of the dioxetane by the metal ion might facilitate decomposition by removing orbital symmetry restrictions or by lending positive character to one or both oxygen atoms, thereby destabilising the peroxy bond:

$$\text{TMD} + M^{2+} \rightleftharpoons \text{complex} \longrightarrow 2\,(CH_3)_2C{=}O + M^{2+} \qquad (7.39)$$

### 7.5.4 Insertion reactions of triphenylphosphine and stannous compounds

Although outside the scope of this review, two very interesting reactions of TMD recently investigated by Bartlett's group should be mentioned here. (a) Triphenylphosphine reacts fast and quantitatively with TMD to form a stable solid phosphorane, which decomposes upon heating to form quantitative amounts of triphenylphosphine oxide and tetramethylene oxide[111]:

$$\text{TMD} + PPh_3 \xrightarrow[6\,^\circ C]{C_6H_6} \text{phosphorane} \xrightarrow{55\,^\circ C} Ph_3\overset{+}{P}{-}O^- + \text{tetramethylethylene oxide} \qquad (7.40)$$

Similar but even more stable phosphoranes were obtained by reaction of TMD with trimethyl or triethyl phosphites or with methyl diphenylphosphinite[112]. (b) Treatment of TMD with stannous chloride in protic solvents results in formation of varying amounts of pinacol and acetone; the yields are solvent dependent. The two products appear to be formed by competing mechanisms[18,89]. For example:

$$\text{TMD} + \underset{\sim 0.4\ \text{equiv.}}{SnCl_2} \xrightarrow{CD_3OD} \underset{37\%}{\text{pinacol-}d_2\ (\text{DO, OD})} + \underset{63\%}{(CH_3)_2C{=}O} + \underset{(\text{no } SnCl_2)}{Sn^{IV}} \qquad (7.41)$$

## 7.6 DIRECT AND SENSITISED PHOTODECOMPOSITION OF DIOXETANES

The photochemistry of dioxetanes presents particularly interesting aspects. For example, the mechanism proposed for the induced decomposition of DED (Section 7.5.1) calls for a surprising step. Indeed, the triplet of DBA is located[29] at ~40 kcal mol$^{-1}$, surely well below any possible spectroscopic triplet of the dioxetane. Some type of 'non-vertical' energy transfer[26] must therefore be invoked, as was suggested in a similar situation involving

$$^{3}\text{DBA}^{*} + \text{(EtO)(O–O)(OEt) dioxetane} \longrightarrow 2\ \text{EtO–C(=O)–H} + \text{DBA} \quad (7.37)$$

benzoyl peroxide[113,114]. Wilson and Schaap's photochemical experiments showed that the decomposition of this dioxetane can clearly be sensitised by DPA or DBA, mostly from their singlet but also from their triplet states, in support of the mechanism proposed to account for the effect of oxygen removal[43].

Although the possibility was discussed, no attempt was made in these preliminary experiments to detect excited formate, or 'recycling' of the excited DBA. In experiments with sensitisers of lower triplet energy, it would be interesting to see if a low cut-off limit can be found and, if so, if it approximates to the activation energy for thermal decomposition ($E_a \approx$ 24 kcal mol$^{-1}$).

Turro and Lechtken[79] investigated the direct photolysis of TMD, with the surprising result that even with exciting light of wavelength as long as 450 nm (for TMD spectrum, see Figure 7.6), the emission was the characteristic emission of acetone fluorescence (~400 nm). Irradiation of TMD causes its decomposition into two molecules of acetone, up to 50% of which is in an excited state. Trapping experiments with *trans*-dicyanoethylene (Section 7.2.6) and comparison of the emission from TMD with pure acetone fluorescence showed that the ratio of triplet to singlet acetone is wavelength dependent: mostly triplets are generated at 366 nm (triplet : singlet = 4 : 1), but up to 35% singlet is generated[115] at 240 nm.

Single-photon counting experiments demonstrated that the rise time of acetone fluorescence derived either from the photolysis of TMD or directly from acetone were the same (~$10^{-10}$ s). Turro *et al.*[115] interpret this result as establishing that no intermediate of lifetime greater than $10^{-10}$ s precedes the formation of singlet acetone from photoexcited TMD.

These authors also studied the decomposition of TMD sensitised by biacetyl[116]. TMD was found to quench strongly the phosphorescence, but not the fluorescence of biacetyl. Moreover, *in the presence of TMD*, triplet biacetyl ($E_T \approx$ 56 kcal mol$^{-1}$) is able to sensitise the Type II reaction of valerophenone ($E_T \approx$ 74 kcal mol$^{-1}$)[73], probably via the following 'anti-Stokes' process:

(7.42)

+ other products (7.43)

One might have expected that triplet acetone ($E_T \approx 78$ kcal mol$^{-1}$), generated from TMD in the same solvent cage as the sensitising biacetyl molecule [equation (7.42)], would immediately be quenched by transfer of energy to this biacetyl. This apparently is not an efficient process.

## 7.7 MECHANISM OF DIOXETANE DECOMPOSITION

A mechanism applicable to all the dioxetanes of Tables 7.1 and 7.2 must first be compatible with the following major observations: (a) excited states are produced efficiently, predominantly with triplet rather than singlet multiplicity, even when the singlet state is accessible; (b) the stability of the dioxetane is little influenced by substituents, although very bulky, rigid substituents (such as adamantane or norbornane) render the dioxetanes unusually stable, whereas α-peroxylactones are much less stable.

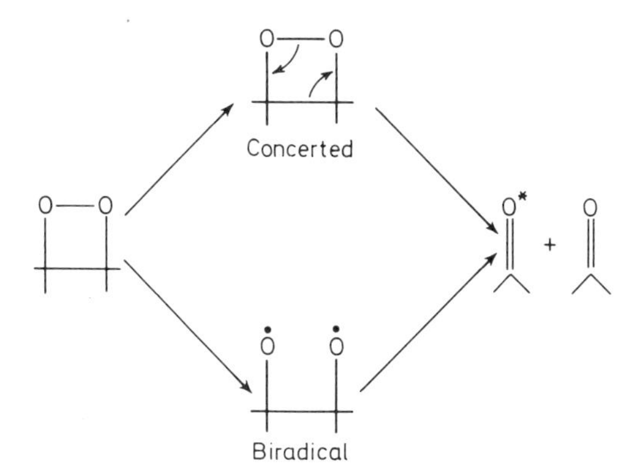


**Figure 7.8** Two extreme mechanisms of dioxetane decomposition: concerted *versus* biradical pathway

Let us now consider two extreme mechanisms (Figure 7.8). First, a truly *concerted mechanism*, in which the O—O and the C—C bonds break and C=O bonds form simultaneously, so that the exothermicity of forming two carbonyl groups helps drive the reaction from dioxetane to products. Second, a *two-step mechanism*, in which the cleavage of the weaker O—O bond occurs first, thus generating a short-lived biradical which cleaves into products in a second step.

McCapra[10] was first to suggest a concerted mechanism, whereby the whole exothermicity of the reaction may be envisaged as channelled into electronic energy of one of the carbonyl fragments, via a simple rearrangement of electrons guided by orbital symmetry rules. This idea is certainly responsible to a large extent for the recent interest in chemiluminescence. Kearns[117] and McCapra[1,118] worked out orbital and state correlation diagrams depicting the cleavage of a dioxetane into two carbonyl fragments (Figure 7.9).

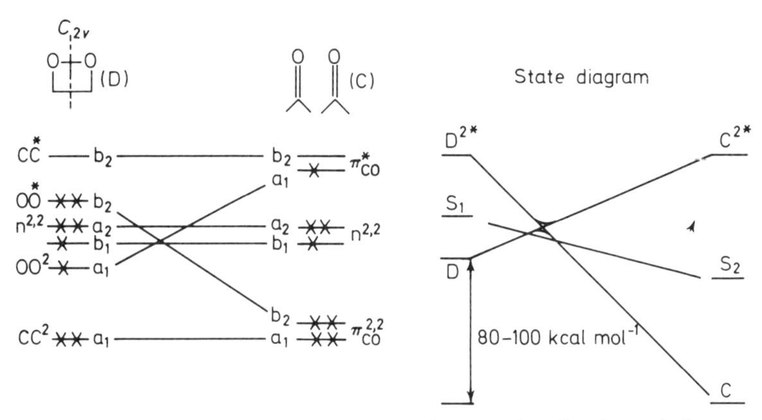


**Figure 7.9** Orbital and state correlation diagrams for the thermal cleavage of a dioxetane. '$S_2$ is an n,π* state of one of the two carbonyl (C) compounds, either singlet or triplet, and $S_1$ is a correlated state of the dioxetane (D). It may be of interest that the n,π* transitions are allowed in this case, and that early breakage of the O—O bond may lead to a lowering of the energy of $S_1$' (From McCapra[1], by courtesy of Wiley)

The ground state electronic configuration of the dioxetane apparently correlates with a highly excited state of the carbonyl products, whereas the doubly excited state of the dioxetane goes down to the ground state of the carbonyl cleavage product. A singly excited state would then be expected to cross both of these, so the ground state reactant, after an initial rise, could cross to the surface which ends as excited product, either singlet or triplet. The activation energy for decomposition would then represent the energy expenditure up to the crossing of the state levels. The main difference between the general aspect of such a state diagram for dioxetane as Kearns' or McCapra's, and the classic Woodward–Hoffmann diagram for the decomposition of cyclobutane into two ethylene molecules[119], is evidently the large exothermicity of the dioxetane reaction, which may bring the excited states of a carbonyl fragment below the transition state (for thermochemical reasons, only one carbonyl can be excited) (Figure 7.10). The details of such diagrams are notoriously difficult to calculate, particularly in this heterocyclic system with the lone electron pairs of oxygen. Kearns[117] has stated that the ground state curve of the dioxetane may be forbidden from crossing the singlet excited state curve of the product, while allowed to cross the triplet curve. This may well be, although simple mindedly one would think that such a spin flip would slow down the decomposition, which is fast ($\log A \approx 13$).

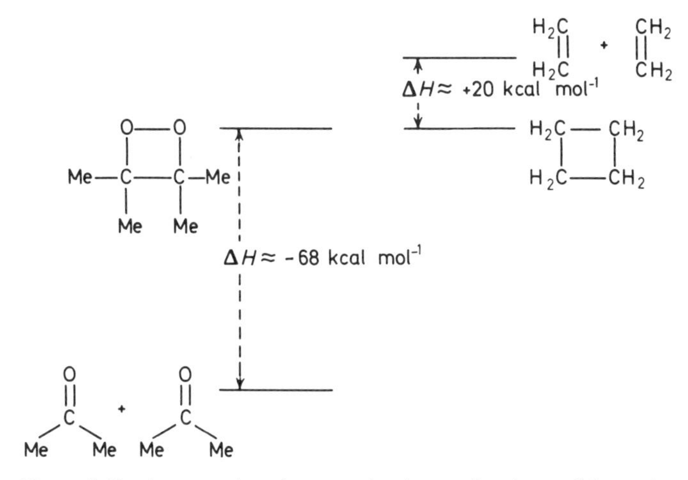


**Figure 7.10** A comparison between the thermochemistry of the cyclobutane–ethylene and dioxetane–carbonyl fragments systems

Nevertheless, it seems to us that the much greater stability of the bis-adamantane- and bis-norbornane-dioxetanes [Table 7.1, (13), and Table 7.2, (26); see also Figure 7.7] is difficult to explain in terms of a truly concerted mechanism. If cleavage of the O—O and C—C bonds were simultaneous, why would these two crowded dioxetanes be so reluctant to come apart? Also, as Richardson *et al.*[88] pointed out, one would expect that in a concerted mechanism the stability of a dioxetane would be reduced by substituent groups which could conjugate with the developing carbonyl group. For example, dioxetane (2) (Table 7.1) should decompose faster than (1), where the phenyl group is replaced by a methyl group; in fact, the rates of these two dioxetanes are the same.

Going on now to the two-step mechanism, let us consider the process of breaking the O—O bond first, by a combination of stretch and twist around the C—C bond:

O—O / C—C ⟶ ·O O· / C—C (7.44)

Breaking the O—O bond of course releases the ring strain and the oxygen atoms fly apart (provided there is no steric hindrance due to the substituents on the ring). The energy of a normal peroxide bond[120] is *ca.* 35 kcal mol$^{-1}$, whereas the ring strain is probably ~26 kcal mol$^{-1}$, its value in cyclobutane[85]. Thus the activation energy can be expected to be less than 35 kcal mol$^{-1}$, not incompatible with the average $E_a \approx$ 23–27 kcal mol$^{-1}$ observed for most dioxetanes. Once the O—O bond is broken, the four-electron rearrangement in the resulting dioxetane should follow immediately. This second step should be very exothermic, since two new carbonyls will be formed at the expense of the one C—C bond.

Using the group additivity method of Benson[85], O'Neal and Richardson[83,84] have calculated the thermochemistry of the dioxetane ring (substituted with up to four methyl groups), as well as that of the corresponding biradical intermediate and carbonyl products. For example, for trimethyldioxetane their calculated value for the reaction enthalpy is $-65.6$ kcal mol$^{-1}$. The *experimental* activation Energy $E_a$ for decomposition of this dioxetane is 23.7 kcal mol$^{-1}$. Hence, according to O'Neal and Richardson, the energy available for electronic excitation of the carbonyl fragment is

$$E_a - \Delta H^\circ_r = 89.3 \text{ kcal mol}^{-1} \tag{7.45}$$

which is sufficient for excitation of either acetone or acetaldehyde singlet states. We will come back later to this result, and cast some doubt on its validity.

The same authors have also calculated the *difference* between the activation energy of trimethyldioxetane and that of the dioxetane derivatives with one more methyl substituent as well as with one or two fewer methyls[83,84]. Because the results of this calculation have often been quoted, we will analyse them at some length here. They are based on the following scheme:

$$\text{dioxetane } (R^1R^2C{-}O{-}O{-}CR^3R^4) \underset{-1}{\overset{1}{\rightleftharpoons}} R^1R^2C(O^\bullet){-}C(O^\bullet)R^3R^4 \xrightarrow[\text{fast}]{2} R^1R^2C{=}O + R^3R^4C{=}O \tag{7.46}$$

The assumption here is that the rate limiting step is 1; therefore that the activation energy of the three steps are such that

$$E_1 > E_{-1} > E_2$$

Figure 7.11 illustrates, in our understanding, their idea of a biradical intermediate sitting in a shallow minimum on the reaction surface between dioxetane and products. We see on this diagram that $E_{-1} = 8.5$ kcal mol$^{-1}$. This activation energy for reclosing the dioxetane ring is considered to be essentially an activation energy for internal rotation around the C—C bond [see equation (7.44)] to get the O atoms to face each other again. O'Neal and Richardson then compare the dioxetane case with the cyclobutane decomposition into two ethylene molecules. On energy grounds one could say that the hydrocarbon biradical wants to go back to cyclobutane, whereas the dioxetane biradical wants to go to carbonyl products. This reasoning leads to an activation energy $E_2$ from biradical to products smaller than the corresponding 6.5 kcal mol$^{-1}$ in the cyclobutane case and smaller than $E_{-1}$ (hence $k_2 \gg k_1$). Finally, they calculated the activation entropy for ring closing, again on the assumption that the reaction coordinate is an internal rotation, taking the value calculated for the analogous hydrocarbons.

The activation parameters calculated on this basis for increasingly methyl substituted dioxetanes agree quite well with the experimental values[83,84]. This is not entirely surprising, since the authors have adopted the experimental parameters for trimethyldioxetane, and only calculated the differences between these and the corresponding values for the other dioxetanes in the series (assuming the same value for $E_{-1} = 8.5$ kcal mol$^{-1}$ and the same ring strain $= 26$ kcal mol$^{-1}$ throughout). These differences are indeed very small, and this truly is the interesting result.

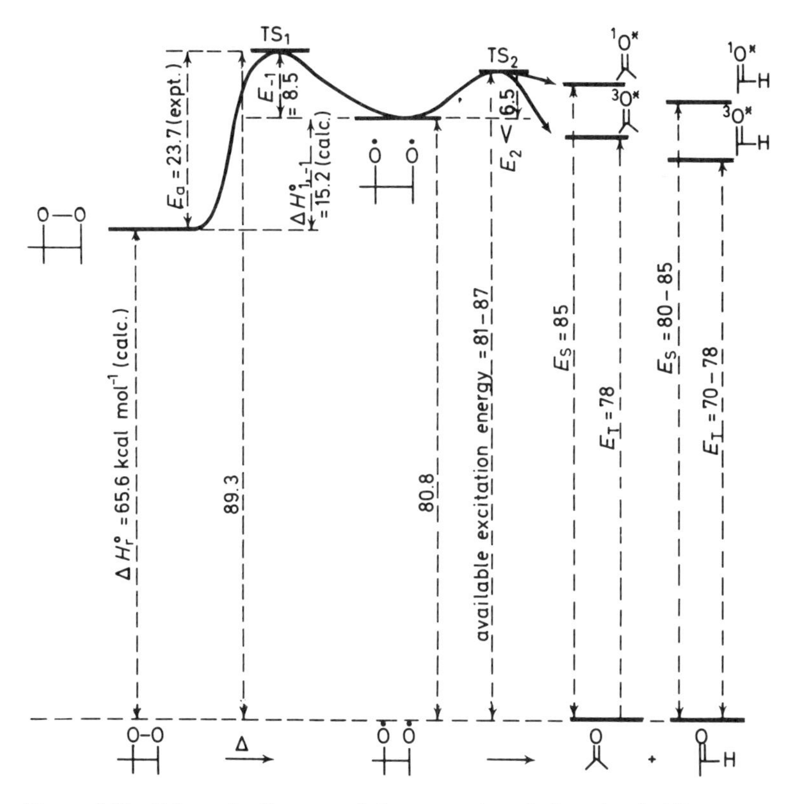


**Figure 7.11** Schematic diagram of the energetics of the trimethyldioxetane decomposition, based on the assumptions and calculations of O'Neal and Richardson[83,84] (energies in kcal mol$^{-1}$)

As mentioned above, the amount of energy available for electronic excitation, as calculated by O'Neal and Richardson, seems indeed sufficient for most singlet and triplet carbonyls, except for the totally unsubstituted dioxetane (giving formaldehyde). This result[83,84] may be in error, however, since it appears that the energy available for excitation was taken as ($E_a - \Delta H°_r$), equation (7.45), whereas it should have been (see Figure 7.11)

$$E_2 - (\Delta H°_r + \Delta H°_{1,\,-1}) \tag{7.47}$$

In this calculation, $E_2$ is only known to be $<6.5$ kcal mol$^{-1}$. Hence the energy available for excitation may be between 2 and 8.5 kcal mol$^{-1}$ *smaller* than given by O'Neal and Richardson. In other words, instead of having 89.3 kcal mol$^{-1}$ at our disposal for electronic excitation of the dioxetane, we may have only anywhere between 81 and 87 kcal mol$^{-1}$, i.e. surely enough for excitation of triplet acetone, but may be not enough to reach its singlet. In view of the arbitrary character of many of the assumptions on which these calculations are based, it is probably not justified to put too much weight on this result. It should be kept in mind, however, that the calculations of O'Neal and Richardson[83,84] appear to be in error by 2–8 kcal $mol^{-1}$ and therefore there is at least a possibility that the biradical mechanism does not allow excitation of the singlet state of some of the carbonyl products, without additional thermal activation.

In any case, the two-step mechanism, like the concerted one, has to be compatible with the observed high yields of triplet fragments. Right after the cleavage of the O—O bond, the unpaired electrons on each O atom are in a singlet state, because of the residual O—O bonding. However, when sufficiently separated, the bonding energy will be negligible and so will be the (antibonding) repulsive interaction in the triplet states. Consequently, the singlet and triplet states will approach one another in energy. This will make it much easier for spin orbit interactions, etc. to cause transitions from singlet to triplet, and *vice versa*. *If statistical equilibrium were attained*, each of three triplets and one singlet state would be equally occupied, so that there would be three times as much triplet as singlet biradical. If the C—C bond breaks, then each singlet biradical will generate two singlet carbonyl fragments (only one of which might, but need not, be excited, for thermochemical reasons; two triplets are ruled out also). Each triplet biradical will form one triplet carbonyl plus one ground state carbonyl. Hence 75% of all carbonyl fragments formed should be in the triplet state.

At present, the highest yields of triplets (0.5 for TMD) do not reach this 75% theoretical limit. However, it is impossible to know if these experimental yields represent the genuine primary yields of carbonyl fragments, or if quenching factors work to reduce these yields before they can be measured. For instance, it is at least conceivable that something like an exciplex process involving the newly born triplet carbonyl and its ground state twin — still within one bond distance of one another — could be an efficient quenching path. On the other hand, a primary yield of triplets higher than 75% is a possibility also, considering that triplet biradicals are more likely to dissociate, since they cannot rebound without first crossing back to singlet.

All this shows that a prompt intersystem crossing from singlet to triplet biradical, before the C—C bond breaks, will increase the yield of triplet products. What are the factors which will favour this intersystem crossing?

Salem[121] has discussed this question in terms of simple orbital pictures and matrix elements with respect to atomic orbitals, particularly the p orbitals. Because a singlet–triplet transition changes the spin angular momentum, yet the total angular momentum is conserved, there must be a rotation of the orbital factor. Salem states that 'the orbital axes must be orthogonal to each other and to the axis around which the angular momentum is being created'. Also some ionic character is required in the wavefunction and there needs to be some orbital overlap (but not too much).

Müller[122] has suggested that these conditions can be met 'by a vibrationally excited dioxetane in which the internuclear distance between the oxygen atoms is sufficiently elongated so that the total wavefunction of the formal biradical may adopt some ionic character, and one-centre LS-coupling terms at oxygen become increasingly important in the evaluation of intersystem crossing rates'. Although it is unlikely that such LS-coupling mechanisms could be efficient enough to induce intersystem crossing during one single O—O dissociative event, one could assume, in addition, that the singlet biradical may close reversibly until intersystem crossing occurs[122] Naturally, the actual calculation of such crossing rates is far off and would probably require elaborate configuration interaction wavefunctions, as well as consideration of Franck–Condon vibrational factors for a vast number of vibrational states.

In truth, there is no evidence that such biradicals, lying between two maxima on the potential energy surface between dioxetane and products, do indeed exist. Their lifetimes may be so short that they have no physical meaning. The truly concerted reaction and the two-step scheme discussed so far bracket a whole array of 'nearly concerted mechanisms', with leading O—O bond cleavage. Such compromises, which would blend the qualities of both extreme mechanisms, may be much closer to the true situation.

This appears to be the position of Turro, who advocates stretching the O—O bond and weakening it first, but getting the C—C bond to weaken then also, before the O—O bond actually breaks. The idea is to benefit immediately from the energy to be released by the formation of the carbonyls. To this effect, Turro *et al.*[48,78] propose a 90° rotation about the C—O axes of the electronic charge on one of the oxygen atoms, the O—O bond being already much elongated. This movement of charge, at 90° from the O—O bond axis, would have the desired favourable effect on intersystem crossing, via spin–orbit coupling, and therefore that particular carbonyl may be formed in a $^3n,\pi^*$ state. Evidently such a near-concerted mechanism would be more exoergic than the biradical pathway. Yet as the O—O bond cleavage would lead the reaction, the greater stability of the bis-adamantane-dioxetane and bis-norbornanedioxetanes may still be understood on the steric grounds discussed earlier.

It is our belief that better and more reliable data on yields and multiplicity of products, as well as accurate thermochemical data, are badly needed before a valid choice is made between these mechanisms or any other.

In any event, the energetics of dioxetanes like TMD or DED put such constraints on the situation that the metal or amine catalysed decompositions (Sections 7.5.2 and 7.5.3), which have activation energies ~10 kcal $mol^{-1}$ lower than the uncatalysed reactions, should not be expected to yield excited singlet states, perhaps not even excited triplets.

## 7.8 CONCLUSIONS

### 7.8.1 Dioxetane chemiluminescence as a tool

Besides the intrinsic interest of the dioxetanes, there is no doubt that the by-products of their study, or their deliberate use as controllable 'dark' sources of excited molecules, will contribute to different areas of kinetics. 'Photochemistry without light', as White realised[59], offers several potential advantages over its light counterpart. These include selective excitation of triplet molecules at a known rate, even in strongly absorbing medium and, in luminescence studies, no complication due to the need to filter out or otherwise exclude the exciting photons, and hence an ideal situation for energy transfer studies.

Recent work of Lechtken and Turro[123] on energy transfer between identical molecules in fluid solution illustrates this point. This problem, evidently difficult to study by conventional methods, was approached in the case of acetone by taking advantage of the long-wavelength photolysis of TMD (Section 7.6), in deuterioacetone (the conditions were such that acetone itself was not photoexcited). Singlet and triplet acetone, from TMD, were trapped selectively by *trans*-dicyanoethylene or by *trans*-diethoxyethylene, respectively. The reaction products were the following oxetanes:

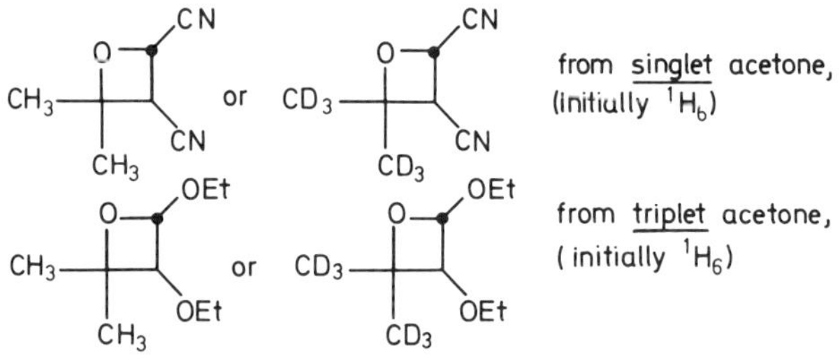

CN, O, $CH_3$, CN, $CH_3$ or CN, O, $CD_3$, CN, $CD_3$ — from singlet acetone, (initially $^1H_6$)

OEt, O, $CH_3$, OEt, $CH_3$ or OEt, O, $CD_3$, OEt, $CD_3$ — from triplet acetone, (initially $^1H_6$)

Determination of the ratio of deuterated to non-deuterated oxetanes, and Stern–Volmer treatment, gave about the same rate constants for singlet–singlet and triplet–triplet energy transfer: $k \approx 3 \times 10^6$ l mol$^{-1}$ s$^{-1}$. Thus this rate of energy hopping between 'identical' acetone molecules is *ca.* $10^3$ times smaller than diffusion controlled. There will be little chance of such a hop during the singlet lifetime, and something like 100 or 1000 hops during the triplet lifetime. (Turro[48,123] arrived at two different numbers, 14 and 1000, without comments.)

Evidently the use of dioxetanes as tools will increase with our understanding of their decomposition mechanism.

### 7.8.2 Dioxetanes in the general context of chemiluminescence in solution

Going back to the beginning of this review, is it fair to say that dioxetanes and α-peroxylactones stand a good chance of fulfilling their promised role of energy-rich precursors in many chemiluminescences involving oxygen? The answer is, at present, a qualified yes. Every dioxetane synthesised so far does indeed generate excited states in fair yields. But, and this was the greatest surprise, these are generally in triplet rather than in singlet states, even when the first singlet seems energetically accessible.

Yet in all the old 'classic' chemiluminescent and bioluminescent systems where dioxetanes or α-peroxylactones appear involved, *singlet* excited states are formed[1–6]. Indeed, the high efficiency of bioluminescence (0.88 *hv* per substrate molecule in fireflies!)[7] requires singlet products: triplet molecules very seldom emit or undergo triplet–triplet annihilation in oxygenated solution.

Could it be, then, that the case for participation of dioxetanes and α-peroxylactones in chemi- and bio-luminescence is not really as strong as we believed? It does rest on quite a considerable amount of circumstantial evidence, but to this day, no dioxetane-like compound has yet been trapped in these

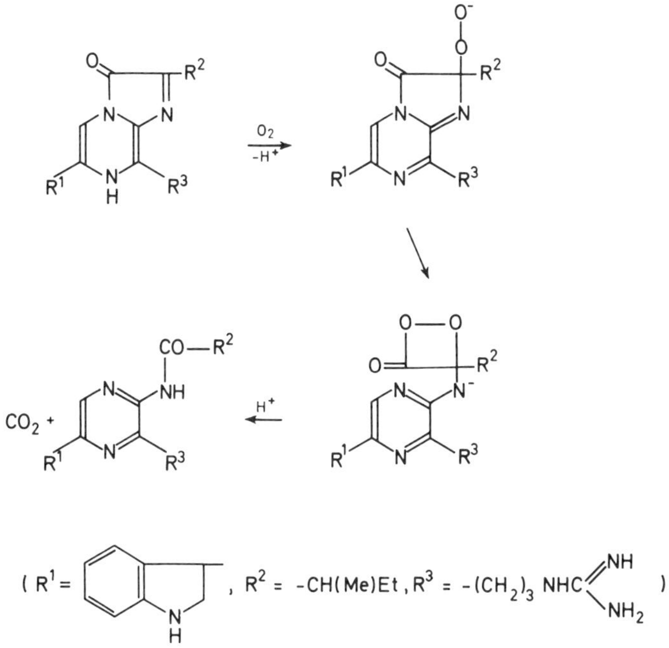

( $R^1$ = [indole ring], $R^2$ = -CH(Me)Et, $R^3$ = -$(CH_2)_3$ NHC(=NH)$NH_2$ )

**Figure 7.12** *Cypridina* luciferin and proposed dioxetane reaction path[124]

examples. Figure 7.12 shows the well-characterised substrate of the shrimp *Cypridina*[124] system (called, in a generic way, its luciferin, all bioluminescence enzymes being called luciferases, although specific luciferins and luciferases differ from one biological system to another).

$$\text{luciferin} \xrightarrow[O_2]{\text{enzyme luciferase}} \text{oxidised products} + h\nu \qquad (7.48)$$

Figure 7.13 represents the luciferin of the firefly system, with its postulated dioxetane intermediate[125]. Both systems generate $CO_2$ and both reactions can readily be effected *in vitro.*

Should the reaction be carried out in an atmosphere of labelled oxygen ($^{18}O$), the dioxetane mechanism evidently calls for incorporation of one atom of $^{18}O$ in $CO_2$, the other in the oxidised substrate. Such experiments have been conducted in the two systems above, as well as in that of the coelenterate *Renilla* where $CO_2$ is also generated and where the luciferin strongly resembles that of *Cypridina*. The results are conflicting: DeLuca and co-workers found no incorporation in the firefly[126] and *Renilla*[127] systems,

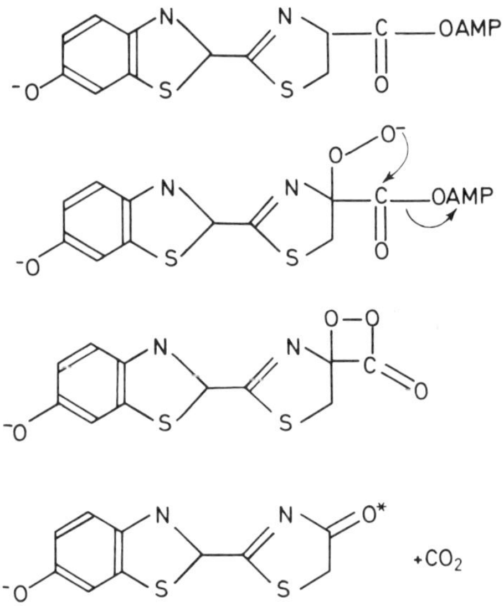

**Figure 7.13** Firefly luciferin and proposed dioxetane reaction path[125]

whereas Shimomura and Johnson did indeed find the labelled oxygen incorporation predicted by the dioxetane hypothesis in the *Cypridina* system[128]. Thus a fair amount of controversy clouded the scene, neither of the opposing parties trying to resolve the conflict by repeating their type of experiments, with their own technique, on the others' system. Shimomura and Johnson[129] appear, however, to have pinpointed the cause of the difficulty by showing that the scale of the oxygen labelling experiments was a decisive factor in the result; with very small amounts of *Cypridina* luciferin ($<1$ μmol), as in the luciferin samples in DeLuca's experiments, the exchange of oxygen between $CO_2$ and water dominates the reaction, resulting in no apparent incorporation of $^{18}O$ in $CO_2$. This seems to be a satisfactory explanation, which proponents of dioxetanes like to consider a convincing argument in favour of their mechanisms.

Not all bioluminescences, it needs be said, lend themselves easily to the suggestion of a dioxetane mechanism. Examples of such heretic systems as that of luminous bacteria[130], where $FMNH_2$ and a long-chain aldehyde are the substrates, or that of Pholas[131,132], where the substrate is a protein and where the superoxide ion $O_2^{\cdot -}$ appears involved in a free-radical mechanism, do not seem to conform to the pattern of a dioxetane intermediate. Could electron transfer mechanisms be involved in these cases?

The mass of known examples of chemiluminescence in solution present an even more battered front than bioluminescence[1–6]. First, and regretfully, a fully convincing mechanism has yet to be proposed for perhaps the most classic example, the luminol reaction, and although dioxetanes have been proposed for a large number of systems (as tentatively pictured in Figure 7.14), proofs are still wanting. We owe to the elegant work of McCapra[1] and others

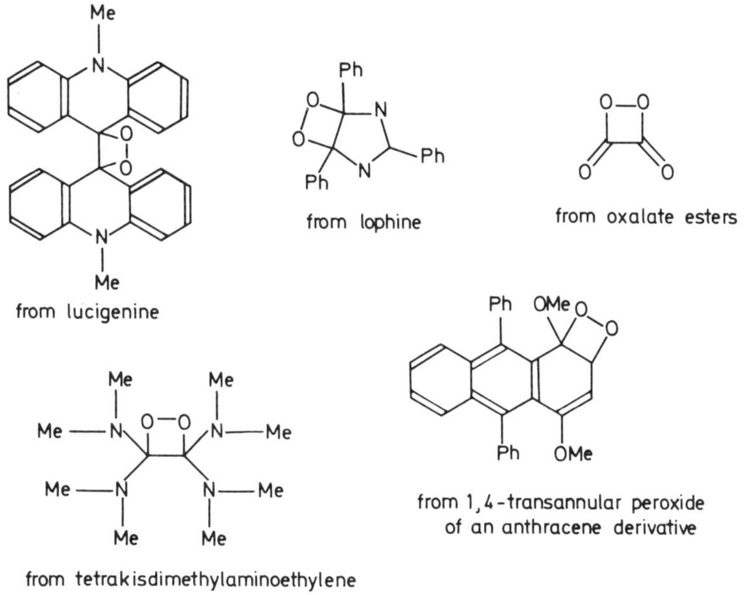

from lucigenine

from lophine

from oxalate esters

from tetrakisdimethylaminoethylene

from 1,4-transannular peroxide of an anthracene derivative

**Figure 7.14** Example of hypothetical dioxetanes proposed as intermediates in 'classical' chemiluminescences

on acridin and acridan derivatives what constitutes probably the best understood complex mechanisms (for example, autoxidation of acridan esters, $\phi = 0.1$, where the emitter is the highly fluorescent *N*-methylacridone) (Figure 7.15).

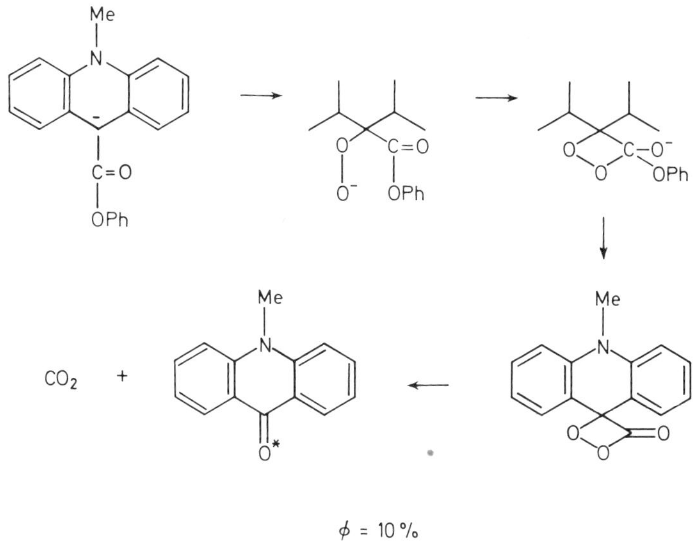

$\phi = 10\,\%$

**Figure 7.15** Autoxidation of acridan esters in strong base (From McCapra[1], by courtesy of Wiley)

What, then, is the connection between the properties of small dioxetanes reviewed here, and the hypothetical dioxetanes postulated in bio- and chemiluminescence? Clearly somewhere along the line from very simple substituents (such as methyl or phenyl groups) to large, more resonant ones (such as in the luciferins), a switch from mostly triplet to mostly singlet products must occur. It is notable that all the dioxetanes listed in Table 7.1 and 7.2 give carbonyl fragments which generally have n,$\pi^*$ first excited singlet and triplet states and are very poor fluorescers. The substituents on the peroxide ring must play a determining role in the partitioning between singlet and triplet products. More synthetic work in this direction, with more drastic substitution, is needed.

Or else, could one play the devil's advocate and argue that if all the dioxetanes encountered in nature appear to generate singlet products, it is because our sampling method is by essence biased? A 'dark' dioxetane source of triplet products could go along unnoticed, yet possibly be active in some unappreciated chemical way[125,133,134].

### 7.8.3 Is there anything sacred about the dioxetane structure?

Is it simply the large exothermicity of the dioxetane decomposition, coupled with the accessibility of carbonyl excited states, or its possible forbiddance in a Woodward–Hoffmann sense, which are the keys to the efficient generation of electronic excitation? Or is the involvement of oxygen important for reasons other than exothermicity?

The recent studies of two very exothermic thermal rearrangements are interesting in that context. Lechtken *et al.*[135] report that Dewar benzenes generate some benzene triplets, in yields $\approx 10^{-3}$–$10^{-4}$ (possibly lower limits), but no benzene singlets:

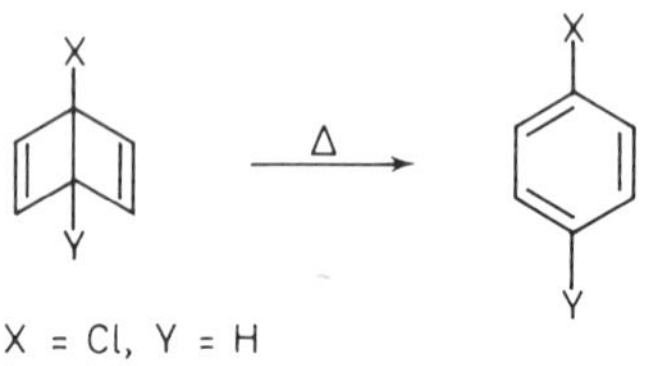

X, Y-substituted Dewar benzene $\xrightarrow{\Delta}$ X, Y-substituted benzene (7.49)

(X = Cl, Y = H
X = Y = H
X = Y = Cl)

Yang *et al.*[136] found no evidence of excited anthracene (singlet or triplet) from 2,3-naphthobicyclo[2.2.0]hexa-2,5-diene:

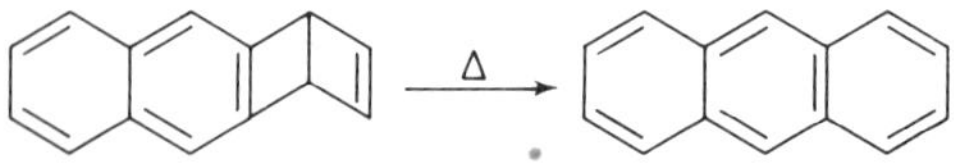

2,3-naphthobicyclo[2.2.0]hexa-2,5-diene $\xrightarrow{\Delta}$ anthracene $\quad E_a = 26.5\ \mathrm{kcal\ mol^{-1}}$ (7.50)

The exothermicity of this latter reaction, 75–80 kcal mol$^{-1}$, would be sufficient to populate either the first singlet or triplet states of anthracene; yet this apparently does not occur. Perhaps the search for triplet anthracene should be pursued further; the only basis for ruling out triplet anthracene is the fact that no fluorescence of anthracene was observed via TT annihilation.

In the case of Dewar benzene, *ca.* 80–90 kcal $mol^{-1}$ are available, probably enough for the generation of triplet excited benzene, certainly not enough for benzene singlets. As in the case of the dioxetanes, a spin flip must take place during the rearrangement path to give triplet products; this does not seem to happen at all as easily as with the dioxetanes.

One aspect which may explain in part the poor (or non-existent) yields of excitation in these two systems could be unfavourable Franck–Condon factors. Dewar benzene is bent out of the plane, but unlike triplet benzene, and therefore vibrationally excited states should be produced. This contrasts with the dioxetane structure, where the non-planarity of the ring matches the configuration of the excited carbonyl.

## Acknowledgements

This work was supported by a grant from the National Science Foundation (GB 31977X) to Professor J. W. Hastings, to whom I am also grateful for many enjoyable and stimulating discussions. I have greatly benefited from conversations with, among others, A. L. Baumstark, L. R. Faulkner and M. E. Landis.

## Note added in proof

Zimmerman and Keck[137] have synthesised dioxetanes (35a–c) by reaction of $^1O_2$ with the corresponding olefins [m.p. (35a) and (35c) $\approx$ 90 °C; rate of decomposition of (35a) (at 71.4 °C) = $4.5 \times 10^{-4}$ $s^{-1}$ in $CCl_4$, similar to TMD].

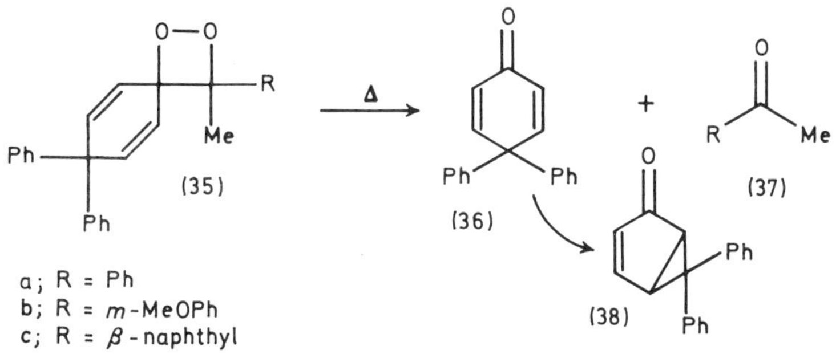

(35) (36) (37) (38)

a; R = Ph
b; R = *m*-MeOPh
c; R = *β*-naphthyl

Excited dienone (36) with $\phi_{isc}$ = 1) undergoes 'photo' rearrangement into (38); the yields of excited (singlet or triplet) products from (35) are determined on that basis, and reported to be 17.1, 14.0 and 12.0% for (35a), (35b) and (35c), respectively. Hence, the nature of group R seems to have little influence on the yields of (36)*, even though, in the case of (35c), the lowest triplet ($\pi$ $\pi^*$) of product (37) (β-acetonaphthone) is lower than that (n–π*) of (36). According to the authors, this shows preferential formation of n–π* carbonyls. However, direct search for excited (37)* is not reported.

Dioxetanes (39) of polyarylfulvenes have been isolated (m.p. 100 °C) as isomerisation products of endoperoxides (40) in acid medium[138].

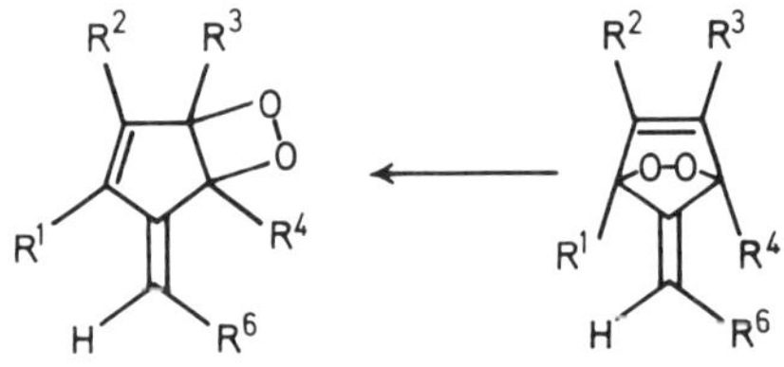

(39) (40)

Tetraethyldioxetane was found quite similar to TMD with respect to yield and multiplicity of excited products, but a little more stable ($E_a$ = 31.6 kcal mol$^{-1}$, log $A$ = 15.6)[139]. The preparation and thermolysis of dioxetanes (4), (5), (7), (8) and (21) (Tables 7.1 and 7.2)[140] and the study[141] of adamantylideneadamantanedioxetane (13) have now been reported in full papers.

A CIDNP experiment with TMD supports the triplet multiplicity of the acetone produced by the dioxetane's thermolysis[142]. The reaction of diphenyl sulphide with some dioxetanes brings about their monodeoxygenation[143]. Simo and Stauff[144] report that the chemiluminescence spectrum of trimethyldioxetane (4) shows three peaks: at 400 nm, acetone or acetaldehyde fluorescence; at 465 nm, acetone phosphorescence; plus a main peak at 430 nm interpreted as the emission from a biradical species formed in the cleavage of (4). Results of quenching experiments are discussed. These authors also reinvestigated the photoluminescence of acetone, with results diverging from those of Turro *et al.*[145]. The photochemistry of TMD was studied by Lechtken and Steinmetzer[146], and Foote and Darling[147] reported on aspects of the photosensitised decomposition of dioxetane (6) (Table 7.1), showing the importance of induced decomposition.

Regarding the intermediacy of a hypothetical dioxetanone in the firefly system (see Section 7.8.2), new experiments[148] on the chemiluminescence of dimethyl-luciferin using labelled oxygen ($^{18}$O) show a level of $^{18}$O incorporation in $CO_2$ consistent with a dioxetane mechanism.

The mechanism of the chemiluminescent thermolysis of dioxetanes has been the object of several recent theoretical papers[149–152]. A two-step mechanism seems generally favoured. One could speculate, however, that a switch from a biradical to a truly concerted mechanism may take place when or if dioxetanes are substituted by large resonant groups. In these cases, in contrast to the excited carbonyls (with n,π* excited states) generated by the 'stable', isolable dioxetanes so far studied, the cleavage products would have delocalised, π,r*, lowest excited states and fluoresce with high efficiency. Such resonant substituents may conceivably reduce the stability of the dioxetanes and increase the proportion of singlet products, as expected from the hypothetical intermediates in bioluminescence.

## References


1. McCapra, F. (1973). *Progress in Organic Chemistry*, Vol. 8, 231 (W. Carruthers and J. K. Sutherland, editors) (New York: Wiley)
2. Vassil'ev, R. F. (1970). *Russ. Chem. Rev.*, **39,** 529
3. White, E. H. and Roswell, D. F. (1970). *Accounts Chem. Res.*, **3,** 54
4. Hercules, D. M. (1969). *Accounts Chem. Res.*, **2,** 301
5. Rauhut, M. M. (1969). *Accounts Chem. Res.*, **2,** 80
6. Gunderman, K. D. (1968). *Chemilumineszenz Organische Verbindungen* (Berlin: Springer-Verlag)
7. Seliger, H. H. and McElroy, W. D. (1960). *Arch. Biochem. Biophys.*, **88,** 136
8. Hastings, J. W. (1968) *Ann. Rev. Biochem.*, **37,** 597
9. Seliger, H. H. and McElroy, W. D. (1965). *Light: Physical and Biological Action* (New York: Academic Press)
10. McCapra, F. (1968). *Chem. Commun.*, 155
11. Herzberg, G. (1966). *Molecular Spectra and Molecular Structure. III. Electronic Spectra and Electronic Structure of Polyatomic Molecules*, 518 (Princeton, N.J.: Van Nostrand)
12. McGlynn, S. P., Azumi, T. and Kinoshita, M. (1969). *Molecular Spectroscopy of the Triplet State*, 166 (Englewoods Cliffs, N.J.: Prentice-Hall)
13. Kopecky, K. R. and Mumford, C. (1968). *Abstracts, 51st Annual Conference of the Chemical Institute of Canada*, June 1968, 41 (Vancouver, B.C.)
14. Kopecky, K. R. and Mumford, C. (1969). *Can. J. Chem.*, **47,** 709
15. Kopecky, K. R., Van de Sande, J. H. and Mumford, C. (1968). *Can. J. Chem.*, **46,** 25
16. White, E. H., Wildes, P. D., Wiecko, J., Doshan, H. and Wei, C.C. (1973). *J. Amer. Chem. Soc.*, **95,** 7050
17. Kopecky, K. R. (1973). Personal communication; Filby, J. E. (1973). *Ph.D. Thesis*, University of Alberta
18. Baumstark, A. L. (1974). *Ph.D. Thesis*, Harvard University
19. Bartlett, P. D., Mendenhall, G. D. and Schaap, A. P. (1970). *International Conference on Singlet Molecular Oxygen and its Role in the Environmental Sciences* (A. M. Trozzolo, editor), *Ann. N.Y. Acad. Sci.*, **171,** 79
20. Bartlett, P. D. and Schaap, A. P. (1970). *J. Amer. Chem. Soc.*, **92,** 3223
21. Mazur, S. and Foote, C. S. (1970). *J. Amer. Chem. Soc.*, **92,** 3225
22. Story, P. R., Whited, E. A. and Alford, J. A. (1972). *J. Amer. Chem. Soc.*, **94,** 2143
23. Bailey, P. S., Carter, T. P., Fisher, C. M. and Thompson, J. A. (1973). *Can. J. Chem.*, **51,** 1278
24. Kopecky, K. R., Lockwood, P. A., Filby, J. E. and Reid, R. W. (1973). *Can. J. Chem.*, **51,** 468
25. Calvert, J. G. and Pitts, J. N. (1966). *Photochemistry* (New York: Wiley) and references therein
26. Wagner, P. J. and Hammond, G. S. (1969). *Advances in Photochemistry*, Vol. 5, 21 (J. N. Pitts, G. S. Hammond and W. A. Noyes, editors) (New York: Wiley) and references therein
27. Turro, N. J. (1967). *Molecular Photochemistry* (New York: Benjamin) and references therein
28. Turro, N. J., Dalton, J. C., Dawes, K., Farrington, G., Hautala, R., Morton, D., Niemczyk, M. and Shore, N. (1972). *Accounts Chem. Res.*, **6,** 92 and references therein
29. Engel, P. S. and Monroe, B. M. (1971). *Advances in Photochemistry*, Vol. 8, 245 (J. N. Pitts, G. S. Hammond and W. A. Noyes, editors) (New York: Wiley-Interscience) and references therein
30. Hatchard, C. G. and Parker, C. A. (1956). *Proc. Roy. Soc.* (*London*), **A235,** 518
31. For example, Lee, J. and Seliger, H. H. (1965). *Photochem. Photobiol.*, **4,** 1015
32. Lee, J., Wesley, A. S., Ferguson, J. F. III, and Seliger, H. H. (1966). *Bioluminescence in Progress*, 35 (F. H. Johnson and Y. Haneda, editors) (Princeton, N.J.: Princeton University Press)
33. Hastings, J. W. and Weber, G. (1965). *Photochem. Photobiol.*, **4,** 1049
34. Hastings, J. W. and Weber, G. (1963). *J. Opt. Soc. Amer.*, **53,** 1410

35. Hastings, J. W. and Reynolds, G. T. (1966). *Bioluminescence in Progress*, 45 (F. H. Johnson and Y. Haneda, editors) (Princeton, N.J.: Princeton University Press)
36. Dunn, D. K., Michaliszyn, G. A., Bogacki, I. G. and Meighen, E. A. (1973). *Biochemistry*, **12,** 4911
37. Papisova, V. E., Shlyapintokh, V. Y. and Vasil'ev, R. V. (1965). *Russ. Chem. Rev.*, **34,** 594
38. Lundeen, G. and Livingston, R. (1965). *Photochem. Photobiol.*, **4,** 1085
39. Turro, N. J. and Lechtken, P. (1972). *J. Amer. Chem. Soc.*, **94,** 2886
40. Turro, N. J., Steinmetzer, H. C. and Yekta, A. (1973). *J. Amer. Chem. Soc.*, **95,** 6468
41. Belyakov, V. A. and Vassil'ev, R. F. (1970). *Photochem. Photobiol.*, **11,** 179
42. Vassil'ev, R. F. (1962). *Nature* (*London*), **196,** 668
43. Wilson, T. and Schaap, A. P. (1971). *J. Amer. Chem. Soc.*, **93,** 4126
44. Berenfel'd, V. M., Chumaevskii, E. V., Grinev, M. P., Kuryatnikov, Yu. I., Artem'ev, E. T. and Dzhagatspanyan, R. V. (1970). *Bull. Acad. Sci. USSR, Ser. Phys.*, **3,** 597
45. Wilson, T. (1974). Unpublished results; also, Turro, N. J. (1974). Personal communication
46. Golan, D. and Wilson, T. (1974). Unpublished results
47. Steinmetzer, H. C., Lechtken, P and Turro, N. J. (1973). *Ann. Chem.*, 1984
48. Turro, N. J., Lechtken, P., Shore, N. E., Schuster, G., Steinmetzer, H. C. and Yekta, A. (1973). *Accounts Chem. Res.*, **7,** 97
49. Turro, N. J. and Steinmetzer, H. C. (1974). *J. Amer. Chem. Soc.*, **96,** 4677, 4679
50. Wildes, P. D. and White, E. H. (1971). *J. Amer. Chem. Soc.*, **93,** 6286
51. Crosby, G. A., Whan, R. E. and Alire, R. M. (1961). *J. Chem. Phys.*, **34,** 743 and references therein
52. El-Sayed, M. A. and Bhaumik, M. L. (1963). *J. Chem. Phys.*, **39,** 2391
53. Wilson, T. (1974). Unpublished results
54. Winston, H., Marsh, O. J., Suzuki, C. K. and Telk, C. L. (1963). *J. Chem. Phys.*, **39,** 267
55. Filipescu, N., Mushrush, G. W., Hurt, C. R. and McAvoy, N. (1966). *Nature* (*London*), **211,** 960
56. Bhaumik, M. L. and Telk, C. L. (1964). *J. Opt. Soc. Amer.*, **54,** 1211
57. Dawson, W. R., Kropp, J. L. and Windsor, M. W. (1966). *J. Chem. Phys.*, **45,** 2410
58. Filipescu, N. and Mushrush, G. W. (1968). *J. Phys. Chem.*, **72,** 3516
59. White, E. H., Wiecko, J. and Roswell, D. R. (1969). *J. Amer. Chem. Soc.*, **91,** 5194
60. White, E. H., Wiecko, J. and Wei, C. C. (1970). *J. Amer. Chem. Soc.*, **92,** 2167
61. Hammond, G. S., Saltiel, J., Lamola, A. A., Turro, N. J., Bradshaw, J. S., Cowan, D. O., Counsell, R. C., Vogt, V. and Dalton, C. (1964). *J. Amer. Chem. Soc.*, **86,** 3197
62. For a discussion of complications in the interpretation of photosensitised reactions, see Ref. 29
63. Lechtken, P., Yekta, A. and Turro, N. J. (1973). *J. Amer. Chem. Soc.*, **95,** 3027
64. Dalton, J. C., Wriede, P. A. and Turro, N. J. (1970). *J. Amer. Chem. Soc.*, **92,** 1318
65. For example (1966). *Organic Photochemistry*, 223 (R. O. Khan, editor) (New York: McGraw-Hill)
66. Scaiano, J. C. (1973). *J. Photochem.*, **2,** 81 and references therein
67. Wagner, P. J. (1966). *J. Amer. Chem. Soc.*, **88,** 5672
68. Wilson, T., Landis, M. E., Baumstark, A. L. and Bartlett, P. D. (1973). *J. Amer. Chem. Soc.*, **95,** 4765
69. Landis, M. E. (1974). *Ph.D. Thesis*, Harvard University
70. Porter, G., Dogra, S. K., Loufty, R. O., Sugamori, S. E. and Yip, R. W. (1973). *J. Chem. Soc. Faraday Trans. I*, **69,** 1462
71. Charney, D. R., Dalton, J. C., Hautula, R. R., Snyder, J. J. and Turro, N. J. (1974). *J. Amer. Chem. Soc.*, **96,** 1407
72. Darling, T. R. and Foote, C. S. (1974). *J. Amer. Chem. Soc.*, **96,** 1625
73. Wagner, P. J. (1971). *Accounts Chem. Res.*, **4,** 168
74. Yang, N. C. and Elliott, S. P. (1969). *J. Amer. Chem. Soc.*, **91,** 7551
75. Richardson, W H., Montgomery, F. C. and Yelvington, M. B. (1972). *J. Amer. Chem. Soc.*, **94,** 9278
76. Engel, P. S. (1970). *J. Amer. Chem. Soc.*, **92,** 6074
77. Cohen, A. D. and Reid, C. (1956). *J. Chem. Phys.*, **24,** 85

78. Turro, N. J. and Lechtken, P. (1973). *J. Amer. Chem. Soc.*, **95**, 264
79. Turro, N. J. and Lechtken, P. (1973). *Pure Appl. Chem.*, **33**, 363
80. Lee, C.-S. and Wilson, T. (1973). *Chemiluminescence and Bioluminiscence*, 265 (M. J. Cormier, D. M. Hercules and J. Lee, editors) (New York: Plenum)
81. Steinmetzer, H. C., Yekta, A. and Turro, N. J. (1974). *J. Amer. Chem. Soc.*, **96**, 282
82. Lechtken, P. and Höhne, G. (1973). *Angew. Chem. Int. Ed. Engl.*, **12**, 772
83. O'Neal, H. E. and Richardson, W. H. (1970). *J. Amer. Chem. Soc.*, **92**, 6553
84. O'Neal, H. E. and Richardson, W. H. (1971). *J. Amer. Chem. Soc.*, **93**, 1828
85. Benson, S. W. (1968). *Thermochemical Kinetics* (New York: Wiley)
86. Adam, W. (1973). *Chemie in unserer Zeit*, **7**, 181
87. Steinmetz, W. E. (1973). Personal communication
88. Richardson, W. H., Yelvington, M. B. and O'Neal, H. E. (1972). *J. Amer. Chem. Soc.*, **94**, 1619
89. Baumstark, A. L. and Landis, M. E. (1974). Personal communication
90. Richardson, W. H., Montgomery, F. C., Yelvington, M. B. and Ranney, G. (1974). *J. Amer. Chem. Soc.*, **96**, 4045
91. Kopecky, K. R. and Mumford, C., quoted by O'Neal, H. E. and Richardson, W. H. (1970). *J. Amer. Chem. Soc.*, **92**, 6553
92. Turro, N. J., Lechtken, P., Schuster, G., Orell, J., Steinmetzer, H. C. and Adam, W. (1974). *J. Amer. Chem. Soc.*, **96**, 1627
93. Kopecky, K. R. and Filby, J. E. (1974). Personal communication
94. Schaap, A. P. (1971). *Tetrahedron Lett.*, 1757
95. Wilson, T., Harris, M. S. and Baumstark, A. L. (1974). Unpublished results
96. Wieringa, J. H., Strating, J., Wynberg, H. and Adam, W. (1972). *Tetrahedron Lett.*, 169
97. Turro, N. J., Schuster, G., Steinmetzer, H. C., Schaap, A. P., Faler, G. R., Adam, W. and Liu, J. C. (1974). To be published
98. Adam, W. and Liu, J. C. (1972). *J. Amer. Chem. Soc.*, **94**, 2894
99. Adam, W. and Steinmetzer, H. C. (1972). *Angew. Chem. Int. Ed. Engl.*, **11**, 540
100. Adam, W. and Liu, J. C. (1972). Unpublished results (cited in Ref. 99)
101. Adam, W., Simpson, G. A. and Yany, F. (1974). *J. Phys. Chem.*, in press
102. Hasty, N. M. and Kearns, D. R. (1973). *J. Amer. Chem. Soc.*, **95**, 3380
103. Schaap, A. P. and Tontapanish, N. (1971). *Symposium on Oxidation by Singlet Oxygen*, Vol. 16, No. 4, A78 (Washington, D.C.: Amer. Chem. Soc.)
104. Schaap, A. P. and Tontapanish, N. (1972). *J. Chem. Soc. Chem. Commun.*, 490
105. Bartlett, P. D. and Ho, M. S. (1974). *J. Amer. Chem. Soc.*, **96**, 627
106. Rio, G. and Berthelot, J. (1971). *Bull. Soc. Chim. Fr.*, 3555
107. Burns, P. A. and Foote, C. S. (1974). *J. Amer. Chem. Soc.*, **96**, 4338
108. Basselier, J. J., Cherton, J. C. and Caille, J. (1971). *C.R. Acad. Sci. Ser. C*, **273**, 514
109. Rigaudy, J., Capdevielle, P. and Maumy, M. (1972). *Tetrahedron Lett.*, 4997
110. Bartlett, P. D., Baumstark, A. L. and Landis, M. E. (1974). *J. Amer. Chem. Soc.*, **96**, 5557
111. Bartlett, P. D., Baumstark, A. L. and Landis, M. (1973). *J. Amer. Chem. Soc.*, **95**, 6486
112. Bartlett, P. D., Baumstark, A. L., Landis, M. E. and Lerman, C. L. (1974). *J. Amer. Chem. Soc.*, **96**, 5267
113. Walling, C. and Gibian, M. J. (1965). *J. Amer. Chem. Soc.*, **87**, 3413
114. Ref. 29, p. 287, and references therein
115. Turro, N. J., Lechtken, P., Lyons, A., Hautala, R. R., Carnahan, E. and Katz, T. J. (1973). *J. Amer. Chem. Soc.*, **95**, 2035
116. Turro, N. J. and Lechtken, P. (1973). *Tetrahedron Lett.*, 565
117. Kearns, D. R. (1971). *Chem. Rev.*, **71**, 395
118. McCapra, F. (1970). *Pure Appl. Chem.*, **24**, 611
119. Woodward, R. B. and Hoffman, R. (1970). *The Conservation of Orbital Symmetry* (New York: Academic Press)
120. Benson, S. W. and Shaw, R. (1970). *Organic Peroxides*, Vol. 1, 105 (D. Swern, editor) (New York: Wiley-Interscience)
121. Salem, L. and Rowland, C. (1972). *Angew. Chem. Int. Ed. Engl.*, **11**, 92
122. Müller, K. (1974). Personal communication
123. Lechtken, P. and Turro, N. J. (1973). *Angew. Chem. Int. Ed. Engl.*, **12**, 314

124. Shimomura, O. and Johnson, F. H. (1973). *Chemiluminescence and Bioluminescence*, 337 (M. J. Cormier, D. M. Hercules and J. Lee, editors) (New York: Plenum) and references therein
125. White, E. H., Rapaport, E., Seliger, H. H. and Hopkins, T. A. (1971). *Bioorg. Chem.*, **1,** 92 and references therein
126. DeLuca, M. and Dempsey, M. E. (1970). *Biochem. Biophys. Res. Commun.*, **40,** 117
127. DeLuca, M., Dempsey, M. E., Hori, K., Wampler, J. E. and Cormier, M. J. (1971). *Proc. Nat. Acad. Sci. USA*, **68,** 1658
128. Shimomura, O. and Johnson, F. H. (1971). *Biochem. Biophys. Res. Commun.*, **44,** 340
129. Shimomura, O. and Johnson, F. H. (1973). *Biochem. Biophys. Res. Commun.*, **51,** 558
130. Hastings, J. W., Eberhard, A., Baldwin, T. O., Nicoli, M. Z., Cline, T. W. and Nealson, K. H. (1973). *Chemiluminescence and Bioluminescence*, 369 (M. J. Cormier, D. M. Hercules and J. Lee, editors) (New York: Plenum) and references therein
131. Henry, J. P. and Michelson, A. M. (1973). *Biochimie*, **55,** 75
132. Michelson, A. M. and Isambert, M. F. (1973). *Biochimie*, **55,** 619
133. For example: Allen, R. C., Stjernholm, R. L. and Steele, R. H. (1972). *Biochem. Biophys. Res. Commun.*, **47,** 679
134. For example: Maugh, T. H. (1973). *Science*, **182,** 44
135. Lechtken, P., Breslow, R., Schmidt, A. and Turro, N. J. (1973). *J. Amer. Chem. Soc.*, **95,** 3025
136. Yang, N. C., Carr, R. V., Li, E., McVey, J. K. and Rice, S. (1974). *J. Amer. Chem. Soc.*, **96,** 2897
137. Zimmerman, H. E. and Keck, G. E. (1975). *J. Amer. Chem. Soc.*, **97,** 3527
138. Le Roux, J. P. and Goasdoue, C. (1975). *Tetrahedron*, **31,** 2761
139. Bechara, E. J. H., Baumstark, A. L. and Wilson, T. (1976). To be published
140. Kopecky, K. R., Filby, J. E., Mumford, C., Lockwood, P. A. and Ding, J.-Y. (1975). *Can. J. Chem.*, **53,** 1103
141. Schuster, G. B., Turro, N. J., Steinmetzer, H.-C., Schaap, A. P., Faler, G., Adam, W. and Liu, J. C. (1975). *J. Amer. Chem. Soc.*, **97,** 7110
142. Bartlett, P. D. and Shimizu, N. (1975). *J. Amer. Chem. Soc.*, **97,** 6253
143. Wasserman, H. H. and Saito, I. (1975). *J. Amer. Chem. Soc.*, **97,** 905
144. Simo, I. and Stauff, J. (1975). *Chem. Phys. Lett.*, **34,** 326
145. Turro, N. J., Steinmetzer, H.-C. and Yekta, A. (1973). *J. Amer. Chem. Soc.*, **95,** 6468
146. Lechtken, P. and Steinmetzer, H.-C. (1975). *Chem. Ber.*, **108,** 3159
147. Foote, C. S. and Darling, T. R. (1975). *Pure Appl. Chem.*, **41,** 495
148. White, E. H., Miano, J. D. and Umbreit, M. (1975). *J. Amer. Chem. Soc.*, **97,** 198
149. Roberts, D. R. (1974). *J. Chem. Soc. Chem. Commun.*, 683–684
150. Dewar, M. J. S. and Kirschner, S. (1974). *J. Amer. Chem. Soc.*, **96,** 7578
151. Turro, N. J. and Devaquet, A. (1975). *J. Amer. Chem. Soc.*, **97,** 3859
152. Barnett, G. (1974). *Can. J. Chem.*, **52,** 3837

# Index